Realism, Science, and Pragmatism

"This is a first-rate collection of essays on the general issue of realism, on the relation of realism to contemporary philosophy of science and epistemology, and on the challenge that has been made to traditional realism by classical pragmatism and neo-pragmatism. The contributors are among the leading scholars in the field, and their essays advance the debates in ways that will provoke response and further inquiry. Anyone interested in the topic of realism, its history and current controversies, will benefit from paying the close attention that these essays deserve."

—John Ryder, American University of Ras al Khaimah, United Arab Emirates

This collection of original essays aims to reinvigorate the debate surrounding philosophical realism in relation to philosophy of science, pragmatism, epistemology, and theory of perception. Questions concerning realism are as current and as ancient as philosophy itself; this volume explores relations between different positions designated as 'realism' by examining specific cases in point, drawn from a broad range of systematic problems and historical views, from ancient Greek philosophy through the present. The first section examines the context of the project; contributions systematically engage the historical background of philosophical realism, re-examining key works of Aristotle, Descartes, Quine, and others. The following two sections epitomize the central tension within current debates: scientific realism and pragmatism. These contributions address contemporary questions of scientific realism and the reality of the objects of science, and consider whether, how or the extent to which realism and pragmatism are compatible. With an editorial introduction by Kenneth R. Westphal, these fourteen original essays provide wide-ranging, salient insights into the status of realism today.

Kenneth R. Westphal is Professorial Fellow in the School of Philosophy at the University of East Anglia, UK

Routledge Studies in Contemporary Philosophy

For a full list of titles in this series, please visit www.routledge.com

24 Philosophical Delusion and its Therapy
Outline of a Philosophical Revolution
Eugen Fischer

25 Epistemology and the Regress Problem
Scott F. Aikin

26 Civil Society in Liberal Democracy
Mark Jensen

27 The Politics of Logic
Badiou, Wittgenstein, and the Consequences of Formalism
Paul M. Livingston

28 Pluralism and Liberal Politics
Robert B. Talisse

29 Kant and Education
Interpretations and Commentary
Edited by Klas Roth and Chris W. Surprenant

30 Feminism, Psychoanalysis, and Maternal Subjectivity
Alison Stone

31 Civility in Politics and Education
Edited by Deborah S. Mower, Wade L. Robison

32 Philosophical Inquiry into Pregnancy, Childbirth, and Mothering
Maternal Subjects
Edited by Sheila Lintott and Maureen Sander-Staudt

33 Authenticity as an Ethical Ideal
Somogy Varga

34 The Philosophy of Curiosity
Ilhan Inan

35 Self-Realization and Justice
A Liberal-Perfectionist Defense of the Right to Freedom from Employment
Julia Maskivker

36 Narrative Identity, Autonomy, and Mortality
From Frankfurt and MacIntyre to Kierkegaard
John J. Davenport

37 Contemporary Feminist Pragmatism
Edited by Maurice Hamington and Celia Bardwell-Jones

38 Morality, Self Knowledge, and Human Suffering
An Essay on The Loss of Confidence in the World
Josep Corbi

39 **Contrastivism in Philosophy**
Edited by Martijn Blaauw

40 **Aesthetics After Metaphysics**
From Mimesis to Metaphor
Miguel de Beistegui

41 **Foundations of Freedom**
Welfare-Based Arguments against Paternalism
Simon R. Clarke

42 **Pittsburgh School of Philosophy**
Sellars, McDowell, Brandom
Chauncey Maher

43 **Reference and Structure in the Philosophy of Language**
A Defense of the Russellian Orthodoxy
Arthur Sullivan

44 **Civic Virtue and the Sovereignty of Evil**
Derek Edyvane

45 **Philosophy of Language and Webs of Information**
Heimir Geirsson

46 **Disagreement and Skepticism**
Edited by Diego E. Machuca

47 **Philosophy in Schools**
An Introduction for Philosophers and Teachers
Edited by Sara Goering, Nicholas J. Shudak, and Thomas E. Wartenberg

48 **A Philosophy of Material Culture**
Action, Function, and Mind
Beth Preston

49 **A Philosophy of the Screenplay**
Ted Nannicelli

50 **Race, Philosophy, and Film**
Edited by Mary K. Bloodsworth-Lugo and Dan Flory

51 **Knowledge, Virtue, and Action**
Essays on Putting Epistemic Virtues to Work
Edited by Tim Henning and David P. Schweikard

52 **The Ontology of Psychology**
Questioning Foundations in the Philosophy of Mind
Linda A.W. Brakel

53 **Pragmatism, Law, and Language**
Edited by Graham Hubbs and Douglas Lind

54 **Contemporary Dualism**
A Defense
Edited by Andrea Lavazza and Howard M. Robinson

55 **Reframing the Intercultural Dialogue on Human Rights**
A Philosophical Approach
Jeffrey Flynn

56 **How History Matters to Philosophy**
Reconsidering Philosophy's Past After Positivism
Robert C. Scharff

57 **The Affordable Care Act Decision**
Philosophical and Legal Implications
Edited by Fritz Allhoff and Mark Hall

58 **Realism, Science, and Pragmatism**
Edited by Kenneth R. Westphal

Realism, Science, and Pragmatism

Edited by Kenneth R. Westphal

Routledge
Taylor & Francis Group
LONDON AND NEW YORK

First published 2014
by Routledge

2 Park Square, Milton Park, Abingdon, Oxfordshire OX14 4RN

711 Third Avenue, New York, NY 10017

Routledge is an imprint of the Taylor & Francis Group, an informa business

First issued in paperback 2017

Library of Congress Cataloging-in-Publication Data

Realism, science, and pragmatism / edited by Kenneth R. Westphal.
pages cm. — (Routledge studies in contemporary philosophy ; 58)
Includes bibliographical references and index.
1. Realism. 2. Science—Philosophy. 3. Pragmatism. I. Westphal, Kenneth R., editor of compilation.
B835.R3265 2014
149′.2—dc23
2013041610

ISBN: 978-1-138-01882-2 (hbk)
ISBN: 978-0-8153-7200-4 (pbk)

Typeset in Sabon
by Apex CoVantage, LLC

Contents

Introduction

Kenneth R. Westphal

Issues about realism are as current and as ancient as philosophy itself. Plato in the *Theatetus* comments on the long-standing battle between philosophical giants and gods about whether, in addition to the physical objects and events we perceive, there are also non-physical, and hence non-perceptible forms or ideas of kinds or characteristics, variously instantiated in physical particulars, but which exist independently both of their instances and of what we may happen to say, think, believe, or know about them. In philosophical usage, the term 'realism' is both basic and polysemic. For example, one can hold realism—in contrast to idealism, irrealism, or agnosticism—if one holds that material objects exist and have various characteristics regardless of what we may say, think, believe, or know about them. One can be a direct realist in the theory of perception by holding that perception is direct awareness of external objects, a moral realist if one believes that there are objective moral values, a scientific realist if one holds that scientific knowledge is about theory-independent phenomena and that such knowledge is possible even about unobservable entities, or a modal realist if one believes that possible worlds are as real as the actual world. In ontology, realism indicates that one grants—in ways which vary from case to case—extra-mental existence to certain kinds of entities, processes, or structures and at least some of their features, such as physical objects, universals, relations, structures, or propositions. Realism about particular objects and about their features or relations became problematic in Twentieth Century philosophy when it became generally recognized that we cannot, as it were, set aside our concepts, theories, beliefs, or, in general, our language to inspect the facts themselves and on that basis assess our beliefs, statements, or theories about them. Realism has remained fraught since.

The fourteen original essays presented here explore the relations that different positions designated as 'realism' may have to each other by examining specific cases in point, drawn from a broad range of systematic problems and historical views from Ancient Greek philosophy up to the present day. Individually and taken together, these essays show how much can be gained by examining issues about realism both systematically and historically. The essays form three groups, Part I: Realism Contextualized, Part II: Scientific Realism, and Part III: Pragmatism and Realism.

Part I contextualizes issues about realism regarding both physical particulars and universals by critically re-examining several major historical and systematic positions on these topics. A hallmark of pragmatism is that both the understanding and the assessment of current views, and the development of improved views, benefit, often centrally, by re-assessing prior views on the same or related issues, whether these prior views be familiar, neglected, under appreciated, or misunderstood. Accordingly Part I attempts neither a historical nor a systematic review of issues about realism; excellent surveys are available elsewhere. Instead, its six chapters re-investigate key historical and systematic issues where new and unexpected insights can be discovered—and discoveries there are in these chapters, which plumb philosophical depths in Ancient, Mediaeval, Modern, and Contemporary philosophy.

The issues are launched, officially yet non-technically, by Jaakko Hintikka in "What Is Real(ism)?" (Chapter 1). Hintikka articulates what is involved in claiming realism about any domain or issue or particular(s), in part by arguing that the use of possible world semantics requires a richer domain of discourse than is provided by any such semantics tailored to any one domain and its attendant possible-worlds model. To use such a logic we must be able to identify individuals across such domains and models, in order to identify any actual individuals within possible worlds, and to identify their merely possible, non-actual existence in some possible worlds, including that possible world which is our actual world. Accordingly, "actuality and existence do not go together," and epistemic logic requires a richer domain of discourse than that of Frege-Russell first-order quantification logic and the possible-worlds semantics built upon it. We need not only the 'is' of identity, the 'is' of predication, and the 'is' of instantiation, but also the 'is' of identifiability, as was implicitly recognized in Aristotle's logic and metaphysics. Epistemic logic requires possible objects as well as actual ones, and epistemic logic is required to use first-order quantification theory and possible-worlds semantics in any actual domain of inquiry.

In "Aristotle's Direct Realism and Some Later Developments" (Chapter 2), Mika Perälä elucidates Aristotle's direct theory of perception within Aristotle's general explanatory project. Aristotle's formal cause is specified in terms of the relevant efficient cause, the activity of which occurs in the activity of the patient. This model implies that, in perception, seeing, *e.g.*, a white object is the type of perception it is because it is caused by a white object, where the object's white color exercises its power of being perceived in the sense of vision as someone's seeing that white of that object. That object's white color is the term to which the activity of the perceiver's sense, as the patient of the object's color's activity, is relative. This analysis enables Aristotle to solve problems in the Megarean and Protagorean anti-realist accounts of perception. Aristotle's view thus contrasts in important regards to those of, *e.g.*, Aquinas and Scotus. Perälä examines two important contrasts. First, he argues that Aristotle, unlike Aquinas—and many of his successors even today—did not resort to the concepts of intentional being and likeness (and their cognates such as representation) to explain why a perception is about its proper object. Second, he points out that Aristotle, unlike Scotus, did not allow that a mental act could be directed at its object even when the object is not the efficient cause of that act (for example, when God or Devil induces such an act in us). Accordingly, Aristotle provides a cogent direct realist theory of perception.

Laurent Cesalli begins his examination of "Key Arguments Supporting Non-semantic Universality" in "Late Mediaeval Realisms" (to invert the sub-title and title of Chapter 3) by noting the highly favorable reception of Scotus's realism about universals by both C. S. Peirce and David Armstrong. Cesalli then critically examines key arguments for the real existence of universals developed by Aquinas, Scotus, Burley, Ockham, Buridan, Richard Brinkley, Nicolas of Autrécourt, Francesco da Prato, John Wyclif, and Dietrich of Freiberg, who argue that (*i*) semantic universality depends upon the existence of metaphysical universals, or that (*ii*) scientific knowledge requires metaphysical universals if it is not to be reduced to psychology or linguistics, or that (*iii*) essences of things exist objectively, no less than do their matter and forms. They analyze relations between universals and particulars in several ways: in terms of a merely formal distinction, mereology, concomitance, partial identity, real identity, or platonic-atomism. Cesalli further elucidates their views by critical comparison with modern forms of realism developed by Bergmann, Armstrong, and Cocchiarella.

In "Descartes on the Formal Reality, Objective Reality, and Material Falsity of Ideas: Realism through Constructivism?" (Chapter 4), Dermot Moran revisits Descartes' account of the formal and objective reality of ideas in order to ascertain more exactly Descartes' commitment to realism. For Descartes, as for the Scotist tradition in general, 'real' means something that can be a '*res*': to be real is to be possible. The reality of some possible thing is expressed by its essence which is reflected in its 'objective reality'. Objective reality comes in degrees. Some 'real' entities also have actuality as a result of being caused. This is their 'formal reality'. For Descartes certain ideas (*e.g.*, of God) have an objective reality so great that it can be accounted for only by those ideas also having formal reality. Other ideas have objective reality but fail to have a formal cause and may even mislead in presenting the kind of objective reality they possess. These are 'materially false' ideas. The received view is that Descartes' employment of these Scholastic notions is confused and that Antoine Arnauld in his Objections makes a number of valid criticisms of Descartes' account. Moran contends that Descartes' distinctions offer powerful insights into the intentionality of the mind and the manner in which the phenomenological character of our experiences can (or cannot) reliably lead us to grasp the nature of reality in itself. Descartes has a complex conception of intentional content that deserves more attention and credit than it has hitherto received.

In "Quine's Conception of Objects: Beyond Realism and Anti-realism" (Chapter 5), Antti Keskinen argues that the apparent tension between

Quine's scientific realism and his epistemological conception of objects as theoretical posits is not resolved by appeal to Quine's naturalism, but instead by the genuinely reciprocal containment between science and epistemology. Quine's scientific realism and his epistemological conception of objects as posits are consistent because the notion of reality is itself always part of a theory; otherwise it is meaningless. The appearance of a tension between those two aspects of Quine's view arises only if it is assumed that objects can be real in some sense other than as posits of a theory included within our best current science. Quine disallows any metaphysical realism that has primacy over epistemology; hence his scientific realism is consistent with his epistemological view of objects as theory-dependent posits. Yet this conception of objects does not entail that the objects talked about in our best current science are less than real, in any admissible, theory-external sense of "real." However, Quine's view further implies that, although *de re* attitude ascriptions have sense, they are rarely true because the conditions necessary for their truthful ascription are rarely satisfied. Though no *reductio,* this implication of Quine's view is highly counter-intuitive.

The answer to Peter Swirski's titular question is no mystery: "Did Sherlock Holmes Inhale Pipe Smoke through a Hole in His Forehead?" (Chapter 6). The question is how we know that Holmes did no such thing, despite his being fictional, and why such knowledge matters. Swirski argues that such knowledge is not based simply upon the text, nor upon possible worlds implicated by the text or the author, not even when guided by a Reality Principle, a Mutual (Shared) Belief Principle, or by the fictional persona of the narrator. None of these proposals properly specify the relevant background beliefs or information required to understand literary texts. Recognizing what is relevant, Swirski argues, requires the reader's reflexive recognition of the *real* author's successfully executed reflexive intentions relevant to a given fiction, facilitated in part by our recognition of an author's use (or abuse) of genre conventions, and by our quintessentially human, natural capacities to understand one another's intentions and acts of directing joint attention. Our understanding of real intentions is required for comprehending one another, and for comprehending fictional truths.

Our reflections on realism began with quantification logic, according to which to be is to be the value of a bound variable, to argue that our understanding of possibility and of possible existence is required to use possible world models of any domain to understand actual features of actual objects or events, thus renewing our appreciation of Aristotle's insights into the 'is' of identification. The systematic and historical trajectory thus launched carries through the essays of Part I to conclude that our understanding of fictional truths requires our understanding of actual intentions and acts of joint attention, without which we could not communicate, and indeed could not be human. Part I thus provides the systematic and historical context of issues about the reality of objects, events, their characteristics, and their relations, within which Parts II and III examine these issues in greater detail.

Part II considers a specific domain of these issues: scientific realism; Part III considers a distinctive approach to these issues: pragmatism.

Part II opens with Panu Raatikainen's "Realism: Metaphysical, Scientific, and Semantic" (Chapter 7). Raatikainen distinguishes and interrelates three influential forms of realism: realism about the external world, construed as a metaphysical doctrine; scientific realism about non-observable entities postulated in science; and semantic realism as defined by Dummett. He first contrasts metaphysical realism about everyday physical objects with idealism and phenomenalism, reviews several potent arguments against these latter views, and argues briefly by induction in support of realism about physical objects. Scientific realism—the idea that natural sciences discover and explain genuine features of natural phenomena—may be commonsense, and may be regarded as the paragon of empirical knowledge, though it has been widely out of philosophical favor, not only since Kuhn's *Structure of Scientific Revolutions* (1962), but throughout the history of empiricism, from Hume to the Logical Positivists, Logical Empiricists and today's Constructive Empiricists. Raatikainen distinguishes three forms of scientific realism: (*i*) scientific theories and their existence postulates should be taken literally; (*ii*) the existence of unobservable entities posited by our most successful scientific theories is justified scientifically; and (*iii*) our best current scientific theories are at least approximately true. Raatikainen argues that only some form of scientific realism can make proper sense of certain episodes in the history of science. He then considers Dummett's influential formulation of semantic issues about realism. Dummett argued that in some cases, the fundamental issue is not about the existence of entities, but rather about whether statements of some specified class (such as mathematics) have an objective truth value, independently of our means of knowing it. Dummett famously argued against such semantic realism and in favor of anti-realism. Raatikainen examines the relation of semantic realism to the metaphysical construal of realism, presents Dummett's main argument against semantic realism, and focuses on Dummett's key premise, that understanding the meaning of a declarative sentence involves knowing the conditions which would make that sentence true. Raatikainen argues against that key premise by appeal to semantic externalism.

Ilkka Niiniluoto's "Scientific Realism: Independence, Causation, and Abduction" (Chapter 8), examines three related criteria of realism within the sciences: mind-independence, causal power, and knowledge by abductive reasoning, by considering what a scientific realist should say about the reality of the past. He argues that realism about the past effectively rules out many anti-realist philosophical positions, such as subjective idealism, phenomenalism, solipsism, positivism, internal realism, social constructivism, and non- or anti-realist varieties of pragmatism. His analysis takes up the theme, announced by Hintikka's opening chapter, of object identification, in connection with Peirce's, Putnam's and with Pihlström's versions of pragmatism.

In "Cognitive Semantics and Newton's Rule 4 of Experimental Philosophy: Scientific Realism without Empiricism" (Chapter 9), Kenneth

Westphal argues that Evans' analysis in "Identity and Predication" (1975) provides the basis for a powerful semantics of singular, specifically *cognitive* reference which directly and strongly supports Newton's Rule 4 of (experimental) Philosophy in ways which support Newton's realism about gravitational force. He first examines Newton's Rule 4 and its role in Newton's justification of realism about gravitational force and then summarizes Evans' account of predication and examines its implications for the semantics of singular cognitive reference. Westphal argues that this semantics of singular cognitive reference is embedded in and strongly supports Newton's Rule 4, and that it rules out Cartesian, infallibilist presumptions about empirical justification generally. He then argues that this semantics of singular cognitive reference reveals a key defect in Bas van Fraassen's main argument for his anti-realist "Constructive Empiricism," and also in many common objections to realism, both commonsense and scientific. More generally, Westphal argues, "realism" has appeared problematic to the extent that, in their focus upon the semantics of conceptual content or linguistic meaning, philosophers have neglected the further requirements for specifically cognitive reference.

In "Naturalism without Metaphysics" (Chapter 10), Jonathan Knowles argues—against wide-spread consensus to the contrary—that scientific naturalism, the thesis that natural science is our unique source of fundamental knowledge and explanation, does not require metaphysical realism, so that a scientific naturalist can reject metaphysics. Drawing on the work of Huw Price, Knowles argues against a naturalistic form of metaphysical realism that builds on a substantive notion of reference, and also argues (contra Devitt and Searle) that one cannot have a substantive realistic position without such a notion. Further, science itself does not militate for a naturalistic metaphysical realism. The correct alternative to realism, Knowles argues, is not Huw Price's "subject naturalism," which purports to explain scientifically the pluralism exhibited by language, including scientific language, by a global "expressivist" theory of content. Knowles argues that Price's approach leads to another, equally problematic kind of metaphysics, and its semantics lacks the scientific credentials claimed for it. The proper middle ground, Knowles contends, recognizes that semantic minimalism need not reduce truth to justification nor to warranted assertability, and that the systematic search for truth, which grapples with what is yet unknown, remains the prerogative of the sciences. These are the keys to a scientific naturalism without metaphysics.

Pragmatist and neo-pragmatist themes are sounded repeatedly in Parts II and III; they are examined in detail in Part III: Pragmatism and Realism. In "Majesty of Truth and the Moral Sentiment: Emerson's and Peirce's Ethico-Ontological Realism" (Chapter 11), Heikki Kovalainen and Douglas Anderson argue that, although they are often logically independent doctrines, realism about physical objects and realism about universals intersect in the religiously influenced interpretation and reception of Plato from the early church Fathers up to Romantics such as Samuel Taylor Coleridge, according to whom Christianity embodies universal truths, nonhuman in origin yet knowable to human reason by intellectual intuition—a faculty Kant famously denied. This Platonic, religiously inclined ontological realism enters American philosophy via Ralph Waldo Emerson. Emerson's notion of the *moral sentiment* is an intellectual faculty of intuiting universal truths—*à la* Coleridge—and a non-human real force operating in reality—anticipating Peirce. Kovalainen and Anderson argue that both Emerson and Peirce advocate *ethico-ontological realism*. Understanding Peirce's theory of inquiry and his ontology requires recognizing their moral and theological aspects. More generally, they contend, understanding the issues of realism about particular objects, their features, and their relations requires grappling with the moral and theological dimensions of these issues.

In "Concepts and the Real in C. I. Lewis' Epistemology" (Chapter 12), Lauri Järvilehto argues that Lewis developed an aspectual realism that avoids relativism. According to Lewis, concepts guide our attention in what we experience. Concepts combine to form conceptual principles, which function as categorial laws by which we classify whatever we experience. If an experience does not conform to our conceptual expectations, we classify that experience as non-veridical. Consequently, our attributions of reality depend in part on the conceptual principles we employ. This seems to result in a very strong relativism: what is real depends upon the concepts and classifications we devise. This apparent relativism arises from a terminological ambiguity: Lewis uses the term 'real' both to designate that to which we attribute reality within our conceptual scheme, and to designate what there actually is, which we encounter, experience, and classify. The fact that a classification works for our purposes, and thus serves to attribute reality to some kinds of particulars, shows that what we so classify is, albeit aspectually, metaphysically real. Thus Lewis advocates perspectivalist or aspectualist realism rather than relativism. Indeed, the very logic of relativity, Lewis argues, undercuts relativism.

In "Pragmatic Realism" (Chapter 13), Sami Pihlström—himself a major exponent of this view—reassesses and further develops his pragmatic realism by re-examining the Kantian roots and character of pragmatic realism and the debates about realism in the classical pragmatism of Peirce, James, and Dewey, and by differentiating and defending his view by critically examining the views of three other contemporary pragmatic realists: Margolis, Westphal, and Vihalemm. Pihlström argues that sustained controversy about and tensions between realism and pragmatism are not a plague, but instead a strength of pragmatism and an important source of its continuing vitality. Pihlström argues that, in both its classical and in its contemporary forms, pragmatic realism is distinct, *e.g.*, from both social constructivist and metaphysical realist accounts of the world and of our knowledge of it, both commonsense and scientific. Recognizing the distinctive virtues of pragmatic realism requires recognizing that whatever we justifiably regard as real must result from inquiry, and those results and the form(s) they take can be neither pre-determined nor presupposed. That is the cardinal mistake of many commonsense, scientific, and metaphysical forms of realism. Pihlström develops a pragmatic realism which combines a (naturalized) transcendental idealism with pragmatic realism and naturalism in a *circular* (though not viciously circular) structure: We transcendentally constitute the world through engaging in worldly (and entirely natural) practices, including practices of inquiry, which themselves are constituted through this same continuing process; there is no Archimedean *fundamentum* of our transcendental world-constitution.

In "McDowell's Pragmatist Anti-anti-realism" (Chapter 14), Eirik Julius Risberg re-examines the debate between Rorty, Davidson, McDowell, and now joining them, the 'New Pragmatists', about the character and status of *objectivity* in a philosophy cleansed of the dualism of conceptual scheme and empirical content. Rorty has long argued that any supposed answerability of our words to a world beyond the linguistic community is fundamentally misguided. The New Pragmatists contend instead that the notion of objectivity is not inimical to pragmatism. In particular, Ramberg argues, against Rorty, that Davidson's insistence upon the irreducibility of the intentional marks a *post-ontological* distinction between the intentional and the non-intentional, which provides for our intentional thought to be answerable to the non-intentional, in a way compatible with pragmatism. Rorty has accepted Ramberg's criticism, but still maintains that McDowell's view of the answerability of thought to the world is metaphysical and beyond the pale of pragmatism. Against Rorty, Risberg argues that Davidson's post-ontological distinction between the intentional and the non-intentional, and Rorty's accepting that distinction, suffice to show that McDowell's "anti-anti-realism" belongs within the pragmatist fold. Although McDowell and Davidson disagree about the boundary between the intentional "space of reasons" and the non-intentional "space of nature," McDowell's distinction is as post-ontological as Davidson's. Consequently, McDowell's view is tantamount to *pragmatist anti-anti-realism*.

Many of these essays originated from the conference, "Realism in Its Multiple Forms: A Case of Mere Homonymy or Identifiable Common Commitments?" (6–9 June, 2011), hosted by the Helsinki Collegium for Advanced Studies. It was sponsored jointly by the Helsinki Collegium for Advanced Studies; by the research project, "The Ethical Grounds of Metaphysics" (Universities of Helsinki and Jyväskylä), funded by the Academy of Finland; by the Nordic Pragmatism Network, funded by NordForsk; and by the Swiss National Science Foundation. All of the contributors, and I in particular, express our gratitude to these sponsors for their manifest confidence in, and concrete support of, our research.

Papers presented there have been substantially revised for the present volume, and several contributions have been specially written for it.

Originally I had proposed to Sami Pihlström that we co-edit this volume. I have consulted him at every step, yet by happy coincidence I handled the editing: I had the time, whereas Sami was busy directing the Helsinki Collegium for Advanced Studies. More significantly, the contributors have all been wonderful collaborators; hence no problems arose which required extra brain-storming. I wish to thank each contributor for his fine contribution and exemplary cooperation; my special thanks are to Sami for his thoughts, advice, and assistance. I believe all the contributors join me in thanking Sami very warmly for having organized the very successful conference which launched this project, and in thanking the sponsors who made that conference possible, and hence this volume too. Last though not at all least, I wish to thank the Helsinki Collegium for Advanced Studies for its warm hospitality and ideal working conditions, where as a (former) member of its Academic Advisory Board I was allowed to spend the final quarter of 2011, when most of the editorial work on this volume was undertaken. The editorial and production staff at Routledge have been thoroughly professional, the very model of integrity in academic publishing, for which we are all very grateful indeed.

Part I

Realism Contextualized

1 What Is Real(ism)?

Jaakko Hintikka

According to a humorous story, a distinguished diplomat was once embarrassed when a lady sitting next to him at a festive dinner asked: "What do you do? What is your profession?" "I am a diplomat." "So what kind of a diploma do you have?" The distinguished diplomat could not think of a diplomatic answer.

In this paper I am trying to help my fellow realists to avoid a similar embarrassment, should a charming lady (or a charming gentleman) ask them at a dinner table, "So how can you tell what is real?" Needless to say, a philosophical realist cannot try to tell what reality consists of or what laws govern it in the same way as a scientist. But we can and should be clear what we are talking about when we speak of reality and of realism.

So, what is real? What does 'real' mean? The first answer that a dictionary is likely to give you is that what exists or what is actual is real. But it is far from clear what the force of the word 'exist' is that is presupposed in the *OED* or some other dictionary. The concept of existence applies most naturally to things or objects, in contrast to properties, relations, functional dependencies, structures, and even sets. What does it mean for instance for a relation to exist? A relation can be said really to obtain between two objects, in contrast to it merely being assumed, believed, or imagined to obtain. But is there a separate use for saying that the relation itself exists as a separate entity? Should we perhaps think that abstract objects like structures and sets exist as separate entities in logically the same sense as particular objects, but only in some non-sensible region of the universe, in order to be able to say that they are real? Or is it that by speaking of the existence of properties and relations we in fact mean their *being* in some wider sense? If so, how are the different senses of being related to each other?

The genesis of modern symbolic logic and of contemporary analytic philosophy involves a resolute attempt to sort out these questions. With his usual modesty, Russell called his enterprise the first real progress of logic since the days of the Greeks. One of its aspects was the interpretation (usually and not unfairly attributed to Frege and Russell) of the different uses of verbs for being as instantiating semantically different meanings of the verb: the 'is' of identity, of existence, of predication and of subsumption.

Verbs for being are ambiguous between these logically different meanings. Although historians do not emphasize the fact, this ambiguity thesis already was a radical novelty. No major thinker before the nineteenth century had maintained it.

The Frege-Russell ambiguity thesis implies that all questions of existence *in any strict sense* must be asked in terms of the existential quantifier. If reality is identified with what exists, the logic of reality—and of the metaphysics and the ontology it presupposes—becomes the logic of the existential quantifier. This vision is epitomized in Quine's question "What is there?" with his one-word answer "Everything." This answer is not a tautology. It is an apt summary of the Frege-Russell metaphysics of their new logic.

On their construal of the logic of existence, all use of the existential quantifier presupposes a universe of discourse, a range of objects from which the values of quantifiers (quantified variables) are chosen. Admittedly, in ordinary discourse all our quantifiers seem to be relative ones, restricted to some specified class of values. This is reflected in the fact that in natural language we normally use quantifier phrases (Russell called them "denoting phrases") like "every woman," "some mother," "any child," rather than absolute unrestricted quantifiers. What Russell tried to do was to replace the logic and metaphysics of denoting phrases by a logic and metaphysics of quantifiers. The manifesto of the new metaphysics was Russell's paper, "On Denoting." Its more accurate title could have been "On Not Denoting," for its message is a paraphrase of all denoting phrases (relativized quantifiers) in terms of absolute quantifiers.

Thus the picture of reality implicit in the Frege-Russell construal of the logic they discovered is like a democratic society consisting of a number of citizens that are at least logically all at the same level. Moreover, the logic developed on the basis of this vision can be complete in the sense of spelling out all conceptual truths only if all the individuals are logically independent of each other. No 'hidden variables' may induce dependencies between them.

This seems to presuppose a metaphysics of exclusive reality of particular objects. An object is real if it exists, and it exists if it is a member of our universe of discourse (or a construct from these members). But what about properties, relations, sets, and all such? The first answer that Frege and Russell gave was simple: they are all objects, too. Wittgenstein followed their example. It was only later that Wittgenstein found that he could not assume such complete independence of basic (simple) objects, even though he had done so in the *Tractatus,* and was gradually led to an altogether different metaphysics.

This notion of object is not obviously reconcilable with the parity and independence of basic objects. Frege's assumption that higher-order objects could be dealt with logically by means of quantifiers for individual variables led to the crash of his system. In order to avoid paradoxes, Russell was led to his theory of types with its hierarchy of logically different objects. In the end the deep and ultimately unsolvable difficulties turned out to be not so much homogeneity of the universe of discourse or the independence of its members, as too narrow a view of the mode of operation of quantifiers.

Meanwhile, indeed much earlier, the Frege-Russell metaphysics (as we may call it) led them (and other philosophers) to a wealth of new questions. The most important was of course: What is the universe of discourse of the language that we actually use? What are the basic objects? The striking preoccupation of the Russell of *Our Knowledge of the External World* and of the Wittgenstein of the *Tractatus* with this question is a direct consequence of their new metaphysics. Frege, with his more limited aims, raised the same question only of the objects of mathematics: What are numbers? It is because of the Frege-Russell metaphysics that the question of realism in mathematics becomes important for philosophers, if not to mathematicians.

All three nonetheless shared basically the same idea. What is real was for them what exists, and what exists is the universe of discourse with its objects. And this universe is the totality of actually existing objects. There is nothing else. Even in logic, we are dealing with the actual world and actually existing objects. In Russell's simile, we are doing so as fully in logic as in zoology, even if in logic the objects in question are more abstract than sundry beasts. There are no other possible worlds. Later the philosophically timid Gödel still shared most of this metaphysics.

This actualism, this restriction of what is real to what exists in the actual world, confused the issue in the foundations of mathematics. Instead of asking what the mode of existence is that mathematical objects enjoy, philosophers thought as if the only questions concerned whether they reside in some Platonic region of the actual world.

Slowly, the limitations of this Frege-Russell metaphysics became apparent. The change is often presented as a revival of the Leibnizian metaphysics of possible worlds. Purely logically (structurally) there is indeed an actual similarity between Leibniz's metaphysics and the possible-worlds semantics (PWS) that was developed in the second half of the twentieth century. However, the real rationale of PWS is different, even though it has not been emphasized in philosophical discussion (some of the pioneers of PWS were motivated by it). PWS spells out the way in which our language and its logic are applied to the world. This way, as Quine has remarked, is not unlike the way in which a scientific theory is actually applied. Outside cosmology, a physical theory is not applied to 'the world'. It is applied to some 'system', which can be some part of the actual universe, some counterfactual nook and corner of it, or maybe a possible scenario. The only requirement is that it has to be so isolated conceptually from the rest of the world so that we can apply the theory in question to it without having to consider what is outside it. This isolation may be physical or it may be reached by abstraction, by disregarding some aspects of the 'system' or, most interestingly, it can be isolated *ad hoc* by specifying its boundary conditions (including initial conditions). Theoretical statisticians and probability theorists are happy to speak of such systems modestly as 'small worlds', but philosophers like to think of these systems as altogether independent 'possible worlds'. This is not necessarily harmful as long as we remember that what is meant is merely one possible application of our language and logic. The point is that such an application has to be considered logically as if it were an entire self-sufficient use of the language and its logic.

A central part of any real possible-worlds semantics would have to be a thorough examination of the logic and epistemology of the nature and the mode of selection of boundary conditions in science. This is an enormously important topic in other respects, too. For instance, no full, general scientific theory has observable consequences except in conjunction with some boundary conditions (including initial conditions). Yet no full-fledged study of that kind seems ever to have been carried out.

In the 'possible-worlds' jargon this self-sufficiency means that such possible scenarios have to be considered epistemologically and semantically as self-contained alternatives to the actual course of events. We can only come to know what the actual world is like, and if we were denizens of another 'world', we could only know come to know what that world is like. For instance, we could not know what other possible worlds it might be an alternative to.

This in turn means that no viable concept that we can significantly use can depend on comparisons between different possible worlds, for we could never know when such a concept is applicable in the actual world. This makes nonsense of much of contemporary modal metaphysics. The only possible uses of that concept and of theories which rely on world-to-world comparisons are in theology, for God presumably could (according to some creeds) survey different possible worlds and compare them freely with each other.

This alone does not seem to require major modifications in a Frege-Russell type metaphysics. Each possible world can have its universe of discourse over which its quantifiers range. We merely have to relativize them to a 'world' (in logician's jargon, to a model).

But this forces us to face a problem. Alternative scenarios can obviously share the same objects. But if we cannot compare them, how can we know who's who in them? How can we as much as speak of the same objects as sometimes recurring in different possible worlds? This conundrum is deepened by the fact that individual objects cannot for logical reasons be defined descriptively (in terms of predicates) without distorting their logical behavior.

The answer is straightforward but seems not to have reached most philosophers' consciousness. Cross-world identification is possible only between worlds that exhibit sufficient structural similarity. That shared structure can on suitable assumptions—Kant would have called them (and perhaps did call them) 'transcendental' assumptions—serve as a framework in which to identify objects, so to speak place them on a conceptual 'map' or 'chart'. This map can be chosen differently for different purposes. In a simple case it can even be a geographical framework, map in the literal sense. I have used as an illustration old, but still widely used, Finnish family names which could identify their bearers as people who lived on a river (Virtanen), an island (Saarinen) or a peninsula (Nieminen), to mention only some of the most common ones. My present assistant's family was once identified as living at the end of the village (Kylänpää). In our present day and age, we of course use a variety of different frameworks of identification.

Thus possible-worlds semantics does not seem to force abandoning or even severely modifying the Frege-Russell semantics. Formal manifestation of this *status quo* might be the possibility—which does not seem to be widely advertised—of developing an epistemic logic simply as a kind of two-sorted first-order logic with an operator. In it "it is known that *S*" would be expressed by *KS* and known references to known (identifiable) objects could be distinguished simply typographically, for instance by using boldface letters. The central novelty would be that this distinction would have to be made also for quantifiers. The result is a perfectly viable simple quantified epistemic logic. There does not seem to be any need to make any changes in our metaphysics.

But this is not the last word on the subject. What does it for instance mean in terms of an epistemic logic that it is known who *b* is? Clearly $(\exists x)\,\mathrm{K}(b = x)$. And what this means is that one can find the slot for *b* in every scenario compatible with what is known. But must that slot be occupied? Must *b* exist in all those possible worlds? No. On the contrary the possibility of being able to say that *b* does not exist in one them presupposes that we can in principle locate the right slot but find it empty. Hence the object that is the value of *x* need not exist in all possible scenarios compatible with what is known.

By the parity of cases, that value of *x* need not exist in the actual world, either. Hence the range of values of *x*, unlike that of *x*, may contain merely possible (epistemically possible) objects. If to exist is to be a value of a bounded variable, we must acknowledge some objects to exist only possibly but not actually. Actuality and existence do not go together. Accordingly, if existence and reality coincide, reality and actuality do not. Thus a drop of logic has exploded into a metaphysical insight.

But is the logic involved here worthy of the name? Is it real logic? Can we even consistently imagine that someone uses a logic with some quantifiers ranging over merely possible objects, or more accurately, over known (identifiable) objects? Yes, not only can we consistently imagine someone doing so. Actual someones have done so in the history of logic and philosophy.

The best-known such someone is Aristotle. He considers in effect (in *Met.* Γ 3) the Frege-Russell ambiguity thesis, but rejects it. The allegedly different Frege-Russell senses of 'is' (or should I say 'εστίν'?) are merely components of one single meaning. Not all of these components need to be present on every occasion when we speak of being. But one of them is indispensable. It is the 'is' of identity, which Aristotle takes to be the 'is' of identifiability, not plain identity. Here he was in fact more perceptive than Frege. Frege's distinction was in fact defective in that he only considered plain-vanilla identities ($a = b$), not the uses of identity for identification. ("You *are* Dr. Livingstone, I presume.") Aristotle spoke of such identities in terms of an object being one, or as we would say, "one and the same." The underlying idea is obvious and at least apparently persuasive. Of course, whatever we say, we must know who or what we are talking about.

Hence when Aristotle in his logic speaks of all *A*'s, he does not mean all actually existing *A*'s, he means all possible identifiable objects that are *A*'s. "All poets are eloquent" implies (together with "Homer is a poet") that "Homer is eloquent" even though Homer does not exist.

From this meaning of Aristotle's quantifiers it follows that he could not express actual existence by means of quantifiers. For him to be was not to be a value of a bound variable. Nor could he have answered the question "what exists?" by answering "everything." I have elsewhere examined how existence is expressed within an axiomatic Aristotelian science. Assumptions of existence enter such a science, not with the help of an existential quantifier, but as a part of the force of certain predicate terms.

This idea was eventually the undoing of Aristotelian logic. When Kant said that "existence is not a predicate," he was not maintaining anything that Aristotle would not have at least verbally agreed upon. What killed Aristotelian logic was Kant's argument that existence cannot even be part of the force of a predicate. This made the all-important notion of (actual) existence homeless in logic until De Morgan, Russell, and others found for it a more-or-less natural place in the membership in a universe of discourse.

This epistemic element in Aristotle's logic and metaphysics should not be surprising in view of his historical circumstances. Aristotle's logic arose from the Socratic *elenchus,* his method of questioning, via the dialectical games that were practiced in Plato's academy. Hence behind Aristotle's logic there lurks historically and systematically the logic of interrogation, the logic of questions and answers. And this 'erotetic' logic is unproblematically nothing but a branch of epistemic logic.

Furthermore, it has turned out that there was a major epistemic element in Greek mathematics and the modes of reasoning employed there. In the old-fashioned axiomatic geometry that was still taught in European schools in the twentieth century, an important part consisted of questions of what was 'given'. Euclid wrote an entire book about what is 'given' or 'data', devoted to questions like "if a so-and-so geometrical object or configuration is given, what else is given?" As is eminently clear and acknowledged later by some translators, this 'given' meant simply 'known'. It did not mean 'existent' and it did mean any person's subjective knowledge. It was a term of epistemic logic.

In a way, a committed Aristotelian metaphysician might even claim to have been indirectly vindicated by the history of analytical philosophy: For when Russell seriously tried to answer the question of what our own universe of discourse consists of, he wound up identifying the ultimate simple objects with the objects of acquaintance. According to Russell's own words, they are the entities of which I know who or what they are. Unlike Aristotle, Russell thought of them as a subclass of actually existing objects; but this does not change their status as known objects in the sense of epistemic logic. Aristotle had his revenge.

All told, we have reached a massive vindication of possible objects as an important aspect of the reality with which our knowledge deals. Reality does not presuppose actuality. This result is thrown into an even sharper profile by the realization that the epistemic objects (objects of knowledge), even though totally 'objective' in the sense of mind-independent and intersubjective, are constituted by frameworks of identification mentioned earlier. In our actual conceptual practice, we operate all the time with the two kinds of epistemic objects which I have called perspectival (*e.g.* perceptual) and public. Some philosophers, Russell included, have mistakenly acknowledged only perspectival objects. Others consider only a public identification as fundamental. They, too, are wrong. What is more, there are yet other ways, so far unexploited, to develop further the metaphysical idea of a possible object. For all their wild and woolly speculations about alien worlds and their denizens, about transworld identities and worldbound individuals, modal metaphysicians have been timid in dealing with the notion of a possible object. The possibilities considered by them are propositional possibilities of states of affairs or courses of events.

Possible objects are typically objects that would be actual in some other possible world. What if there were possible but unrealized objects happily residing in our actual world? Logically, this would not be any stranger than assuming in epistemic logic that there are identifiable objects of which we do not know that they happen not to exist. Surely it might conceivably turn out that there never was a Homer even if we speak of him legitimately as a definite person.

What happens if we liberate our speculations about possible objects from the framework of possible worlds? What if there are in the actual world virtual objects that are just like the actual ones except they do not exist? Their behavior might conceivably even have observable consequences. That would finalize the divorce of reality and actuality.

Is the idea of full-fledged metaphysically possible but nonfactual objects completely unrealistic speculation? Suppose we are as foolhardy as to drop the prefix 'meta' from our speculations. What would physical counterparts to the metaphysical speculations presented here look like?

Now, possible-worlds theories do in fact exist in quantum physics in the form of the so-called many-worlds interpretations. But what could a possible-object quantum theory be like? Again such theories have in fact been proposed, even though they have not played a major role. If Bohr had not discouraged Kramers and Slater around 1924, they might have presented a full-fledged interpretation of an electron as a collection of virtual oscillators, one for each possible state of an electron.

I strongly suspect that such possible-object interpretations could have a great future, not only in metaphysics, but in physics. Perhaps we already have experimental evidence for possible but unactualized objects. In a Bohm-Aharonov experiment, an actually observed electron is influenced by a magnetic field from which it is isolated by an impenetrable copper tube. How come? An actual electron cannot penetrate the boundary, but a virtual one can. Hence if the observed electron is 'really' a bunch of virtual ones, some of them can be located inside the tube. Hence the observed electron can 'feel' a field, even from the other side. Is the Bohm-Aharonov phenomenon thus perhaps an exercise in experimental metaphysics?

Whatever the answer ultimately should turn out to be, we have here an intriguing guess as to where we should look for an answer to the question: "What is real?" And isn't that what a realist should be asking?

POSTSCRIPT

The aim of this paper is not to convince the reader once and for all that the ideas and interpretations put forward here are the last word on the subject. What I have tried to do is to summarize some recent results and to bring out connections between several different lines of thought and interpretations that I have pursued. The relevant writings of mine are included below.

REFERENCES

Hintikka, Jaakko, 1981. "Kant on Existence, Predication and the Ontological Argument." *Dialectica* 35.1: 127–146.

———, and Merrill B. Hintikka, 1986. *Investigating Wittgenstein.* Blackwell, London.

———, 1999. "On Aristotle's Notion of Existence." *The Review of Metaphysics,* 52.4: 779–805.

———, 2005. "On the Different Identities of Identity: A Historical and Critical Essay." *Language, Meaning, Interpretation* 2: 117–139.

———, 2006. "Ta Meta Ta Metaphysika: The Argumentative Structure of Aristotle's *Metaphysics.*" In: V. Hirvonen, *et al,* eds., *Mind and Modality: Studies in the History of Philosophy in Honour of Simo Knuuttila* (Leiden, Brill), 41–55.

———, and Risto Vilkko, 2006. "Kant's Influence on the Development of Modern Logic." In: H. Lenk and R. Wiehl, eds., *Kant Today—Kant Aujourd'hui—Kant Heute.* (Münster, LIT Verlag), 112–126.

———, 2007. "It All Depends on What 'Is' Is: A Brief History (and Theory) of Being." In: Jon Burmeister and M. Sentesy, eds., *On Language: Analytic, Continental and Historical Contributions* (Newcastle upon Tyne, Cambridge Scholars Publishing), 51–64.

———, and Risto Vilkko, 2007. "Existence and Predication from Aristotle to Frege." *Philosophy and Phenomenological Research* 73.2: 359–377.

———, (forthcoming). "On the Ensemble Interpretation of Quantum Mechanics."

2 Aristotle's Direct Realism and Some Later Developments

Mika Perälä

1 INTRODUCTION

A student of Aristotle's *De Anima* faces various challenges in making sense of its discussion of perception: How is this discussion related to Aristotle's general explanatory project concerning the affections of the soul? What, exactly, is the account of perception, and how far does it extend? Is it an attempt to explain all types of perception, including misperception and even hallucination? How could we perspicuously contrast it with later theories? And why should one take it seriously in the first place?

Many scholarly disagreements are fundamental, and perhaps irreconcilable, because exegetical and philosophical puzzles are tightly intertwined.[1] First and foremost, Aristotle's characterization of perception as the reception of perceptible form without matter (*De Anima* 2.12, 424^{a}17–19) has given rise to different interpretations that are based on rather different intuitions about what makes an account of perception a plausible philosophical theory worth attributing to Aristotle. Many interpreters take Aristotle to be committed to direct realism, the claim that we are directly aware of real, mind-independent objects. In their interpretation, the received form is to be understood as a sort of likeness, or, as some say, representation, that links us to the external object: the likeness in question is not an object of awareness, not at least in typical cases of perception, but something by which we perceive an external object (Modrak 1987; Wedin 1988; Burnyeat 1992, 2001; Shields 1997). However, there are others who consider Aristotle as an indirect realist: according to them, Aristotle would take us to perceive external objects by perceiving and being aware of their effects upon our sense organs. In this interpretation, the sense organs take on the form of an object 'literally', *e.g.* by turning red or smelly, and we perceive an external object by being aware of this change in them (Sorabji 1991, 2001; Everson 1997).

Against this background, it is rather difficult to see a way forward. In search of resolution, some scholars have tried to take a position that steers clear of the excesses of earlier interpretations (*e.g.*, Caston 2005, Stein 2009, Lorenz 2007). My approach in this paper is somewhat different. I shall rethink Aristotle's theory of perception as part of his explanatory project in general, and set it into a historical perspective, including the philosophers of perception Aristotle explicitly criticizes. I shall also contrast Aristotle's view with two interpretations that developed some of Aristotle's insights to entirely new directions: one by Thomas Aquinas and the other by Duns Scotus. An aim of this paper is to clarify and illuminate the distinctive features of Aristotle's position as opposed to these other theories.

As I see it, the most fundamental question about perception Aristotle addressed was this: what makes a perception the perception it is? This question asks for the definition of perception, which, in Aristotle's theory, requires a reference to the four different types of cause as distinguished in the *Metaphysics* and *Physics,* but also in the *De Anima* (see Bolton 2005): material, formal, efficient, and final. The material cause of a perception is to be identified with the sense organ and its capacity for perceiving, and the final cause with self-preservation (*De Sensu* 1, $436^{b}19$–$437^{a}1$). What particularly interests us for the present are the efficient and formal causes, which are generally taken to constitute the cause and content of perception. How are the two related?

An answer to this question, I shall argue, can be found in Aristotle's peculiar model of causation introduced in *Generation and Corruption* 1.7 and *Physics* 3.3, and applied to perception in the *De Anima.* This model suggests that the content of perception is essentially dependent upon the cause of perception. I shall draw attention to two characteristics of the model. First, when the sense is acted upon by the perceptible object, *i.e.,* it receives the perceptible form, two different types of activity occur in the sense: the activity of the perceptible object and the activity of the sense. Alternatively, and using Aristotle's own distinction between being and number, the two could be seen as two different aspects, or beings, of one and the same activity because they are correlatives: the activity of the perceptible object consists in the realization of the object's perceptibility to the sense, while the activity of the sense amounts to the sense's being sensitive to what is perceptible. In contrast to what many interpreters assume, this model need not commit Aristotle to any specific view about the received form. In particular, Aristotle, unlike Aquinas, did not postulate two different ways of being a form, one material and the other spiritual or intentional, in order to explain the difference between ordinary qualitative change such as becoming hot, and perceiving hot. Second, in Aristotle's view, perception's being relative to what is perceptible is, as I shall argue, explained by the fact that perception is by definition caused by the perceptible object. On this point, Aristotle can be seen to be in agreement with many later philosophers, including Thomas Aquinas, but it is instructive to contrast this view with Duns Scotus's denial of it, which was based on an interpretation of Aristotle's account of the different types of relative in *Metaphysics* 5.15. Scotus's position is significant because it separated the cause of perception (and thought) from the content of perception (and thought) in a way that offered an explanation of the possibility of certain radical scenarios: God's or the Devil's intervention in the causal chain between an object and our perception (or thought) of it.

I shall argue that Aristotle is a direct realist about sense perception. In contrast to the other interpreters with this tendency, however, I do not assume that Aristotle took the received form as some sort of representation of a mind-independent object. Instead, I interpret the received form simply as the realization, *i.e.* activity, of a mind-independent perceptible form in the perceptual capacity, claiming that for Aristotle, there is no perception of a mind-independent object beyond the receptivity of this activity. My focus is on perceptions concerning what Aristotle refers to as the proper perceptibles, *i.e.* perceptibles that are specific to each single sense: color, sound, smell, taste, and the objects of touch. Aristotle extended his account to perceptions concerning the common perceptibles, including shape, size, number, movement, and rest, which are equally perceived by any one of the single senses, but this extension raises further questions that need not concern us for the present. I shall chiefly discuss veridical perception because Aristotle assumes that perceptions of the proper perceptibles are always, or for the most part, true. I shall briefly remark on the cases of misperceiving the proper perceptibles Aristotle examined in the *De Sensu,* but ignore others that seem to require a more complex account.

This paper is divided into three main parts. First, I give an overview of Aristotle's explanatory project as it is introduced in *De Anima* 1.1. Second, I present the main elements of Aristotle's account of perception in the light of *De Anima* 2.5 and 2.12, contrasting it with Aquinas's understanding of the text. Third, I elaborate on Aristotle's account by examining his discussion of perception as a relative in the light of *De Anima* 2.6 and *Metaphysics* 4.5 and 5.15. In this connection, I shall refer to Scotus's interpretation of the distinctions between the different types of relative as presented in *Metaphysics* 5.15 and point out that Aristotle did not distinguish between causal relatives and terminal relatives in the same way as Scotus. I conclude by raising an objection to Aristotle's view and giving a tentative response.

2 DEFINING THE AFFECTIONS OF THE SOUL

Aristotle sets out in *De Anima* 1.1 to define affections of the soul. By 'affection' (πάθος), he refers to any kind of states or activities of the soul such as anger, desire, and perception. He suggests that the definitions to be given should be of the same general type as the definition of becoming angry:

> Becoming angry is [*i*] a kind of change [*ii*] belonging to such-and-such a body or part of the body or capacity [*iii*] due to a certain thing [*iv*] for the sake of something. (*De An.* 403^{a}26–27)

This definition consists of four parts that express the four different types of cause we are familiar with from *Physics* 2.3 and *Metaphysics* 5.2: [*i*] formal,

[*ii*] material, [*iii*] efficient, and [*iv*] final. In analogy with becoming angry, seeing would be defined as follows: seeing an x is [*i*] a kind of change (*i.e.* reception of a form) [*ii*] belonging to the eyes' capacity for seeing [*iii*] due to y [*iv*] for the preservation of animal. Setting aside the material and final causes, the definition can be reformulated thus: seeing an x is [*i*] the type of change [*iii*] that is caused by y (*cf.* Charles 2010, 287).

This formulation makes it clear that the type of change is specified with reference to its cause y. We cannot give a definition of seeing purely in terms of form and matter, or [*i*] and [*ii*], because the form in question is indeterminate without reference to the cause that gives rise to the form. Aristotle gave a very helpful example in *Metaphysics* 8.4:

> Regarding the form, the cause is the account, but the account is unclear unless it is accompanied by its cause. What is an eclipse, for example? 'Loss of light'. However, if one adds 'due to the Earth intervening', this is the account together with the cause. (1044^{b}12–15)

Aristotle here defines lunar eclipse and states that its formal cause is given by an account (λόγος). The account is a linguistic description of eclipse, and the idea is that it helps us to separate eclipse from every other entity. "Loss of light" is a good first approximation, but the problem is that the phenomenon referred to can be due to a variety of causes, for example, due to the moon's surface turning black and thus becoming non-reflective. Therefore it is a substantial addition that the loss of light is related to the Earth. Without reference to the Earth, the eclipse remains indeterminate, and the corresponding account of it is unclear (ἄδηλος) (see Charles 2000, 37 and 293; Bolton 2005, 212–3). I suppose that this is true of all cases (*cf. Met.* 7.17, 1041^{a}26–30).

This enables us to distinguish between the different types of seeing in terms of their causes. Now one might think, and many have thought, that one of Aristotle's fundamental theses was that in optimal conditions, the object of seeing is its cause, *i.e.* $x = y$. For example, seeing the white is [*i*] the type of change [*iii*] that is caused by the white.

This is not, however, a complete definition, because we need to specify in more detail what constitutes the efficient causal agent here. In Aristotle's view, the white can be either determinate (ὡρισμένον) or indeterminate (ἀόριστον) depending on where it occurs. For example, when the white is in the air or water, it is indeterminate. This means that it looks different when one perceives it close by or from a distance. By contrast, the white that is inherent in a solid body is determinate. So the environment or context (τὸ περιέχον) determines which type of color is present (*De Sensu* 3, 439^{b}1–6). This raises a further question: what happens to the identity of the white inhering in the milk when it acts upon the perceptual capacity through the medium? Does it lose something of its determinacy, for example, in poor lightning? I shall not elaborate on this question for the present though I wish to emphasize that, in Aristotle's view, no quality exerts causal power on its own: it is only on account of its being a property of a substance that it is causally effective. Thus understood, the proper efficient cause of seeing the white is that which is white (*i.e.* the white thing) and, as we shall see, it possesses this power, not just in virtue of being white, but in virtue of being visible, a capacity that is based on its being white.

Here Aristotle seems to be referring by 'environment' primarily to the medium, but the way in which he proceeded suggests that the environment may include another object that lies close by. This enabled him to explain a variety of cases that we would refer to as visual illusions. For example:

> It is possible that the white and the black are juxtaposed so closely that either separately would be invisible due to their minor distance, though they would be jointly visible. This [compound] could appear neither white nor black, and, as it must have some color, and can have neither of these, this color must be something mixed—in fact, a species of color different from either. (*De Sensu* 3, 439^b19–25)

Aristotle considered this as a reason for taking colors as different mixtures of white and black. In this case, the invisibility is due to small size. Further on (440^a21–23), he referred to a case in which the invisibility arises from a very small temporal distance between two movements or changes (as involved, for example, in the so-called Wagon Wheel Effect). What interests us is the fact that there may be two different types of seeing the grey that differ in causal terms:

> [A] Seeing the grey is [*i*] the type of change [*iii*] that is caused by a grey color patch [*v*] under optimal circumstances (*i.e.,* at an optimal distance, the medium being entirely transparent *etc.*)

> [B] Seeing the grey is [*i*] the type of change [*iii*] that is jointly caused by white and black color patches [*v*] under special circumstances (*e.g.,* the patches being far from the perceiver, and close to one another).

It is worth noting that [A] and [B] need not differ in subjective terms: they are not necessarily subjectively distinguishable. Condition [*v*] determines the circumstances in which our perceptual capacity is supposed to be acted upon. It is evident that a corresponding condition concerning circumstantial factors needs to be added to all definitions of perceiving.

I shall not extend my discussion to more complex types of seeing such as seeing the white and square, or seeing the white and black, or seeing the white Socrates. These would require a separate study (on such an extension, see *e.g.,* Charles 2000, 124–129). At this point, however, it is worthwhile to make two clarificatory remarks.

First, the type of efficient cause that is required in a definition is what Aristotle refers to as a *per se* cause. Only a *per se* cause can explain what makes a certain thing the thing it is. Each thing has innumerable accidental causes which do not have explanatory power irrespective of a *per se* cause (*Physics* 2.5, 196^{b}24–29). In the case of seeing, *per se* causes include colors, shapes, *etc.*, that have the power of being perceived (*De An.* 2.6, 418^{a}8–11).

Second, a philosophically informed reader of Aristotle might ask whether the perceptual contents in [A] and [B] are the same or different. This is an intriguing question. In order to give an adequate answer, we should determine first what would constitute perceptual content for Aristotle. Is it the received form? Is it something that is conceptually different from the activity of the sense? Philosophers of perception are accustomed to distinguishing between a mental state and its content, but Aristotle worked in a different conceptual framework. Without committing myself to any specific view about perceptual content in Aristotle, let me state something that I think would help us clarify Aristotle's view in contemporary terms, and it is this: Perception for Aristotle is an experience (or mental state) whose intrinsic character is ontologically dependent upon the states of its physical environment external to the senses, including perceptible objects. Thus understood, definitions [A] and [B] single out two different types of seeing whose intrinsic character varies because they are ontologically dependent upon different states of physical environment. If this is correct, Aristotle need not admit that [A] and [B] require something which could be referred to as the common content or common factor (see also Charles 2000, 120). In contemporary philosophy of perception, this notion of content is used to explain what is common to veridical and non-veridical mental states (*e.g.*, Lowe 2008, 99), but Aristotle was presumably unfamiliar with it. In his view, as I see it, a difference in the types of cause implies a difference in the types of change involved in perception, *i.e.*, the reception of form, regardless of whether we can discern the difference. This follows from his idea that a definition of perception consists in an explanation of why the perception in question arises, and this explanation must include causal factors that are external to the perceiver.

3 PERCEPTION AS THE EXERCISE OF THE CAPACITY FOR PERCEIVING

Let us examine the relationship between the efficient and formal causes in more detail. In *De Anima* 2.5 Aristotle introduced a general model to give an account of this relation. Before going into the text, let me mention as a terminological point that Aristotle used the Greek term αἴσθησις for both the act of perceiving, in short, perception, and the capacity for perceiving, *i.e.*, the sense. The context determines which is signified in each case. Another term for the sense (and occasionally for the sense organ) is τὸ αἰσθητικόν, 'that which is capable of perceiving', and the term for the perceptible object τὸ αἰσθητόν, 'that which is perceptible'.

3.1 *De Anima* 2.5

Aristotle agreed with his predecessors that perception (αἴσθησις) seems to be a type of change, but he criticized them for failing to make the necessary distinctions. They had argued that perceiving is to be acted upon by the like (416^b35). As was typical, Aristotle admitted that this is true in one sense, but false in another one. His objection was that what is to be acted upon is unlike, but when it is acted upon, it is like what acts upon it. His predecessors had missed this point, Aristotle argued, because they had failed to make an important distinction between the sense's capacity or power (δύναμις) for perceiving, and the exercise or activity (ἐνέργεια) of this power, in short, the sense-in-capacity and the sense-in-activity (417^a12–13). The same distinction, according to Aristotle, applies to the perceptible objects (417^a13–14). His point was this: it is one thing for the object to have a power for being perceived, and another thing for it to exercise this power. When the object has the power but does not exercise it, the object is perceptible but not perceived, *i.e.,* the perceptible-in-capacity, whereas when the object uses its power on the sense, the object *is* perceived by the sense and it is a perceptible-in-activity.

Aristotle raised another objection to his predecessors, and it was this: perception is not a change of quality such as the cold espresso pot becoming hot when it is put on a hot stove. This type of change entails that a property is replaced by its contrary (417^b2–3). By contrast, when an object acts upon our capacity for perception, there is nothing in the capacity that would change with respect to quality: no exchange of quality occurs. What happens is a transition from inactivity to activity (417^b3–7). Aristotle likens this type of change to exercising the knowledge of grammar (417^a32–$^b1$). When I am asleep, my knowledge of grammar is out of use. But when I wake up and start using my knowledge, no change in my grammatical skills needs to have happened. I merely begin to articulate myself.

The distinctions Aristotle made were a response to some problems he saw in his predecessors' work. This is important for the understanding of the nature of Aristotle's theory. A major part of his project was to revise the anti-realist account of perceptible objects that he attributed to the Megarians and to Protagoras. On this account, "neither cold nor hot nor sweet nor in general any perceptible object will be when it is not being perceived" (*Met.* 9.3, 1047^a4; see also *De An.* 3.2, 426^a20–22). The Megarians had based their view on a certain account of modality: only that which is actual is possible. So objects are perceptible only when they are actually perceived. Protagoras's view was somewhat different as it arose from his doctrine that "man is the measure of all things." In addition to this doctrine, Plato attributed in the *Theaetetus* to Protagoras another doctrine, the so-called Secret Doctrine:

> . . . everything is really change, and there is nothing but change. Change has two forms, each an infinite multitude, but the one having the capacity for acting upon, and the other for being acted upon. And through the intercourse and mutual friction of these two there comes to be an offspring infinite in multitude but always twin births, on the one hand that which is perceptible, on the other, the perception, the perception in every case being generated together with what is perceptible and emerging along with it. (156a5–b2)

Aristotle could not accept the idea that a perceptible object only exists in relation to a perception. His criticism was that this view fails to distinguish between the object's power for being perceived and the object's exercising this power when it is perceived. Only the latter is dependent upon the exercise of the sense in the way in which the Megareans and Protagoras claimed.

Although Aristotle was highly critical of his predecessors' views, he did not entirely reject the Protagorean understanding of perception, the idea that perception stands in relation to its object. I would even suggest that his understanding of the perception as a relative was a revision of the Protagorean view. Thus understood, Aristotle's view did not arise out of thin air, and there was a model to build upon in his tutor's work.

It is worth noting that in this connection, Aristotle confined himself to talking about the capacity for perceiving. No reference is made to the sense organs that are required for the exercise of this capacity. There is extensive debate about whether the organs need to be qualitatively changed. In some cases at least, namely smell and taste, Aristotle posits such a change: in order to smell something, the organ of smell must become slightly dryer, and in order to taste anything, the tongue has to be moistened (for smell, see *De An.* 2.9, 422a6–7 and *De Sensu* 5; for taste, see *De An.* 2.10, 422^a34-$^b5$; see also Lorenz 2007, 191). This issue about change in sensory organs may be set aside for the present because, as it seems, Aristotle focused on the relationship between the perceptible form and the sense when giving his causal model. This is obvious from the way in which he elaborated the model in *De Anima* 2.12.

3.2 *De Anima* 2.12

In *De Anima* 2.12 Aristotle makes two points about his causal model: first, this model is supposed to explain not only perception of proper perceptibles, but also perceptions concerning more than one proper perceptible; second, the nature of perceiving is characterized in terms of the reception of perceptible forms. As stated in the Introduction, the first point would require a separate study, and I shall not go into it here. Let us focus on the second. Here is the text:

> The sense is that which receives the perceptible forms without matter. Just as the wax receives the design of the signet ring without the iron

> or gold, taking on the gold or bronze design, but not as gold or bronze, so the sense, in each case, is acted upon by that which has color, flavor, or sound, though not insofar as each of these objects is called what it is, but insofar as each is of the aforementioned kind and in accordance with the relevant account. (424ᵃ17–24)

The sense's receiving of the perceptible forms is thus likened to the wax's receiving of the design of the signet ring. As a terminological remark, it is worth noting that what the sense receives are the perceptible forms without matter (τὰ αἰσθητὰ εἴδη ἄνευ τῆς ὕλης). The expression "perceptible" highlights that the sense is receptive of the forms without matter in a qualified sense: it is sensitive to the perceptibility of these forms. This is confirmed by the way in which Aristotle expounds the analogy: the sense is affected not by the object's quality as such, but by its perceptibility. This is how I understand the contrast: "not insofar as each of these qualities is called what it is, but insofar as each is of the aforementioned kind." Thus understood, "the aforementioned kind" (τοιονδί) refers to the power of being perceived. For example, the sense of sight is acted upon by an object's color not insofar as it is a color, but insofar as it exercises its power of being seen. Many interpreters construe Aristotle's contrast in different terms (see *e.g.*, Caston 2005, 305–306; Polansky 2007, 343–344), but I think the suggested interpretation helps us to smoothly link this contrast with some other commitments Aristotle makes, and in particular with his idea of the sense's becoming like the perceptible quality's exercise of its power. I shall discuss this point below (§3.4). It is also worth noting that Aristotle refers to that which receives the forms as the sense (αἴσθησις), not as the sense organ. This shows that he is not discussing whether a perception consists in, or is accompanied with, a change in the sense organ; this is a different question, and though an important one, we may ignore it. Let us first state what, if anything, is uncontroversial, namely the general explanatory model which is supposed to justify the analogy in the first place.

The model is as follows: A change, whether a qualitative change or something else, necessarily requires an agent and a patient. The agent is that which changes something towards having some property *F*, whereas the patient is that which undergoes the change to having *F*. This implies, in Aristotle's view, that the agent has the capacity for acting upon the patient in respect of *F*, while the patient has the capacity for being acted upon by the agent in respect of *F*. In other words, the patient needs to have the capacity for being *F*. So the idea is that what is *F* causes the state being *F* in another thing (see Coope 2007). This is Aristotle's understanding of the "like causes like" principle. Let me note that he places no restriction on how many properties the causal agent may have. This enables him to explain several changes with reference to one agent that acts upon the patient insofar as it has each of those properties. I take it that the agent is an accidental compound (*e.g.*, that which is white, square, *etc.*), though I cannot defend this view in detail here.

A reader of Aristotle asks what explains the difference between ordinary changes such as a change of color, and whatever change, or exercise, constitutes a perception. In response, the difference is not that all ordinary changes are receptions of form *with* matter (*e.g.* changing color through being painted red), because there are changes such as being stamped by a signet ring that do not submit matter to the recipient. What is distinctive of perception as opposed to other receptions of form without matter, I argue, is receptivity of the object's perceptibility. This is a special kind of being informed, and that which is informed here is the sense and not the sense organ.

The qualification "in accordance with the relevant account" (κατὰ τὸν λόγον) is most conveniently rephrased as "in accordance with the account of the various types of perception." Thus understood, it would refer to the fact that each perceptible quality acts upon the perceptual capacity in different ways: colors on sight, sounds on hearing, and so forth. This interpretation would be neutral about the question of whether, and if so, how Aristotle's idea of the qualities as ratios between the extremes (*De Sensu* 6, 445b20–27; 3, 439b25–440a6), and his idea of the sense as a corresponding ratio (*De An.* 2.12, 424a30; 3.2, 426a27–426b7), would enter into the general causal model. This is a challenge for any interpretation of the text, but need not be addressed here.

Commentators have had different ideas about the reception of the perceptible form without matter. It would be beside the point to go into details of different interpretations; what is relevant here is to contrast Aristotle's view with an interpretation that takes a step further than what is explicitly stated in his text.

3.3 Thomas Aquinas on the Reception of Form without Matter

Aquinas expounds *De Anima* 2.12, 424^{a}17–24 as follows:

> . . . [*i*] sometimes the form that the thing affected receives from its agent has the same manner of being in the thing affected and in the agent. This happens when the thing affected has the same disposition for the form that the agent has. For everything is received in another according to the manner of the recipient. So if the thing affected is disposed in the same manner in which the agent is, then the form is received in the thing affected in the same manner as it existed in the agent. In that case, form is not received without matter. . . . [*ii*] Sometimes, on the other hand, a form is received in the thing affected according to a manner of being different from the agent's, because the thing affected's material disposition for receiving is not like the agent's material disposition. In that case, a form is received in the thing affected without matter, insofar as the thing affected is made like the agent with respect to form and not matter. And this is how sense receives form without matter. For the form has a different manner of being in the sense and in the sense object: for in the

sense object it has natural being, whereas in the sense it has intentional and spiritual being. (*In De An.* §§552–553; trans. R. Pasnau)

It is noteworthy that while Aristotle discussed the reception of *perceptible* forms without matter, Aquinas merely refers to the reception of forms without matter. Supposing that he did not neglect to mention 'perceptible' by accident, his concept of intentional being is most aptly seen as an attempt to explain the concept of perceptibility in terms of a certain mode of being a form. In this attempt, Aquinas followed Averroes who had introduced this concept in his great commentary on the *De Anima* (p. 221, 277, 317 Crawford). The motivation behind this attempt was that there must be some explanation for how the object's perceptibility is conveyed from the external object to the sense organ through the medium. Averroes's solution was that the form of the object is present both in the medium and the sense organ in an intentional way (Knuuttila 2008, 13). This was also Aquinas's view: "A spiritual alteration occurs in virtue of a species' being received in a sense organ or the medium in an intentional way" (*De An.* §418 on $418^{b}20$–26). He restated the distinction between the natural and intentional ways of being a form in commenting at §589 on another reference to the reception of form without matter in *De Anima* 3.2, $425^{b}23$–24 thus: "Color has a double being: a natural one in a sensible thing, and a spiritual one in the sense."

In the above citation, Aquinas draws attention to two different types of causal exchange. First [*i*], there are cases in which the patient has the same capacity for having a form as the agent. In those cases, the patient takes on the form with matter. This means that the patient undergoes a qualitative change, for example, cold air is warmed up by hot sunshine. Second [*ii*], there are cases in which the patient lacks the capacity that the agent has. In those cases, the patient receives the form without matter, for example, the sight receives a color without being colored.

Aquinas was usually very modest in his comments on the text. He rarely introduced distinctions that have no clear textual grounds. In this case, the distinction between the two ways of being a form clearly arises from his understanding that the two different types of causation require two different types of being a form. Thus understood, the form's intentional being plays an important explanatory role in the account of perception. The presence of a given form in our soul is clearly that which makes a perception the perception it is, and the presence of another form would give us a different perception. In a different vocabulary, the presence of form gives us the content of perception. And since the presence of form is determined by its cause, it would be reasonable to posit that in Aquinas's interpretation, the cause of perception determines the content of perception.

Aquinas did not refer here to another distinction that he made later on in his commentary (§718) and in his systematic works, and which was commonplace in late mediaeval epistemology, namely the distinction between that which (*id quod*) is perceived or thought and that by which (*id quo*) it is perceived or thought (see *e.g.*, Spruit 1994, Tachau 1988). The significance of the distinction was to stress that the objects of perception and thought need to be distinguished from the means by which we perceive and cognize them. What is perceived is in a body, and what is thought is in the things, whereas the means by which the various objects are grasped are in the soul. Aquinas referred to them as the sensible and intelligible species. It is reasonable to suppose that the idea of species as the perceptual means was at work already here. This can be judged from the following: When glossing Aristotle's distinction between the perceptible and intelligible objects at 2.5, 417^{b}19–29, Aquinas used another term, *similitudo* or likeness, which presumably took over species at this point. He states that the senses "corporeally receive a likeness of the thing being sensed," and that, in contrast, the intellect "incorporeally and immaterially receives a likeness of what it cognizes" (§377; trans. R. Pasnau). Likeness is something which stands for the object of perception and thought: "A thing's likeness, received in the senses, represents that thing as it is singular," and "A likeness received in intellect . . . represents that thing as the defining character of a universal nature" (*ibid.*). Thus understood, likeness is something by which we perceive or think of an object. As a species of the object's form, it represents the form of the object. Since *similitudo,* or likeness, can be given various interpretations, some involving picture, others blueprint or another type of representation (see *e.g.* King 2007), and it is not the task of this paper to determine which fits best Aquinas's understanding, let us refer to it generally as a representational item or, in short, representation.

There is a controversy about what Aquinas precisely meant by likeness and species. A 'standard' interpretation is that it is merely a means by which we apprehend the form of object (*e.g.* Kenny 1993, 35–36; Perler 2000). Another interpretation is that it is the immediate object of awareness: it is by being aware of this item that we perceive and think of objects that are not immediately present to the soul (Pasnau 1997, 195–219, and 2008; Panaccio 2001). In either interpretation, Aquinas's comments seem to be based on the assumption that a perceptual or intellectual act is about its mind-independent object because the perceiver or thinker has a representation of the object within him or her, and the representation is a representation of the object because it is caused by the object, and because it is in some sense isomorphic with the object. For ease of reference, let me call this interpretation the Representation Dependency Thesis: a perception of an *x* is the perception it is because the perceiver has within him a representation of *x* that is ultimately caused by *x*. According to this thesis, representation is assigned an explanatory task. The crucial question for now is whether the thesis succeeds in making sense of Aristotle's view.

3.4 Rejection of Aquinas's Interpretation

My understanding is that Aristotle was not committed to the Representation Dependency Thesis and the underlying idea of the intentional being of form.

I should like to raise two textual concerns about the thesis. First, Aristotle did not state that qualities such as colors would have two kinds of being. His distinctions, as I see them, are rather one between being (*e.g.*) colored and being perceptible (*Met.* 11.9, 1065b32–33), and another between being perceptible and being perceived. An object's being colored is different from its being visible, even if the latter is based on colors or something else, such as fire or fluorescent substance. Second, when Aristotle talks about becoming like, or assimilation, and being like, he refers to the sense, not to the received form. He concludes *De Anima* 2.5 thus:

> The capacity for perceiving, as is stated, is in-capacity (δυνάμει) like the perceptible object is already in-fulfillment (ἐντελεχείᾳ). Therefore, when it is being acted upon, it is not like the object, but having been acted upon, it has been made like it, and it is like it. (418a3–6)

This suggests that Aristotle's understanding of perception is more straightforward than Aquinas's interpretation assumes. It appears that the capacity for perceiving could be assimilated to the perceptible object without any likeness of the object in the soul. To put it differently, the sense seems to assimilate the object directly, using no further means or intermediaries. In the light of this passage, we could understand the reception of a perceptible form without matter as the sense's assimilating the perceptible form in-fulfillment. Here representation plays no alleged explanatory role. What plays that role is the idea that perception is to be understood in relative terms. Perception is about its object because it stands in relation to the perceptible form in-fulfillment. This form, being the term for the perception, is in a metaphysical sense a constituent of perception, and gives that which is perceived.

This is difficult, so let us restate Aristotle's view in terms of his causal model. Consider a colored object. The object's color is perceptible because the object has the power for causing a perception in the capacity for seeing. When the object exercises this capacity, it has a certain property *F*, which changes the capacity for seeing to the state being *F*. Property *F* is the color's being perceptible-in-activity. This is all that perception requires for Aristotle. The object has given the sense the color's being perceived, and this is what the sense has received: there is only one thing. So it would be misleading to state that when we perceive we have an internal likeness that represents an external form by being formally identical with the latter. In addition to this causal model, Aristotle's view can also be restated using his notion of relative thus: when we perceive a color, we are in a state that is relative to the color's being perceptible-in-activity. The term of this relative, "the color's being perceptible-in-activity," refers to something which occurs in the capacity for seeing. It is therefore something internal. On my interpretation, the perceptible object (αἰσθητόν, 'that which is perceptible'), is external only in the sense that the object's power for causing perception is grounded in an external body (*De An.* 2.5, 417b20–21). Thus understood, perception is a certain accident of perceiver. One might think that this is not a realist position, but it is worth stressing that the color's being perceptible-in-activity is due to the object's action on our sense of sight: this is the way in which the object reveals itself to us. As I understand it, this position is a distinctive variant of direct realism.

I shall explicate the idea of perception as a relative more carefully below (§4), but before that, I need to adduce further evidence concerning Aristotle's causal model. First, I reject a possible objection that is motivated by a certain understanding of Aristotle's distinction between that which is perceptible (αἰσθητόν) and that which is perceived (αἴσθημα) (§3.5). Then I turn to more general considerations about causality (§3.6). This discussion reveals, I think, that Aristotle did not postulate a representation to explain how perceptual and intellectual acts can be about their objects.

3.5 An Objection: How Different Persons Can Perceive the Same Object

One might contend that Aquinas's postulation of likeness, as opposed to the object of perception, is well-founded because Aristotle himself introduced a distinction between perceptible object (αἰσθητόν) and what is perceived, *i.e.*, percept (αἴσθημα), and the latter seems, as is argued, to serve the same function as Aquinas's likeness. Aristotle rarely uses αἴσθημα in the *De Anima* (only twice, at 431^{a}15 and 432^{a}9), but since he takes it on these occasions to be on a par with φάντασμα, 'that which appears', a key term in the account of *phantasia,* memory, desire and thought, it must play some significant role in his account of perception as well. Now one might suggest that Aristotle's motivation for distinguishing between perceptible object and percept was to explain what distinguishes A's perceiving an *x* from B's perceiving that *x* without giving up the assumption that *x* is the same for A and B. The concept of percept would give this explanation thus: A's and B's perceptions of *x* are numerically different in that each perceiver has his or her own percept as a causal result of *x,* but their perceptions concern the same object because the percepts they have respectively are formally the same as that one object. Thus understood, the percept's likeness to the perceptible object is given in terms of formal sameness, and this explains why a perception is of its object (Wedin 1988, 170; Spruit 1994, 101).

This interpretation is taken to gain support from *De Sensu* 6 in which Aristotle discusses whether different persons can perceive the same object. Aristotle's answer is this:

> In perceiving that which is the primary cause of change—*e.g.* bell, frankincense, or fire—all perceive something that is one and the same in number. What is peculiar to each is different in number, but the same in form, and this is why many people see, smell, and hear [the same object] at the same time. (446^{b}21–25)

In the suggested interpretation, what is peculiar to each is the percept (and the perception as an affection of the soul), but because the percept is the same in form as the object, perceptions by different people concern the same object. This is an ingenious interpretation; however, it is not conclusive because there is a simpler alternative. It would be very reasonable to suppose that what is peculiar to each is a perception which is an affection of the soul. Given that, perceptions by different people can be claimed to be of the same object, because they are the same in form. Aristotle is very cursory here, but I think we should gloss his point in the light of his explanatory project in general. This suggests that the sameness in form is explained with reference to the sameness in efficient cause, which is the ultimate level of explanation for perception. A's hearing of a bell's sound is the same in form (*i.e.*, is of the same type) as B's hearing of that bell's sound because A's hearing is caused by the same sound. If this is correct, the sameness in form, or likeness, does not play an independent explanatory role. A perception's being like to its object is explained in terms of its being caused by that object. As I see it, the interpretation of this passage does not require a reference to a percept. In any case, talk of formal sameness raises some difficult questions, for example, how much variation a given form admits in terms of clarity and distinction: Does a faint token constitute the same form as another, more distinct one? Aristotle does not discuss this question, but one might think that also here he could refer to the efficient cause as the criterion.

3.6 The Causal Model in *De Anima* 3.2 and *Physics* 3.3

Let us begin with *De Anima* 3.2 which spells out the relations between the activity of the perceptible object and the activity of the sense:

> [*i*] The activity of the perceptible object and the activity of the sense are one and the same, but [*ii*] their being is not the same. [*iii*] I refer, for example, to the sound's activity and to the hearing's activity. One may have hearing, but not hear, and that which has a sound is not always sounding. [*iv*] However, when that which is capable of hearing is engaged in hearing and that which is capable of sounding is sounding, then the activity of hearing and the activity of sound come about at the same time (and one might refer to the one as hearing and to the other as sounding). [*v*] If indeed it is true that the change and the acting [and the affection] take place in that which is acted upon, it is necessary that both the sound and the hearing insofar as their activity is concerned occur in that which has the capacity; for the activity of that which is capable of acting and inducing a change is realized in that which is acted upon. (425^b26–426^a5)

Aristotle starts by making two general claims regarding the perceptible object and the sense: [*i*] the activity of the perceptible object and the activity of the sense are one and the same, and [*ii*] their being is different. He does not tell us in general what these claims mean, but points out some of the implications of claim [*i*] in the case of sounding and hearing [*iii*]. Aristotle's choice of example is deliberate because it enables him to draw attention to the difference between the activity and the capacity of the perceptible object. Unlike in the case of the other proper perceptible objects of the single senses, there is a word for the activity of sound, namely ψόφησις, 'sounding'. So ψόφος with the qualification κατ' ἐνέργειαν, 'in accordance with activity', yields sounding: thus the activity of sound takes place in that which hears. Aristotle makes this point against the anti-realists who argued that perceptible objects exist only when they are perceived. In Aristotle's view, their failure was to overlook that perceptible objects can be both in activity and in capacity. For example, sound is perceptible-in-activity when it is heard, and in that case, it is called sounding. But it can also be perceptible-in-capacity. As I see it, this leaves room for Aristotle to suppose that a sound *qua* sound can act upon the air (and the ears as bodily organs) even when it is not heard, and this means that the sound induces some sort of change (*e.g.* a wave reflection) in the air (and the ears). One might think that this is in fact necessary for the perceptible object to act upon the *capacity for perceiving,* for it is difficult to see how the sound *qua* perceptible could be in contact with the capacity unless it is present to the organ of this capacity insofar as it is a sound. However, once the sound *qua* perceptible acts upon the capacity for hearing, it is heard, and the corresponding activity of the capacity for hearing is hearing. The same holds, *mutatis mutandis,* for the other sense modalities. (See also Johansen 1998, 252–267; for a different view, see Irwin 1988, 313–314).

Aristotle points out two implications of his view. One is that sounding and hearing are simultaneous [*iv*]; the other is that sounding takes place in that which is capable of hearing [*v*]. It also seems that Aristotle takes sounding and hearing to constitute a single change, a view that is implied in *Physics* 3.3, though this remains at best implicit here. Since he makes no attempt to clarify and to justify claim [*ii*] in this context, it is advisable to turn directly to *Physics* 3.3 where both claims are clarified and substantiated in more general terms and in great detail. Towards the end of this chapter Aristotle poses the following complex question:

> Or can it be that: [*a*] it is *not* absurd that the activity of one thing should be in another (for teaching is the activity of that which is capable of teaching, but it is *on* something, and not cut off, but is of this on this); and [*b*] there is, also, nothing to prevent the activity of two things being one and the same, not as the same in being, but in the way that what has a capacity is related to what is acting, and [*c*] it is also not necessary that the teacher learns, even if to act upon and to be acted upon are the same thing, provided they are not the same in the sense that the definition that

> gives the 'what it was to be' is one (as with 'raiment' and 'clothing'), but in the sense in which the road from Thebes to Athens is the same as the road from Athens to Thebes, as was said earlier? (202^{b}5–14; trans. E. Hussey; modified)

Aristotle's subsequent discussion reveals that he endorses claims [*a*] to [*c*]. Claim [*a*] states that acting upon and being acted upon take place in the patient, and claim [*b*] goes further by saying that they constitute one and the same activity or change. Aristotle is careful to rule out what this type of sameness does not consist in. Claim [*c*] states that the sameness in question is not the sameness in definition. Thus understood, what it is to teach is not the same as what it is to learn. In general, what it is to act upon is not the same as what it is to be acted upon. However, his suggestion about what the sameness consists in remains somewhat unclear. The two activities should be the same in a sense somewhat similar to that in which the road from Thebes to Athens is the same as the road from Athens to Thebes. It is obvious that the two roads constitute numerically one and the same entity, but it is not clear how this would be analogous with teaching and learning. Is it because the two happen in the same thing, namely the learner? Or is it because what is taught is what is learnt (in optimal conditions)? A suggestion is that they are the same in that teaching stands in relation to learning as form to matter: the action of teaching is "imprinting" the knowledge of the teacher on the learner (Charles 1984, 15). Claim [*c*] makes the further statement that the teacher need not learn. This point would require a detailed explanation that is not, however, relevant here.

As a general account of causation, *Physics* 3.3 helps us to elucidate *De Anima* 3.2, but we need to make certain qualifications. Aristotle states in *De Anima* 2.5 that learning, unlike perceiving, involves a change (or a number of changes) from one condition to another (417^{a}31–32). This makes teaching and learning a process that takes time, and that can be interrupted. By contrast, once a perceptible quality acts upon the sense, and nothing prevents it, we perceive the quality completely. The same applies to thinking.

3.7 Another Objection: The Indirectness of the Object's Causal Power

An objection against the suggested interpretation would be that Aristotle allowed the object to act upon the sense only indirectly: it acts upon the medium and the medium acts upon the sense. Regarding color, Aristotle argued as follows:

> Seeing arises when the perceptual capacity is acted upon by something. However, it is impossible that it be acted upon by the seen color itself. It remains then that it is acted upon by the medium. Hence it is necessary that there is something in between. (*De An.* 2.7, 419^{a}17–20)

In view of this passage, it would be odd to claim, as I do, that the seen color, *i.e.* the perceptible color in-activity, is a constituent of perception. This is possibly a more serious objection, but I respond that it is based on a misunderstanding concerning the role of the medium. An appropriate sort of medium is an important factor insofar as through it the perceptible form of the object acts upon the sense. This is why the medium needs to be transparent, as Aristotle insists. It is also a causal link between the object and the sense insofar as it provides continuity between the object and the sense, so there is no gap between them. Aristotle believes that no perceptible object except for touch and taste can act upon the sense by direct contact. However, this does not require the medium to be a causal agent in its own right. Aristotle's point is, as I see it, that the perceptible object is the causal agent all the way through the medium to the sense (see also Johansen 1998, 133–136). In this way, the medium is one of the standing conditions for perception, and as such it is comparable to sufficient lighting, healthy sense organs, and the perceiver's being sober, fit, and free of strong passions, which are all conditions determining whether the object can act upon the sense as it should under optimal circumstances. If any one of these conditions fails to be satisfied, the object cannot act upon our sense as it should, and the result is an illusion, or a hallucination, or the object's failure to affect the sense at all. However, if the objection concerning the indirectness of the object's acting were germane, it would be difficult to see how Aristotle could argue that the acting-upon and the being-acted-upon constitute only one event.

4 PERCEPTION (AND SENSE) AS A TYPE OF RELATIVE

I have referred previously to the fact that for Aristotle, perception is to be understood as a 'relative' (πρός τι). He concludes his discussion of the perceptible objects in *De Anima* 2.6 by stating, "Of those objects that are perceptible in their own right, the proper perceptibles are primarily perceptibles, and it is with respect to these that the essence of each sense is given" (418^{a}24–25). This is an important claim, and deserves further consideration. As a general point of clarification, let me first note that the expression "with respect to these" translates the Greek πρὸς ἅ, which is a technical term for Aristotle. In general, πρός τι signifies a relative. In Aristotle's view, a relative is an accident that inheres in a subject, and that stands in relation to its term. The relative is to be contrasted with the modern notion of relation which does not imply that the relation be grounded on one of its relata (see Henninger 1989, 4–10).

One way in which Aristotle uses the notion of relative is to make a claim about one thing's being prior to another. This is the way in which he uses the relative above: he makes a claim about the sense's being defined with reference to its proper object. For example, sight is the sense it is because it sees a color, and hearing is the sense it is because it hears a sound. This means that the object of perception is prior to the perception in definition.

One might be puzzled at what this priority more precisely consists in. Why did Aristotle define perception in terms of its object? His answer, I argue, was that the object makes a perception the perception it is, and this is because the object is the efficient cause of the perception of it. In this sense, the object is prior to the perception also in causal terms. Aristotle states this point in *Metaphysics* 4.5:

> For perception is surely not the perception of itself, but there is something beyond the perception, which must be prior to the perception; for that which moves is prior in nature to that which is moved, and if they are said in relation to one another, this is no less the case. (1010^{b}35–1011^{a}2)

4.1 Three Types of Relative: *Metaphysics* 5.15

Aristotle distinguishes in *Metaphysics* 5.15 between three types of relative: mathematical, causal, and, to use Duns Scotus's term, terminal.

> Things are relative [1] as double to half and treble to a third, and in general that which contains something else many times to that which is contained many times in something else, and that which exceeds to that which is exceeded; [2] as that which can heat to that which can be heated, and that which can cut to that which can be cut, and in general the active to the passive; [3] as the measurable to the measure and the knowable to knowledge and the perceptible to perception. (1020^{b}26–32; trans. W. D. Ross. In: Barnes 1984, 2)

Aristotle's subsequent discussion is puzzling in many details. What concerns us here is how the causal relatives [2] are related to the terminal ones [3]. A crucial question is whether the two are independent. The question is, for example, whether an object can be perceived without its being the cause of that perception. I have argued above that this should not be possible in Aristotle's view, and in fact, there is nothing in *Metaphysics* 5.15 that would count against this argument. Aquinas shared Aristotle's view when it comes to embodied souls. In the case of disembodied souls, however, he conceded that God can illuminate the intellect of a separate substance without there being a causal relation between the substance and the intellect through an intelligible species (*Summa Theologiae* 1 q89 a1 ad3).

But matters are complex. Consider the measurable and the measure. It seems that an object can be measured without its being the efficient cause of measurement. This is because that which is measured is the object of action, whereas what is perceived or known is not an object of action: it is the cause of perception and knowledge, respectively. This shows that it is not a general truth that causal relatives and terminal relatives co-occur.

4.2 Duns Scotus on the Real Distinction between Causal and Terminal Relatives

It is illuminating to contrast Aristotle's account with Duns Scotus's. He took occasion to argue that, not even in the case of perception and knowledge, does a causal relative necessarily co-occur with a terminal one. Scotus elaborated upon Aristotle's distinction between the two, using thinking as an example (see Pini 2011):

> Notice that between the act of thinking and its object, as that object moves the intellect, or more properly as that object causes the act of thinking, there is a relation belonging to the second mode of relatives, just like the relation holding between a son and his father or between the heat produced and the heat producing [it]. The relation between the intellect as something that can be moved and its object as what moves it, too, belongs to the second mode of [relatives], just as the relation of what can be heated to what heats it. But in addition to these relations belonging to the second mode [of relatives], there is another relation between the act of thinking and its object as between something terminated and what terminates it [*i.e.*, as between something directed at something else and what is directed at]. For the act of thinking is not only [caused] by its object as by an efficient cause, whether total or partial; [rather, the act of thinking] is [also] [directed] at its object as at what terminates it or as that about which [that act] is. (Duns Scotus *Ord.*, I, d. 3, p. 3, q. 2, nn. 477–478 (Vat. 3: 285–86); trans. G. Pini. In: Pini 2011.)

The second mode Scotus refers to concerns relatives that are required in acting and being acted upon. These include the act of thinking and its object and, as Aristotle notes, the act of perceiving and its object. Scotus characterizes the third mode in terms of what is terminated (*terminatum*) and what terminates it (*terminans*). Aristotle merely speaks of one thing standing in relation to another, for example, seeing is seeing a color. The reason why Scotus refers to the two modes is that he wants to argue that they are independent of each other:

> The difference between these [two] relations is evident, because the one exists without the other. The former relation exists without the latter in an instance of heat produced [by something heating]. The latter relation exists without the former in the act of thinking about a stone, if [that act] were brought about by God in me immediately. (Duns Scotus, *ibid.* n. 479 (Vat. 3: 286); trans. G. Pini. In: Pini 2011.)

This is a significant argument. Scotus does not suppose that in the natural order of things, we could think about a stone without the stone being the cause of our thinking of it. His idea is rather that it is not logically necessary that thinking about a stone would require the stone being the cause of my thought about it. It follows that the causal relation between object and thought is contingent. A further point by Scotus is that we can think about an object even when the object is non-existent, such as a mermaid, which is composed of a woman and a fish, both perceptible items, respectively. In this case, the thought stands not in a real relation to the mermaid, but in a potential real relation.

This is not the occasion to consider further details of Scotus's philosophy. The point that I want to make is this: Aristotle did not separate these two modes of being a relative in the way in which Scotus did. For Aristotle the cause of thought and the cause of perception are by definition the object of thought and the object of perception, respectively. This means that perception and thought, according to Aristotle, are factive mental states. As the case of seeing the grey in [B] (§2) indicates, Aristotle has resources to account for some cases of illusion in causal terms. He also shows interest in some cases of hallucination in the *De Insomniis* (see 458b25–29). His explanation of these cases in analogy with dreams seems to require a more extended causal model with *phantasia* as a further explanatory capacity, and *phantasma* as its contents. This is evidence that he is inclined to give a somewhat different account of perceptual and hallucinatory experiences, being presumably opposed to the assumption that these experiences have content common to both. If this is correct, Aristotle's view would be in line with what recent disjunctivists claim about perceptual content. In my view, this observation is very significant, but it would require another study to determine in which sense Aristotle could be regarded as a disjunctivist.

5 CONCLUSION: AN OBJECTION TO ARISTOTLE'S VIEW AND A RESPONSE

Aristotle's theory certainly strikes many philosophers of perception as odd and unintuitive, especially those who are sympathetic with the common content assumption, which derives from Descartes and early modern empiricists, and even further from John Duns Scotus who, as noted in §4, separated causal and intentional (terminal) relations (see *e.g.* Moran 1996 and King 2004). The most obvious concern of these later thinkers is whether Aristotle can give a satisfactory account of illusions and hallucinations: how, exactly, do accounts of these experiences differ from the account of veridical perception? Even a more fundamental concern relates to the very nature of Aristotle's causal theory, a concern that can be expressed as follows.

In Aristotle's theory, the efficient cause is taken to constitute the same activity (event) as the result of the activity, although the two are defined in different terms: one being the cause and the other its result (Coope 2007). Teaching, for example, is seen to take place in the pupil's activity of learning, and the raising of a hand is claimed to constitute the same event as the rising of the hand. By analogy, on this model, the perceptible item's activity takes place in the perceiver's activity of perceiving, which means that the perceptible item's activity is not just a cause but also a term (in Aristotle's sense of this term) of the activity of perceiving: in other words, it is something that is part of the activity of perception, giving its contents. Now, a common content theorist might object that this cannot be quite right: set aside Aristotle's theory, if we consider our perceptual experiences as they present themselves, it appears that what we are directly aware of are merely our own mental contents, not the external objects or their activity in us. This is, it is argued, especially apparent in cases in which our perceptual experience misrepresents the object, or where there is no external object at all, and these cases strongly suggest that the situation is no different in the case of veridical perception.

If this is correct, it would be unwarranted to claim that the perceptible item is part of veridical perceptual content. The alternative view proposed by the critics, then, is to make some distinction between the way in which perceptual experiences are determined by the perceptible items and the way in which these experiences represent those items. As Dermot Moran's paper in this volume makes clear, some such distinction was drawn by Descartes when he distinguished between the idea understood as 'materially' (or 'formally') and the idea understood as 'objectively'.

This is not the right place to defend Aristotle against criticism such as this, but his view has a major advantage over the alternative view. As I see it, Aristotle's view is very reasonably motivated by what could be taken as the commonsense intuition about sense perception. Common sense tells us that sense perception is for the most part a reliable source of knowledge about the external world, and if this is correct, this should also be taken into account in all theorizing about the nature of sense perception. Given that, the sheer logical possibility that the perceptual experiences we take to be veridical sense perceptions would in fact be mere hallucinations caused by the Devil or an experimenting scientist does not give an adequate and privileged starting point for the study of the nature of sense perception. In my interpretation, Aristotle's causal account of sense perception is congenial to this commonsense intuition, and his account can be seen as an attempt to vindicate such an intuition.[2]

NOTES

1. All translations are mine unless otherwise indicated. The abbreviations of the titles of Aristotle's works are given in the bibliography. *De Sensu* refers to the *De Sensu et Sensibilia* that is part of the collection entitled *Parva Naturalia*. References to Aristotle's texts are given to the book and/or chapter, and the Bekker edition page number, column, and line; for example, *De Sensu* 1, 436^b19–437^a1. For convenience, references to Aquinas's commentary on *De*

Anima are to the paragraphs of the widely used Marietti edition, which is, however, superseded by the more recent Leonine edition. As to Scotus, *Ord.* is the abbreviation of *Ordinatio*. *Ord.* I, d. 3, p. 3, q. 2, nn. 477–478 (Vat. 3: 285–86) refers to *Ordinatio*, Book 1, distinction 3, part 3, question 2, paragraphs 477–478, in *Opera omnia* 3: 285–286.

2. Early drafts of this paper were presented in the conference "Realism: Mere Homonymy or Common Commitment(s)?" (Helsinki, June 2011) and in the History of Philosophy Research Seminar (Helsinki, February 2012). I was also able to test some of my main ideas in the Workshop in Ancient Philosophy (Oxford, November 2011) and in the conference "Early Modern Metaphysics of Causality and Representation" (Turku, December 2011). I am very grateful to the organizers and participants of these meetings for their help in shaping my views on this topic. In addition, I would like to thank Peter Lautner for his invaluable written comments on the penultimate version, and the editor of this volume for his admirable patience, generous suggestions for the improvement of argument, and for revising my English. The remaining shortcomings are, of course, my responsibility.

REFERENCES

Primary Sources

Aristotle, 1956. W. D. Ross, ed., *De Anima* (= *De An.*). Oxford, The Clarendon Press.

———, 1957. W. Jaeger, ed., *Metaphysica* (= *Met.*). Oxford, The Clarendon Press, 1957.

———, 1955. W. D. Ross, ed., *Parva Naturalia. A Revised Text with Introduction and Commentary.* Oxford, The Clarendon Press.

———, 1950. *Physica* (= *Phys.*) W. D. Ross, ed., Oxford, The Clarendon Press.

Avicenna, 1953. *Commentarium magnum in Aristotelis De Anima libros.* F. S. Crawford, ed., Corpus Commentariorum Averrois in Aristotelem Versionum Latinarum, V.I.1. Cambridge, Mass., The Medieval Academy of America.

Duns Scotus, 1954. *Opera omnia*, vol. 3. Civis Vaticana, Typis Polyglottis Vaticanis (= Vat.).

Plato, 1995. E. A. Duke, W. F. Hicken, W. S. M. Nicoll, D. B. Robinson, and J. C. G. Strachan, eds., *Opera I. Euthyphro, Apologia, Crito, Phaedo, Cratylus, Theaetetus, Sophista, Politicus.* Oxford, The Clarendon Press.

Thomas Aquinas, 1948–50. P. Caramello, ed., *Summa theologiae*, 3 vols. Turin, Marietti.

———, 1948. A.M. Pirotta, ed., *In Aristotelis librum De anima commentarium*, 3rd ed. Rome, Marietti.

———, 1984. R.-A. Gauthier, ed., *Sentencia libri De anima.* Sancti Thomae de Aquino doctoris angelici Opera omnia iussu Leonis XIII P.M. edita 45.1. Rome, Commissio Leonina; Paris, Vrin.

Translations

Aristotle's Physics. Books III and IV. E. Hussey, tr., Oxford: The Clarendon Press; New York, Oxford University Press, 1983.

The Complete Works of Aristotle. The Revised Oxford Translation, 2 vols. J. Barnes, ed., Princeton, New York: Princeton University Press, 1984.

Thomas Aquinas, 1999. *A Commentary on Aristotle's De Anima.* R. Pasnau, tr., New Haven and London, Yale University Press.

Secondary Literature

Bolton, R., 2005. "Perception Naturalized in Aristotle's *De Anima*." In: R. Salles, ed., *Metaphysics, Soul, and Ethics in Ancient Thought: Themes from the Work of Richard Sorabji* (Oxford, The Clarendon Press), 209–244.

Burnyeat, M. F., 1992. "Is an Aristotelian Philosophy of Mind Still Credible? A Draft." In: M. C. Nussbaum and A. O. Rorty, eds., *Essays on Aristotle's* De Anima (Oxford, The Clarendon Press), 15–26.

———, 2001. "Aquinas on 'Spiritual Change' in Perception." In: D. Perler, ed., *Ancient and Medieval Theories of Intentionality* (Leiden, Boston, and Köln, Brill), 129–153.

Caston, V., 2005. "The Spirit and the Letter: Aristotle on Perception." In: R. Salles, ed., *Metaphysics, Soul, and Ethics in Ancient Thought: Themes from the Work of Richard Sorabji* (Oxford, The Clarendon Press), 245–320.

Charles, D., 1984. *Aristotle's Philosophy of Action*. Ithaca, New York, Cornell University Press.

———, 2000. *Aristotle on Meaning and Essence*. Oxford, Oxford University Press.

———, 2010. "Definition and Explanation in the *Posterior Analytics* and *Metaphysics*." In: D. Charles, ed., *Definition in Greek Philosophy* (Oxford, Oxford University Press), 286–328.

Coope, U., 2007. "Aristotle on Action." *Proceedings of the Aristotelian Society Supplementary Volume* 81.1: 109–138.

Henninger, M. G., 1989. *Relations: Medieval Theories 1250–1325*. Oxford, The Clarendon Press.

Irwin, T. H., 1988. *Aristotle's First Principles*. Oxford, The Clarendon Press.

Johansen, T. K., 1998. *Aristotle on the Sense-Organs*. Cambridge, Cambridge University Press.

Kenny, A., 1993. *Aquinas on Mind*. London and New York, Routledge.

King, P., 2004. "Duns Scotus on Mental Content." In: O. Boulnois, E. Karger, J.-L. Solère and G. Sondag, eds., *Duns Scot à Paris, 1302–2002*. Textes et Etudes du Moyen Age No. 26 (Turnhout, Brepols Publishers), 65–88.

———, 2007. "Rethinking Representation in the Middle Ages: A Vade-Mecum to Medieval Theories of Mental Representation." In: H. Lagerlund, ed., *Representation and Objects of Thought in Medieval Philosophy* (Aldershot, Ashgate), 81–100.

Knuuttila, S., 2008. "Aristotle's Theory of Perception and Medieval Aristotelianism." In: S. Knuuttila and P. Kärkkäinen, eds., *Theories of Perception in Medieval and Early Modern Philosophy* (Sine loco, Springer), 1–22.

Lorenz, H., 2007. "The Assimilation of Sense to Sense-Object in Aristotle." *Oxford Studies in Ancient Philosophy* 33: 179–220.

Lowe, E. J., 2008. "Against Disjunctivism." In: A. Haddock and F. Macpherson, eds., *Disjunctivism: Perception, Action, Knowledge* (Oxford, Oxford University Press), 95–111.

Modrak, D., 1987. *Aristotle: The Power of Perception*. Chicago and London, The University of Chicago Press.

Moran, D., 1996. "The Inaugural Address: Brentano's Thesis." *Proceedings of the Aristotelian Society*, supp. vol.: 1–27.

Panaccio, C., 2001. "Aquinas on Intellectual Representation." In: D. Perler, ed., *Ancient and Medieval Theories of Intentionality* (Leiden, Boston, and Köln, Brill), 185–201.

Pasnau, R., 1997. *Theories of Cognition in the Later Middle Ages*. Cambridge and New York, Cambridge University Press.

———, 2008. "Id quo cognoscimus." In: S. Knuuttila and P. Kärkkäinen, eds., *Theories of Perception in Medieval and Early Modern Philosophy* (Sine loco, Springer), 131–149.
Perler, D., 2000. "Essentialism and Direct Realism: Some Late Medieval Perspectives." *Topoi* 19: 111–122.
Pini, G., 2011. "Can God Create My Thoughts? Scotus's Case against the Causal Account of Intentionality." *Journal of the History of Philosophy* 49.1: 39–63.
———, (forthcoming). "Two Models of Thinking: Thomas Aquinas and John Duns Scotus." In: G. Klima, ed., *Intentionality, Cognition and Representation in the Middle Ages* (New York, Fordham University Press).
Polansky, R., 2007. *Aristotle's* De Anima. Cambridge, Cambridge University Press.
Shields, Ch., 1995. "Intentionality and Isomorphism in Aristotle." *Proceedings of the Boston Area Colloquium in Ancient Philosophy* 11: 307–330.
———, 2007. *Aristotle.* Oxford, New York, Routledge.
Sorabji, R., 1991. "From Aristotle to Brentano: The Development of the Concept of Intentionality." *Oxford Studies in Ancient Philosophy,* supp. vol.: 227–259.
———, 2001. "Aristotle on Sensory Processes and Intentionality. A Reply to Myles Burnyeat." In: D. Perler, ed., *Ancient and Medieval Theories of Intentionality* (Leiden, Boston, and Köln, Brill), 49–61.
Spruit, L., 1994. *Species Intelligibilis from Perception to Knowledge. Volume One: Classical Roots and Medieval Discussions.* Leiden, New York and Köln, Brill.
Stein, N., 2009. "After Literalism and Spiritualism: The Plasticity of Aristotelian Perception." In: G. V. Riel and P. Destrée, eds., *Ancient Perspectives on Aristotle's* De Anima (Leuven, Leuven University Press), 17–33.
Tachau, K. H., 1988. *Vision and Certitude in the Age of Ockham: Optics, Epistemology and the Foundations of Semantics 1250–1345.* Leiden, New York, København, and Köln, Brill.
Wedin, M., 1988. *Mind and Imagination in Aristotle.* New Haven and London, Yale University Press.

3 Late Mediaeval Realisms

Key Arguments Supporting Non-semantic Universality

Laurent Cesalli

1 INTRODUCTION

In the following sense, universality is a fact: human languages comprise terms equally predicable of many, and those terms express concepts under which fall many individuals of the same kinds. How is this to be explained? Late Mediaeval logicians and metaphysicians distinguish roughly two types of explanation. On the one hand, the phenomenon of universality is reduced to an ontological fact: universality is explained by the existence of universals, *i.e.* non-singular entities; on the other, it is presented as a mere semantic phenomenon: universality boils down to the under-determination of semantic (or cognitive) relations, and its elucidation does not require the existence of anything over and above signs (or concepts) and the singular entities they are signs (or concepts) of. In good scholastic Latin, the *in essendo* explanation faces the *in significando* one. To be sure, no realist would deny that there is universality *in significando,* for this is the very starting point of the whole (philosophical) story; rather, he would insist that we experience semantic universality only *because* there is another, non-semantic type of universality.[1]

In what follows, I shall introduce and discuss some late Mediaeval arguments in favor of such a realist account of universality. More precisely, four main issues will be addressed: (*i*) the nature of universality, (*ii*) arguments for non-semantic universality, (*iii*) the relation of universals to particulars, (*iv*) propositional realism. Also, a possible case of Mediaeval idealism will be briefly considered, and some retrospective, contemporary views on Mediaeval realism will be presented (right at the beginning) and briefly discussed (at the very end).

This survey aims to show that late Mediaeval realists explored many different ways to state and to defend their positions. As we shall see, there have been attempts to demonstrate the non-semantic nature of universality by means of arguments from language and meaning, from scientific knowledge, and from ontology. In addition, the tricky issue of the relation between universals and particulars has also been addressed in many competing ways, by using concepts of identity and distinction, and of parts and wholes. Furthermore, the two extreme philosophical positions of Platonism and idealism (the other form of anti-realism) are not absent from the late Mediaeval landscape, though they may be qualified as marginal positions. Finally, a systematic convergence between Mediaeval propositional realism and contemporary forms of realism will be suggested.

2 RETROSPECTIVE VIEWS ON MEDIAEVAL REALISM

According to himself, Charles Sanders Peirce was sympathetic with scholastic realism.[2] More precisely, he favored a form of so-called Scotist realism, going back to John Duns Scotus (1265–1308), the rightly called *Doctor subtilis,* famous for (among other things) his extensive use of formal distinction and his ontology of *formalitates*.[3] In "Some Consequences of Four Incapacities," Peirce writes, taking a stance midway between naivety and common sense:

> . . . scholastic realism is usually set down as a belief in metaphysical fictions. But, in fact, a realist is simply one who knows no more recondite reality than that which is represented in a true representation. Since, therefore, the word 'man' is true of something, that which 'man' means is real. . . . The great argument of nominalism is that there is no man unless there is some particular man. That, however, does not affect the realism of Scotus; for although there is no man of whom all further determination can be denied, yet there is a man, abstraction being made of all further determination. (Peirce 1868, *CP* 5:312)

In contemporary terms, one could say that Scotus defends a kind of immanent realism: universals exist independently of thought and language, but they do not have a separate mode of existence. Universals exist first *in re* and only secondarily *in mente*—and hence, they are all but fictions.[4] Scotism also finds a favorable reception by David Armstrong, who sees the view as a remarkable anticipation of (his) scientific realism:

> . . . although particularity and universality are inseparable aspects of all existence, they are neither reducible to each other nor are they related. Though distinct, their union is closer than relation. Scotus talked of a mere 'formal distinction' between the thisness and the nature of particulars. The situation is admittedly profoundly puzzling, but, it is suggested, the Scotist view is the most satisfactory one which can be found.[5]

Although Scotus is (partially, but essentially) followed by a major figure of Mediaeval realism like Walter Burley (1275–1344), and mercilessly criticized by the prince of the nominalists, namely William of Ockham (1287–1347), a contemporary commentator does not hesitate to gather them under the single label of 'conceptual realism':

> Ockham and Buridan are generally described as nominalists, but on our account [*i.e.* on conceptual realism], they really were conceptualists. Burley is said to be a realist, but his realism was really a form of conceptual natural realism [as opposed to conceptual intensional realism]. (Cocchiarella 2007, 170)

Contrary to Peirce's and Armstrong's positions, such an ecumenical view tends to grind the dissent and thus, to lose sight of the specificity of realism as opposed to nominalism. To the eyes of Mediaeval philosophers at least, such an open attitude is not an option: while no one would deny the existence of general concepts, one has either to accept, or reject, non-singular entities—and there seems to be no go-between.

3 MEDIAEVAL REALISMS (ANTI-NOMINALISMS)

Although the opposition between nominalism and realism assumes different forms during the Middle Age—in the twelfth century *reales* fight *nominales* about the status of the objects of faith,[6] whereas the later opposition exemplified by Burley and Ockham concerns primarily universals and, secondarily, states of affairs[7]—I shall concentrate on realism as opposed to nominalism, *i.e.* realism about universals and propositions. Before considering the main arguments against the reduction of universality to a merely semantic phenomenon, however, I shall set the stage by sketching three standard, cardinal positions about the nature of universality.

3.1 The Nature of Universality

First to be mentioned is the uncontroversial view that universals come in three states: *in re, post rem* and *ante rem*. Such is the originally Neoplatonic opinion defended among others by Albert the Great (*ca.* 1200–1280) and Giles of Rome (d. 1316), but most famously by Thomas Aquinas (1225–1274). The three states are respectively that of the universal nature as real, metaphysical constituent of any individual thing (as such, though, the nature is not fully, or 'actually' universal); the universal concept acquired by abstraction from perceptual data (such a concept is fully or 'actually' universal); and the divine idea which is exemplified by any creature of a given type:

> . . . one has to say that the universal is threefold. One is in the thing, that is: the nature itself which is in the particulars, although it does not have actual universality in them; there is also a universal which is taken from the thing by abstraction, and that one is posterior to the thing . . . and there is also a universal with respect to the thing, which is prior to it, just as the form of a house is in the mind of its builder.[8]

Less consensual is the nominalist position advocated by authors such as William of Ockham (*ca.* 1287–1347) and John Buridan (d. 1358). They definitely go for the *post rem* option, insisting on the semantic/cognitive nature of universality—so Ockham, for example, flirts with paradox, when he writes: "every universal is one, singular thing, and thus, it is not universal but in virtue of signification, and because it is a sign of several things"[9]—and indeed, since every sign (conceptual or verbal) is a singular thing, every universal is singular. Buridan, for his part, pleading against any hypostatization of the predicate, plays Aristotle against Plato:

> . . . if Socrates is a man, then Socrates and man are the same thing absolutely. Therefore, one should not posit distinct universals in the singulars outside of the soul. Thus, Aristotle was right to say in the prologue of the *De anima* that either a universal is nothing, or it is posterior with respect to things existing outside of the soul. Nothing is a universal in the way Plato wanted it, but for sure, a universal is a term or a concept in the mind through which we grasp several things at once. . . .[10]

At the other end of the spectrum, the staunch realist John Wyclif (1330–1384) argues for the primacy of metaphysical universals over logical and grammatical ones. Note that if this doctrine recalls that of the three states, it diverges from it on a crucial point: whereas Aquinas pointed to the universals' lack of actuality *in re,* Wyclif, by contrast, underscores that universals as metaphysical parts of extra-mental things are universals in the most actual sense of the word:

> . . . there are three ways to consider universals. The first, quite a rough one, is that of the grammarians who conceive of things in their signs only; the second is that of the logicians who know universals in their significates. . . . The third . . ., which I prefer, is the consideration of the metaphysicians who say that the divine mind produces the most actual universality in things regardless of the consideration of any creature, so that the 'being human' common to every human being is the common humanity which is the species or nature of every single human being.[11]

Summing up, the late Mediaeval landscape of theories of universals is delimited on the one side by a radical *in significando* position (Ockham, Buridan), and, on the other, by a radical *in essendo* one (Wyclif). As for the doctrine of the three states of the universal, strictly speaking, it would have to be characterized as an *in cognoscendo* position, since, faithful to Aristotle's word in the beginning of the *De anima,* it seems to acknowledge full universality only for concepts acquired by abstraction.

3.2 Arguments for Non-semantic Universality

A first type of arguments put forward in support of the non-reducibility of universality to a merely semantic phenomenon takes language and its functioning as its starting point. Such a position is representatively held by the Franciscan Richard Brinkley (*fl.* 1350). Brinkley's idea is the following: according to the logicians, a universal is first and foremost a term predicable of many. Such a predicability, however, depends on the existence of a general concept expressed or signified by the general term, a concept which is itself a natural sign of something—call it *x*—in the sense that it naturally derives from *x*. Now *x*, according to Brinkley, is nothing but a real, metaphysical universal, *i.e.* a non-singular entity existing independently of any linguistic or cognitive act:

> Aristotle . . . says that in cognizing, the intellect follows the thing. Thus, because it is such in reality, the intellect cognizes that way, and not vice versa. Therefore, since the intellect has a universal concept naturally predicable of many, it is required a priori that what this concept first signifies be by itself indifferent to the many things thus cognized. . . . One concludes therefrom that . . . a real and metaphysical universal corresponds to the logical universal as its first term [*i.e.*, as its fundament].[12]

In what is probably his last work, Walter Burley develops another line of argumentation. According to him, there would be no science without universals, for to know is always to know something, and there is no science but of universals. Such an epistemic (or 'scientific') defense of realism is (rhetorically) presented as a kind of *opinio communis*:

> Not only a few doubts can be raised about universals. The first is whether universals exist in the nature of things or not. . . . Concerning the solution of the first doubt . . ., nearly all agree in saying that because the sciences are about universals, and because one has to know in advance of the subject matter of a science that it exists—since there is no science of the non-existent, as it is said in the first chapter of the *Posterior Analytics*—, universals must exist in the nature of things.[13]

Ockham, for example, would plainly disagree: science does not require the extra-mental existence of universals, for its object (*obiectum*) is nothing but a mental proposition (*i.e.* the conclusion of a demonstrative syllogism), and its subject (*subiectum*), nothing but the subject term of that mental proposition.[14]

A third type of arguments, which one finds for example in Aquinas' *De ente et essentia,* is of ontological nature. Roughly, the idea is that a universal is the essence (or quiddity) of a thing. Contrary to later and more robust realists such as Burley or Wyclif, Aquinas insists that essence is a universal properly speaking only insofar as it is thought, and not insofar as it exists in singular things. The Avicennian background obviously plays an important role here: essence in itself is neither singular nor universal, such determinations being accidental with respect to pure essence.[15] The fact that universality ultimately depends on thought, however, does not reduce it to a mere mental phenomenon: the universal essence in the mind only exists because individuals of the same species do 'share' (or 'have') the same singularized essence which, with respect to its mental counterpart, functions as a *fundamentum in re*:

> But this nature [*i.e.* human nature] exists in two ways: in the singulars, and in the soul, and in both it becomes accidentally determined; in the singulars, it has a further multiplicity of being according to the diversity of the singulars. . . . It appears, therefore, that human nature considered absolutely makes abstraction of any kind of being, but in such a way that it does not exclude any of them; and nature, considered that way, is what is predicated of all individuals. However, one cannot say that universality pertains to nature thus understood, because unity and community essentially belong to the universal, and neither of them pertains to human nature insofar as it is considered absolutely. . . . Similarly, one cannot say that the essence of genus or species [*i.e.* universality] occurs to nature as it exists in individuals, because human nature is not found in individuals with such a unity as to be one single thing convenient to all, something ⟨, however,⟩ which is required by the essence of the universal. Therefore, it remains that the essence of species occurs to human nature according to the being it has in the intellect.[16]

Just as the pure human nature (*humanitas tantum*) is in potency with respect to singularity or universality, a singularized human nature—that is: man's essence as it exists in each singular man—is in potency with respect to universality, *i.e.* the concept 'man' under which all members of the species equally fall. In other words, the ontological constitution of individual things (whatever exists is a singularized essence) is mirrored by the cognitive constitution of concepts (every universal is a generalized essence)—and this explains the apparently trivial fact that general terms can be truly predicated of singulars, as in '*Socrates est homo*'.

3.3 The Relation between Universals and Particulars

By accepting the existence of mind- and language-independent universals, one not only gets an answer to intriguing questions—how are we to account for the objectivity of knowledge, the validity of concepts, the meaning of general terms?—one also faces a series of not insignificant difficulties, one of them consisting in the explanation of how universals are related to their particulars. As we shall now see, several competing models were developed by Mediaeval realists, ranging from the formal realism of Duns Scotus to the Platonic realism of Nicolas of Autrécourt (*ca.* 1300–1369).

According to Scotus, universals exist *in re* as formal constituents of particular things. A formal constituent is a kind of non-integral part (a hand, for example, is an integral part of a human body, but humanity is only a formal constituent of Socrates). Roughly speaking, one can say that *A* and *B* are (only) formally distinct iff *A* and *B* are really identical (they belong to one and the same real thing *T*) but *A* and *B* are responsible for the fact that *T* falls under different concepts—for example: Socrates falls under the concepts 'animal' and 'rational' in virtue of two of its formal parts:[17]

> . . . for that reality from which the notion of genus is taken is naturally prior to the one from which the notion of difference is taken; and that reality from which the genus is taken is determinable and contractable by the reality of the difference. . . . Thus, just as in the same thing, there are different . . . formal entities (as in whiteness) from one of which the notion of genus is taken (here the notion of colour), and another formal entity from which the notion of difference (of whiteness) is taken . . ., similarly, there is in the same thing a positive entity from which the specific nature is taken, and a formally distinct entity from which the ultimate individual difference is taken. . . .[18]

The different metaphysical constituents of things are also called *formalitates* (or *rationes*). They stand in a precise order of priority and posteriority, the more general being prior with respect to the more specific, from the highest genus to the last, individual difference or *haecceitas*. Similar to what we noted about Aquinas, Avicenna's metaphysics also constitutes the background of Scotus' ontology: there are pure essences, which in themselves are neither universal nor particular, but from which every existing entity is derived by a (multiple and complex) process of 'contraction' to the effect that the ultimate, individual difference (or *haecceitas*) is also the last in a series of successive contractions. Although formal parts can only be grasped by the intellect and not by the senses, they nonetheless provide predication with an objective ground, since subject and predicate typically correspond to only formally distinct parts of individuals, as for example in "*Socrates est homo*," or "*Socrates est albus*."

Walter Burley, a pupil of Scotus with a certain tendency for radicalization, endorses successively two theories, both of which claim that universals are parts of individuals, and are really distinct from them. According to the first theory—call it the Burley *A* theory—universals are integral parts (or parts proper) of individuals:

> In reply to the argument proving that a universal is not a part of a particular substance, one has to say that a universal is a part of a determinate substance, and that Socrates is not only composed of this matter and this form, but rather of this matter, and this form, and of the universals essentially predicated of him. Thus, that which is precisely composed of this matter and this form is not Socrates, but a part of Socrates.[19]

The difficulties of such a position are quite obvious: if humanity, for example, is a part of Socrates and of Plato, how can Plato pass away without causing the loss of the entire species? Furthermore, if two members of the same species bear incompatible properties, should one say that one and the same humanity is, say, at the same time white and non-white?[20]

Accordingly, Burley abandons his first account for a refined theory—call it the Burley *B* theory—according to which universals are not integral, but something like 'concomitant' parts of individuals: ". . . I say that although a universal is an extra-mental thing, nevertheless, it is not a part of an individual. . . ."[21] The intellective soul of Socrates, for example, is a part proper of Socrates, but it is not universal—it is a singular form perfecting a singular matter; Socrates' humanity, by contrast, is universal, but it is not a part of him properly speaking, but only something which essentially accompanies Socrates' essence—something like a 'concomitant part' which 'declares' or 'manifests' his essence:

> Just as form is twofold, . . . the quiddity is twofold, because one is the quiddity which is the form perfecting matter, and the other is the form declaring the quiddity. The quiddity perfecting the matter is a part of the individual . . .; but the quiddity declaring the quiddity is not a part of the individual . . ., nor does it belong to the essence of such an individual, but is essentially concomitant to its essence.[22]

Provided that the Burley *B* theory is able to avoid the mereological paradoxes generated by the Burley *A* theory, it remains vague on the status of such 'concomitant parts' without which no individual can exist, but which are not integral parts of them.

The Dominican master Francesco da Prato (*fl. c.* 1340) explores another way, pushing the concept of identity to (and perhaps, even beyond) its outer limits. Universals, he claims, are numerically distinct, but essentially identical in their particulars:

> . . . I say that the case of Peter's soul existing in his hand and foot is not similar to the case of the human nature existing in Christ and Juda. For since Peter's soul is numerically one . . ., it exists in his hand and foot as numerically the same entity; but common human nature . . . is not in [Christ and Juda] as numerically the same entity; rather it exists in them as numerically distinct entities, although it exists in them as essentially the same real entity.[23]

This amounts to a theory of something like 'partial identity', for a universal—say humanity—is at the same time *different* in many (humanity is numerically distinct in Socrates and Plato) and *the same* in many (the numerically distinct humanity in Socrates and Plato is specifically the same). Thus, universals turn out to be only partially identical with their particulars.

John Wyclif will go one step further in contending that universals and particulars are really identical—every universal is a particular and vice versa—although universals and particulars remain formally distinct:

> . . . some say that every substance is singular and that, when it is universally apprehended, it is universal. . . . A second opinion says that a universal is not something of its singulars, since it is distinct . . . from the singular by opposition. . . . I take an intermediate position . . . and concede to the first opinion that every universal is a singular and vice versa, although the universal and the singular are formally distinct.[24]

Contrary to what someone like Scotus, for example, would say, particular entities do not contain universals as 'contracted' essences so that universals *qua* universals only exist in the mind. Wyclif insists that universals as such are identical with particulars. The formal distinction between the two grounds the objectivity of the distinction between the presentation of Socrates, and that of humanity, although there is no real difference between them: ". . . this [particular] man is really and essentially the human nature," as Wyclif puts it in his *Tractatus de logica*.[25]

The last model for the relationship between universals and particulars I shall consider is one of the very few cases of Mediaeval Platonism, namely the theory developed by Nicolas of Autrécourt:

> And if you ask whether, just like there is an objective universal in the intellect, similarly there exists a subjective [*i.e.* transcendent] universal corresponding to it, one should say that it is the case. . . . And one should say of this subjective universal that it is the quiddity of the singulars, the principle of their being, and [one should add that those subjective universals are] the principles posited by Plato according to which the singulars and many other things are cognized by conformity.[26]

Such a use of the categories of 'subjective' and 'objective', although perfectly standard in the late Middle Ages, deserves a short explanation. 'Objective' means 'in the mind as cognitive content', whereas 'subjective' means 'existing independently of any mental act'. Thus, according to Nicolas, quiddities exist out there, and they are nothing but Platonic Ideas functioning as ontological and epistemological principles (when we know of a given thing what it is, we do so by grasping its conformity with its principle of being, *i.e.* with the Idea it instantiates).

3.4 Propositional Realism

Besides the classical universals' realism, another form of late Mediaeval anti-nominalism—namely: propositional realism—can be mentioned in the frame of this short study. Propositional realism is the view according to which statements (*propositiones,* in Mediaeval terminology) possess special significates (*significata*) which are distinct both from mental propositions (acts of judging) and from the mere things signified by simpler expressions such as names, verbs or adjectives.[27] To make a long story short, I shall only say something about the views of Burley and Wyclif, the most strenuous defenders of this type of realism (Cesalli 2007).

Burley gives several formulations of his view, and one can legitimately wonder if we have to do with just one, or rather with several theories. Without trying to decide this tricky point here, I shall simply quote, and comment on, two representative passages. In the first one, Burley (strangely enough!) introduces the notion of *propositio in re* as a kind of "objective *mental* proposition":

> . . . one has to say that there is a twofold proposition in the intellect: one which is made by the intellect and has a subjective being in the intellect. . . . The other proposition only has an objective being in the intellect. . . . In other words, we can say that one is a real proposition ⟨*propositio in re*⟩, and one a proposition in the intellect, calling that proposition which only has an objective being in the intellect 'real proposition', and the other, which has a subjective being in the intellect, 'proposition in the intellect'.[28]

Burley puts his readers in an uncomfortable situation: as a proposition in the intellect, the *propositio in re* is a mental entity; but insofar as it is *in re,* it belongs to the extra-mental world. So, how is this passage to be understood? A possible—and, in my view, also plausible—reading consists in understanding Burley's apparently ambiguous way of speaking as a means to express the following idea: a *propositio in re* as described above is the mental content one has in mind when grasping an extra-mental fact or situation. The following passage, I believe, supports such an interpretation:

> It is clear that a certain complex thing is signified by a vocal proposition, as well as by a mental proposition, [a complex thing] which is neither the thing precisely signified by the subject, nor the thing signified by the predicate, but an aggregate of these things; and this [complex] thing, which is the ultimate and adequate significate of the vocal and the mental proposition is a certain coupled being which, therefore, can be called a real proposition.[29]

If one also takes into account that the very notion of *propositio in re* is introduced in an extended discussion about the nature of truth, one can easily see that Burley's propositional realism is a theory of the objective foundation of propositional truth: when one truly judges that *p*, one has in mind the cognitive content p^* (the *propositio in re* as presented in the first quote above), which precisely corresponds to the fact p^{**} (the *propositio in re* as presented in the second quote above).

Developing Burley's fundamental insight, Wyclif gives a more articulated theory of non-linguistic propositions distinguishing explicitly between facts or situations (this is the fifth kind of propositions described below) and individuals as real propositions (the fourth kind):

> . . . note that there are five kinds of propositions: mental, vocal and written; a real proposition [*propositio realis*], like any thing motivating the intellect to compose truly or falsely; and the fifth proposition is 'to be such as is signified by the proposition' [*sic esse sicut propositio significat*]. . . . A real proposition is like this man, this stone, etc., because in this man, there is this person, which is like the subject; and similarly, there is the human nature, which is like the predicate; and there is the essence of this man, which is the real copula coupling this man with his nature. . . . The fifth proposition is the truth signified on the side of the things. . . .[30]

Remarkably, the propositional realism defended by Wyclif entails that individuals are described more adequately as facts than as things—an idea which, as we shall see later, is comparable to the way a contemporary author, namely Gustav Bergmann, conceives of ordinary objects (below, §5). Such a view is not excluded by Burley's theory, but it is certainly not asserted with the same clarity as it is in Wyclif's.

4 A CASE OF MEDIAEVAL IDEALISM?

So far, we have briefly considered two kinds of late Mediaeval realism, about universals and about propositions, both of which are forms of anti-nominalism. As is well known, realism also exists as a form of anti-idealism. In such a sense, however, all Mediaeval thinkers are definitely realists: the world is how it is independently of the way we perceive it, and think of it. A possible exception, in the person of Dietrich of Freiberg (1250–1310), has been mentioned by historians of Mediaeval thought, so that the question could be raised—and answered positively—as to whether Kant had a Mediaeval forerunner (Flasch 1972). The reason to see a Mediaeval Kant in Dietrich is his claim that the intellect has the power to "constitute things," and in particular, to constitute the categories other than substance, quality and quantity.[31] If there is a tendency to idealism in Dietrich, then it is at most a tendency to a partial, and thus, moderate, idealism. For if the intellect has indeed a constitutive power, whatever it can bring about (and which is not a pure being "of second intention")[32] is always "ontologically based" on a being which is not (at all) constituted by the intellect:

> . . . some beings are purely of reason . . . and such ⟨beings⟩ are things of second intention absolutely. But other beings are purely of nature . . .

> like a substance, a quality, and others. But others occupy a kind of middle position: with respect to that which is formally and principally signified by [their] name, they depend on the operation of the intellect; however, with respect to a certain nature which they convey in their concept—be this nature a substance or something concrete belonging to a substance— . . . such beings are things of first intention as well. . . . [33]

Human being (*homo*) is an example of a thing which is constituted by the intellect without being a thing of second intention: The thing named by '*homo*' is constituted by the intellect on the bases of the thing named by '*animal*' and of the thing named by '*rationale*'.[34] At first sight, a position like the one of Dietrich seems to be comparable to Putnam's "realism with a small *r*" (as opposed to metaphysical realism, with a big *R*):

> Locke held that the great metaphysical problem of realism, the problem of the relation of our concepts to their objects, would be solved just by natural scientific investigation . . .; Kant held that Locke was wrong. . . . I am suggesting that on this subject, Kant was right and Locke was wrong. (Putnam 1990, 162)

> What I am saying, then, is that elements of what we call 'language' or 'mind' penetrate so deeply into what we call 'reality' that the very project of representing ourselves as being 'mappers' of something 'language-independent' is fatally compromised from the very start. (Putnam 1990, 28)

Whereas the constitutive function of the intellect could plausibly be invoked as a *tertium comparationis,* the fact that between Dietrich and Putnam, there is Kant, makes a difference: Dietrich's intellect *is* mapping—creatively, true enough!—something which exists independently of it; Putnam's mind (or language), by contrast, makes much more than that, for the very existence of what is to be mapped seems to depend upon mental activity.

5 THREE CONTEMPORARY TYPES OF REALISM

Before concluding this study, I shall briefly come 'back' to some contemporary views which, at least for two of them, seem to be quite close to what some Mediaeval realists had in mind in their polemics against nominalists. The first position is that of Gustav Bergmann who, in his critique of Brentano and Meinong, counts universals as things, and ordinary objects as facts:

> Many philosophers divide the entities . . . into two kinds. One kind is simple; the other, complex. Yet they do not have a clear notion of a complex. . . . In [my ontology], all complexes are facts and all facts are complexes; all simples are things and all things are simple. Things, though, are not the only constituents of facts. Each fact has at least one

> constituent of the ontological kind I call nexus, which, in turn is a subkind of the ontological kind I call subsistents. . . . Take a world, such as mine, in which there are universals. The latter are things, not facts. . . . There is a categorical difference between things and ordinary things. The latter are facts (complexes), not things. (Bergmann 2004, 3:4–5)

Bergmann's realism is of a mereological type and includes a relational entity (the nexus) which is in charge of bundling simple things into ordinary things, *i.e.* complexes or facts. The complexity described here is infra-individual: Socrates (full stop) is a fact composed of several things among which one necessarily finds the universal proper to human beings (and many others).

The second position is that of David Armstrong who also considers perceived individuals as metaphysical complexes. Contrary to Bergmann, though, Armstrong's facts (or states of affairs) are not relational complexes—something like a nexus does not belong to the ontology of the Australian philosopher—but instantiated properties. Socrates (full stop) is nothing but a particular instantiating certain properties:

> In the Universals theory we are examining, particulars instantiate properties. . . . Suppose that *a* is *F*, with *F* a universal. . . . It appears that we are required to recognize *a*'s being *F* . . . as [an] item in our ontology. I will speak of these items as states of affairs. Others have called them facts. Why do we need to recognize states of affairs? . . . If *a* is *F*, it entails that *a* exists, and that the universal *F* exists. However, *a* could exist, and *F* could exist, and yet it fail to be the case that *a* is *F*. . . . *a*'s being *F* involves something more than *a* and *F*. (Armstrong 1989, 88–89)

Beyond their differences, what Bergmann's and Armstrong's views have in common with several of the forms of Mediaeval realism considered above is their acknowledgment of the existence of universals independently of mind and language; also, they defend a kind of realism according to which universals are something like 'parts' or constituents of ordinary things, and in doing this, they clearly side with Aristotle against Plato.

The third position I shall consider—Nino Cocchiarella's 'conceptual realism', already mentioned in the first part of this paper—attempts to reconcile the ancestral opposition between Plato and Aristotle by identifying, within conceptual realism, an Aristotelian *and* a Platonic component:

> Conceptual realism is a formal ontology framed within the context of a naturalistic epistemology and a naturalistic approach to the relations between language, thought, and reality as based on our scientific knowledge of the world. . . . It is directed . . . primarily upon the structure of thought. . . . Conceptual realism contains both a natural realism and an intensional realism, each of which can be developed as separate subsystems that are compatible within the larger framework, one containing

> a modern form of Aristotelian essentialism, and the other containing a modern counterpart of Platonism based on the intensional contents of our speech and mental acts. (Cocchiarella 2007, 24)

The advertised compatibility of (modern forms of) Platonism and Aristotelianism is quite intriguing. It might perhaps be explained by considering that the realism at stake is *conceptual,* that is: "primarily directed upon the structure of thought," so that one can legitimately ask if such a realism is not an epistemological theory, rather than a metaphysical one; but if this is true, then Cocchiarella's mention of Mediaeval authors such as Burley and Ockham in this context[35] is somewhat inappropriate, for it seems beyond doubt that they approached the problem of universals *more metaphysico.*

6 CONCLUDING REMARKS

Late-Mediaeval realists reject a purely semantic (or cognitive) account of universality: universals are neither terms nor concepts, but real, mind- and language-independent entities. Different argumentative strategies are developed in favor of non-semantic universality:

From language and meaning: Semantic universality depends upon the existence of metaphysical universals.

From scientific knowledge: If science is to be taken seriously—*i.e.,* if it is not reduced to psychology or linguistics—metaphysical universals must exist.

From ontology: Essence is a metaphysical constituent of things, objectively given in the extra-mental world.

Faced with the problem of the 'relation' between universals and particulars, late-Mediaeval realists adopted different attitudes in order to explain how one thing can exist in many others:

Formal distinction	Scotus
Mereology	Burley *A*
Concomitance	Burley *B*
Partial identity	Da Prato
Real identity and formal distinction	Wyclif
Platonic-atomism	Autrécourt

There are striking affinities between late Mediaeval forms of realism and contemporary ones—more precisely: Bergmann's idea that universals ("things") are constituents of ordinary objects ("facts"), as well as Armstrong's immanent realism and his idea that whatever exists is a particular instantiating at least one universal (or property). These ideas come close not only to the realisms of universals of Scotus, Burley, Da Prato, and Wyclif, but also and foremost to the propositional realism of Burley and Wyclif.

Are late Mediaeval realisms forms of scientific or *a posteriori* realism à la Armstrong? This question calls for a balanced answer. On one hand, one would say that *they are not,* for as Armstrong himself suggests—and even if the term *scientia* appears here and there in their arguments—one cannot adequately say that late Mediaeval realists consider that 'science' alone determines which universals exist; on the other hand, one would say that *they are,* for insofar as Armstrong introduces scientific realism by distinguishing it from nominalism *and* Platonic realism, late Mediaeval thinkers—with the possible exception of Autrécourt—shared this central conviction with their remote Australian epigone.

NOTES

1. On Mediaeval theories of universals in general (and realism in particular), see Kaluza (1988), Conti (1990), (2008), Wöhler (1992), de Libera (1996), (1999), and Erismann (2011).
2. See Peirce (1931–58), *CP* 1:6, 1:16–17, 5:312, 6:175, 6:328. On Peirce's Scotism, see Tiercelin (1999).
3. On Scotus' ontology, see McCord Adams (1982), Marmo (1989), Noone (1999), and Pini (2002), (2007). The *distinctio formalis* must be distinguished from two other main types of distinction, the *distinctio realis,* and the *distinctio rationis.* The *distinctio realis* and *distinctio formalis* are *ex natura rei, i.e.,* what they distinguish is distinct regardless of any intellectual operation; *e.g.,* the bricks in a wall are really distinct; the genus (animal) and the species (*homo*) in Socrates are formally distinct in him because they have different definitions, although in him they are really identical. The *distinctio rationis* is not *ex natura rei:* what is distinguished in the sentence 'Socrates is identical with himself' is distinct only in virtue of an intellectual operation. Note that the later scholastic tradition (Suarez, *Disputationes metaphysicae,* VII.15) rejects the formal distinction on the ground that it is reducible to a *distinctio rationis cum fundamento in re* (= *distinctio rationis ratiocinatae*), while the mediaeval *distinctio rationis* is identified with a *distinctio rationis sine fundamento in re* (= *distinctio rationis ratiocinantis*); see Knebel (2002).
4. Scotus, *Quaestiones in librum Porphyrii Isagoge* (1999), q. 4, n. 12, p. 28: ". . . the universal comes from the mind. And when it is objected: 'therefore, it is a fiction', I say that this does not follow, because nothing *in re* corresponds to a fiction; but something *in re* corresponds to the universal, ⟨something⟩ by which the intellect is motivated to form such a ⟨universal⟩ concept. . . . Therefore I say that ⟨the universal⟩ effectively comes from the mind, but that materially and originally or occasionally, ⟨it comes⟩ from a property *in re.*" (". . . universale est ab intellectu. Cum dicitur 'igitur est figmentum', dico quod non sequitur, quia figmento nihil correspondet extra; universali autem aliquid correspondet extra a quo movetur intellectus ad causandum talem intentionem. . . . Dico igitur quod effective est ab intellectu, sed materialiter sive originaliter sive occasionaliter a proprietate in re.")

5. Armstrong (1978), 2:3, *cf.* 1:*xiv–xv*: "... contemporary philosophers... have largely ignored the possibility of developing a theory of objective universals, where the particular universals admitted are determined on the basis of scientific rather than semantic considerations. It might perhaps be argued that... Aristotle and the Scholastic realists were ahead of contemporary philosophy in this matter, although handicapped by the relative backwardness of the science and the scientific methodology of their day. . . ." For another contemporary defense of Scotism, see Baxter (2001).
6. In order to save the unity of the object of faith (and thus, of faith itself) in spite of the temporal relativity of believers with respect to what they believe, the *reales* claim that the object of faith is a thing (*res*), whereas the *nominales* maintained that it is an *enuntiabile* (or *dictum*) which, beyond its temporal variations, conserves a unity similar to the one of a name under its casual variations. On that topic, see Chenu (1923), Courtenay (1991), Iwakuma (1992), (1997).
7. See Cesalli (2012).
8. Thomas Aquinas (2002), *Super II Sententiarum,* d. 3, q. 3, a. 2, ad. 1: "Ad primum ergo dicendum, quod est triplex universale. Quoddam quod est in re, scilicet natura ipsa, quae est in particularibus, quamvis in eis non sit secundum rationem universalitatis in actu. Est etiam quoddam universale quod est a re acceptum per abstractionem, et hoc posterius est re; et hoc modo formae Angelorum non sunt universales. Est etiam quoddam universale ad rem, quod est prius re ipsa, sicut forma domus in mente aedificatoris; et per hunc modum sunt universales formae rerum in mente angelica existentes, non ita quod sint operativae, sed quia sunt operativis similes, sicut aliquis speculative scientiam operativam habet." On this opinion and its origins, see de Libera (1996), 245–262, and Erismann (2011), 46–59.
9. William of Ockham, *Summa logicae,* I, 14 (1974, 48): "Dicendum est igitur quod quodlibet universale est una res singularis, et ideo non est universale nisi per significationem, quia est signum plurium."
10. Buridan (1509), *Quaestiones super octo librum Physicorum Aristotelis,* I, 7: "... si Sortes est homo, eadem res penitus est Sortes et homo. Ideo sic non sunt ponenda universalia preter animam a singularibus distincta. Ideo bene dicit Aristoteles in prohemio *De anima* quod universale nichil est aut posterius est rebus singulariter extra animam existentibus. Nihil enim est universale secundum istum modum quem ponebat Plato, sed bene est universale terminus sive conceptus in mente quo simul indifferenter concipimus plures res. . . ." (*Cf.* Aristotle, *De anima,* I, 1, 402b8).
11. Wyclif (1985, 80–81), *De universalibus,* c. 3: ". . . triplex est consideratio de universalibus. Prima grammaticorum satis grossa, qui concipiunt rem in suis signis tantum. Secunda logicorum, qui cognoscunt universalia in suis signatis sed secundum habitudinem ad intellectum eorum. . . . Tertia . . . autem maneries loquendi sanctorum doctorum Ecclesiae, cui ego innitor, est consideratio metaphysicorum dicentium quod intellectus divinus facit universalitatem actualissimam in rebus, ut eo ipso quod commune est omni homini esse hominem, sive creatura hoc consideret sive non, aeque actualiter per intellectum divinus est communis humanitas, quae est species vel natura cuiuslibet hominis singularis!"
12. Brinkley (2008), *De universalibus,* §§5, 10: "Dicit enim Aristoteles in 4° *Metaphysice* quod intellectus sequitur rem in intelligendo. Unde quia sic est in re, ideo intellectus sic intelligit et non e converso. Ergo cum intellectus habeat conceptum universalem naturaliter predicabilem de pluribus, requiritur a priori quod illud quod primo significat iste conceptus sit de se indifferens ad illa plura. . . . Ex isto processu concluditur quod . . . universali loyco tanquam eius

primus terminus universale reale et metaphysicum correspondet" (quoted in Cesalli 2008, 301, 303). (*Cf.* Aristotle, *Metaphysica,* IV, 7, 1011b26–27).

13. Burley (1999, 10), *Tractatus de universalibus,* c.1: "Circa universalia sunt dubitationes non paucae: prima est utrum universalia existant in rerum natura vel non. . . . In solucione primi dubii . . . concordant quasi omnes, quia dicunt quod ex quo scientie sunt de universalibus et de subiecto scientie debet precognosci quod est, quoniam de non existente non est scientia, ut dicitur in primo Posteriorum, oportet quod universalia habeant esse in rerum natura." (*Cf.* Aristotle, *Analytica posteriora,* I, 1, 71a12–13).
14. William of Ockham (1967, 266), *Ordinatio,* prol., q. 9: ". . . differentia est inter subiectum scientiae et obiectum scientiae, quia subiectum scientiae est subiectum conclusionis, sed obiectum scientiae est illud quod scitur et terminat actum sciendi. Huiusmodi autem est ipsa conclusio scita. Et ita subiectum est pars obiecti; et si sit obiectum, non est nisi obiectum partiale."
15. See Avicenna, *Liber de philosophia prima* (= *Metaphysica*), V, 1. On that topic, see de Libera (1996, 277–283), and de Libera and Michon (1996), 47–51.
16. Thomas Aquinas, *De ente et essentia,* c. 3: "Haec autem natura habet duplex esse: unum in singularibus et aliud in anima, et secundum utrumque consequuntur dictam naturam accidentia; in singularibus etiam habet multiplex esse secundum singularium diversitatem. . . . Ergo patet quod natura hominis absolute considerata abstrahit a quolibet esse, ita tamen quod non fiat praecisio alicuius eorum. Et haec natura sic considerata est quae praedicatur de individuis omnibus. Non tamen potest dici quod ratio universalis conveniat naturae sic acceptae, quia de ratione universalis est unitas et communitas; naturae autem humanae neutrum horum convenit secundum absolutam suam considerationem. . . . Similiter etiam non potest dici quod ratio generis vel speciei accidat naturae humanae secundum esse quod habet in individuis, quia non invenitur in individuis natura humana secundum unitatem ut sit unum quid omnibus conveniens, quod ratio universalis exigit. Relinquitur ergo quod ratio speciei accidat naturae humanae secundum illud esse quod habet in intellectu."
17. *Cf., e.g.,* Scotus, *Lectura,* I, d. 2, pars 2, q. 1–4, n° 275.
18. Scotus (1982, 282–283), *Lectura,* II, d. 3, pars 1, q. 5–6, n° 171: ". . . sed illa realitas a qua accipitur ratio generis est prior naturaliter illa realitate a qua sumitur intentio differentiae, et illa realitas a qua sumitur genus est determinabilis et contrahibilis pre realitatem differentiae. . . . Unde sicut in eadem re sunt diversae perfectiones formales sive entitates formales (ut in albedine), a quarum una accipitur intentio generis (ut intentio coloris), et alia entitas formalis a qua accipitur intentio differentiae (albedinis) . . ., sic est entitas positiva in eadem re a qua accipitur natura specifica, et entitats formaliter alia a qua accipitur differentia ultima individualis. . . ."
19. Burley, *Expositio libri De anima* (quoted in Conti 2000, 323): "Ad aliam quae probat quod universale non est pars substantiae singularis, dicendum quod universale est pars substantiae determinatae, nec componitur Sortes sufficienter ex hac materia et ex hac forma; immo ex hac materia, et ex hac forma et ex universalibus praedicatis in quid de eo. Unde illud quod praecise componitur ex hac materia et hac forma non est Sortes, sed pars Sortis."
20. Such are the arguments William of Ockham opposed to Burley's first mereological account of the relation between universals and particulars—see Ockham (1974, c. 14, 15).
21. Burley (1501), *In Physicam Aristotelis expositio,* prologus, fol. 9r^{b}: ". . . dico quod quamvis universale sit res extra animam, tamen non est pars individui . . . ."
22. Burley (1497), *Super artem veterem Porphyrii et Aristotelis. Liber Praedicamentorum,* c. 2; fol. d3r^{b}: ". . . sicut forma est duplex . . ., sic quidditas est duplex : quia quaedam est quidditas quae est forma perficiens materiam et

quaedam est forma declarans quidditatem. Quidditas quae est forma perficiens materiam est pars individui . . .; sed quidditas declarans quidditatem non est pars individui . . ., nec est de essentia talis individui, sed est essentialiter concomitans essentiam eius."

23. Francesco da Prato (2003, 110), *Tractatus de universalibus,* art. 5: ". . . dico quod non est simile de anima Petri existente in manu et in pede, et de natura humana communi existente in Christo et in Iuda. Nam anima Petri, cum sit una anima numero . . . existit in manu et in pede Petri secundum eandem realitatem numeralem; sed natura humana communis . . . non est in eis secundum eandem realitatem sive entitatem numeralem, immo est in eis secundum diversas entitates numerales, quamvis sit in eis secundum eandem realem entitatem specificam."
24. Wyclif (1985, c. 4: 86–90): ". . . aliqui dicunt quod omnis substantia est singularis et, ut universaliter apprehenditur, est universalis . . . Secunda via dicit quod universale non est aliquod suorum singularium, cum . . . distinguatur ex opposito a singulari. . . . Ego autem per medium incedo concordando extrema, et concedo cum prima opinione quod omne universale est singulare et econtra, licet distinguantur formaliter ab invicem."
25. Wyclif (1893, 15), *Tractatus de logica,* c. 5: ". . . iste homo est realiter et essentialiter natura humana."
26. Nicolas of Autrécourt (1939, 266), *Tractatus utilis ad videndum an sermones peritatheticorum fuerint demonstrativi* [= *Exigit ordo*]: "Et si quaeras utrum, sicut est universale objectivum apud intellectum, sit aliquod universale subjectivum correspondens ; et diceretur quod sic. . . . Et de isto esse universali subjectivo dicecretur quod est quidditas singularium, principium essendi eis et principia intelligendi ea et multa conformiter quae ponebat Plato."
27. Just as the nominalist negates the existence of non-singular entities in matter of universals (there are no special things signified by general terms), he negates the existence of special significates in matter of propositions (there are no special entities to be signified by propositions and which would be distinct from the entities signified by simple expressions).
28. Burley (1973, 61), *Commentarius in librum Perihermeneias,* 1.26: ". . . dicendum quod in intellectu est duplex propositio: una quae efficitur ab intellectu et habet esse subiectivum in intellectu. . . . Alia est propositio quae solum habet esse obiectivum in intellectu. . . . Sub aliis verbis: possumus dicere quod quaedam est propositio in re et quaedam in intellectu, appellando illam propositionem quae solum habet esse obiectivum in intellectu propositio in re et aliam quae habet esse subiectivum in intellectu propositio in intellectu."
29. Burley (1497), fol. g6r^{b}: "Et ex hoc patet quod per propositionem in voce et etiam in conceptu significatur aliqua res complexa que non est proprie aliqua res precise significata per subiectum nec res significata per predicatum sed aggregatum ex his et illa res que est ultimum et adequatum significatum propositionis in voce et in conceptu est quoddam ens copulatum et propter hoc potest dici propositio in re."
30. Wyclif (1893), c. 5: 14–15: ". . . notandum quod quintuplex est propositio, scilicet: mentalis, vocalis et scripta propositio; realis, ut quelibet res movens ad componendum vere vel false; et quinta propositio est sic esse sicut propositio significat. . . . Propositio realis est, ut iste homo, iste lapis, etc. quia sicut in alia propositione est subiectum et predicatum et copula, sic in isto homine est dare istam personam, que est pars subiectiva speciei humane, que est tamquam subiectum ; et est dare similiter naturam humanam, que essentialiter inest isti homini tamquam predicatum, et realiter predicatur de isto homine. Et est dare essentiam istius hominis, que est realis copula copulans istum hominem cum sua natura. . . . Quinta propositio est veritas significata a parte rei. . . ."

31. See Dietrich of Freiberg (1983, esp. 181–201), *Tractatus de origine rerum praedicamentalium.*
32. Things "of first intention" are natural kinds: men, trees, flowers, *etc.;* things "of second intention" are construed by the mind in its logical operations: subjects, predicates, syllogisms, species, genera, *etc.*
33. Dietrich of Freiberg (1983, c. 5: 199): ". . . quaedam sunt entia, quae pure sunt res rationis . . . et huiusmodi sunt simpliciter res secundae intentionis. Quaedam auten sunt entia, quae sunt pure res naturae . . . ut substantia, qualitas et cetera. Quaedam autem quasi medio modo se habent: quantum enim ad id, quod formaliter et principaliter significatur per nomen, sunt ab operatione intellectus, secundum rationem tamen alicuius naturae quam important in suo intellecetu, sive haec natura sit substantia, sive aliquid substantiae concertum . . . huiusmodi sunt etiam res primae intentionis. . . ."
34. See Dietrich of Freiberg (1983, c. 5: 198).
35. See Cocchiarella (2007, 170), quoted above.

REFERENCES

Armstrong, D., 1978. *Universals and Scientific Realism.* 2 vols. Cambridge, Cambridge University Press.

———, 1989. *Universals: An Opinionated Introduction.* Boulder, Col., Westview Press.

Baxter, D.L.M., 2001. "Instantiation as partial identity." *Australasian Journal of Philosophy,* 79.4: 449–464.

Bergmann, G., 2004. *Realism: A Critique of Brentano and Meinong.* In: *idem.*, E. Tegtmeier, ed., *Collected Works,* (Frankfurt, Ontos), vol 3.

Brinkley, Richard, 2008. L. Cesalli, ed., *De universalibus,* §§ 5 and 10. In: L. Cesalli, "Richard Brinkley '*contra dialecticae haereticos*' (*Summa logicae* II). Une conception métaphysico-logique de l'universel." *Documenti e studi sulla tradizione filosofica medievale* 19: 277–333.

Buridan, John, 1509. *Quaestiones super octo librum Physicorum Aristotelis.* Parisiis.

Burley, Walter, 1497. *Super artem veterem Porphyrii et Aristotelis. Liber Praedicamentorum.* Venetiis.

———, 1501. *In Physicam Aristotelis expositio.* Venetiis. Rpt: Hildesheim: Olms, 1972.

———, 1973. *Commentarius in librum Perihermeneias.* In: S. F. Brown, "Walter Burley's Middle Commentary on Aristotle's Perihermeneias (*Commentarius medius*)." *Franciscan Studies* 33: 45–134.

———, 1999. *Tractatus de universalibus.* H.-U. Wöhler, ed., Leipzig, Verlag der Sächsischen Akademie der Wissenschaften zu Leipzig.

———, 2000. *Expositio libri De anima.* In: A. Conti, "Significato e verità in Walter Burley." *Documenti e Studi sulla Tradizione Filosofica Medievale,* 11: 317–350.

Cesalli, L., 2007. *Le réalisme propositionnel. Sémantique et ontologie des propositions chez Jean Duns Scot, Gauthier Burley, Richard Brinkley et Jean Wyclif.* Paris, Vrin.

———, 2012. "States of Affairs." In: J. Marenbon, ed., *The Oxford Handbook of Medieval Philosophy* (Oxford, Oxford University Press), 442–444.

Chenu, M. D., 1923. "Contribution à l'histoire du traité de la foi." In: *idem.*, *Mélanges thomistes* (Le Saulchoir, Kain), 123–140.

Cocchiarella, N., 2007. *Formal Ontology and Conceptual Realism.* Dordrecht, Springer.

Conti, A., 1990. "Studio Storico-critico." In: A. Conti, ed., Johannes Sharpe, *Quaestio super universalia* (Firenze, Olschki), 209–336.

———, 2008. "Categories and Universals in the Later Middle Ages." In: L. Newton, ed., *Medieval commentaries on Aristotle's Categories* (Leiden, Brill), 369–409.
Courtenay, W., 1991. "Nominales and Nominalism in the Twelfth Century." In: Z. Kaluza and A. de Libera, eds., *Lectionum varietates. Hommage à Paul Vignaux* (Paris, Vrin), 11–48.
de Libera, A., 1996. *La querelle des universaux. De Platon à la fin du Moyen Âge.* Paris, Seuil.
———, 1999. *L'art des généralités. Théories de l'abstraction.* Paris, Aubier.
de Libera, A. and C. Michon, 1996. *Thomas d'Aquin, Dietrich de Freiberg. L'Être et l'essence. Le vocabulaire médiéval de l'ontologie.* Paris, Seuil.
Dietrich of Freiberg, 1983. *Tractatus de origine rerum praedicamentalium.* In: L. Sturlese, ed., Dietrich von Freiberg, *Opera omnia, III, Schriften zur Naturphilosophie und Metaphysik.* Mit einer Einleitung von K. Flasch, J.-D. Cavigioli, R. Imbach, B. Mojsisch, M.-R. Pagnoni-Sturlese, R. Rehn und L. Sturlese, eds. Hamburg, Meiner.
Erismann, C., 2011. *L'homme commun. La genèse du realisme ontologique durant le haut Moyen Âge.* Paris, Vrin.
Flasch, K., 1972. "Kennt die mittelalteriche Philosophie die konstitutive Funktion des menschlichen Denkens? Eine Untersuchung zu Dietrich von Freiberg." *Kant Studien* 53: 182–206.
Francesco da Prato, 2003. *Tractatus de universalibus.* In: F. Amerini, ed., *I trattati* De universalibus *di Francesco da Prato e Stefano da Rieti (secolo XIV).* Spoleto, Centro Italiano di Studi sull'Alto Medioevo.
Iwakuma, Y., 1992. "Twelfth-Century Nominales. The Posthumous School of Peter Abelard." *Vivarium,* 30.1: 97–109.
———, 1997. "*Enuntiabilia* in Twelth-Century Logic and Theology." In: C. Marmo, ed., *Vestigia, Imagines, Verba: Semiotics and Logic in Medieval Theological Texts (12th–14th Century)* (Turnhout, Brepols), 19–35.
Kaluza, Z., 1988. *Les querelles doctrinales à Paris. Nominalistes et réalistes aux confins du xiv*[e] *et du xv*[e] *siècles.* Bergamo, Pierluigi Lubrina Editore.
Knebel, S., 2002. "Entre logique mentaliste et métaphysique conceptualiste. La *distinctio rationis ratiocinantis.*" *Les Études philosophiques* 2: 145–168.
Marmo, C., 1989. "Ontology and Semantics in the Logic of Duns Scotus." In: U. Eco and C. Marmo, eds., *On the Medieval Theory of Signs* (Amsterdam, John Benjamins), 143–193.
McCord Adams, M., 1982. "Universals In: the Early Fourteenth Century." In: A. Kenny, N. Kretzmann, J. Pinborg, and E. Stump, eds., *The Cambridge History of Later Medieval Philosophy* (Cambridge, Cambridge University Press), 413–439.
Nicolas of Autrécourt, 1939. *Tractatus utilis ad videndum an sermones peritatheticorum fuerint demonstrativi* [= *Exigit ordo*]. In: J. R. O'Donnell, "Nicholas of Autrécourt." *Medieval Studies* 1: 179–266.
Noone, T., 1999. "La distinction formelle dans l'école scotiste." *Revue des sciences philosophiques et théologiques,* 83: 53–72.
Peirce, C. S., 1868. "Some consequences of four incapacities." Rpt. in: *CP* 5:264–317.
———, 1931–1958. P. Hartshorn, P. Weis, and A. Burks, eds., *The Collected Papers of Charles Sanders Peirce,* 8 vols. Cambridge Mass., Harvard University Press; cited as '*CP*', by volume: paragraph numbers.
Pini, G., 2002. *Categories and logic in Duns Scotus. An interpretation of Aristotle's* Categories *in the Late Thirteenth Century.* Leiden, Brill.
———, 2007. "Scotus on Universals: A Reconsideration." *Documenti e studi sulla tradizione filosofica medievale,* 18: 395–409.
Putnam, H., 1990. J. Conant, ed., *Realism with a Human Face.* Cambridge Mass., Cambridge University Press.

Scotus, John Duns, 1982. L. Modrić, S. Bušelić, B. Hechich, I. Jurić, I. Percan, R. Rosini, S. Ruiz de Loizaga, and C. Saco Alarcón, eds., *Lectura. Prologus, dist. 1–7*. Città del Vaticano, Typis Polyglottis Vaticanis.
———, 1999. R. Andrews *et al.*, eds., *Quaestiones in librum Porphyrii Isagoge*. Saint Bonaventure, NY, The Franciscan Institute.
Thomas Aquinas, 2002. *Super II Sententiarum*. www.corpusthomisticum.org/snp 2002.html#3805.
———, 1933. *De ente et essentia*. www.corpusthomisticum.org/oee.html.
Tiercelin, C., 1999. "L'influence scotiste dans le projet peircien d'une métaphysique scientifique." *Revue des sciences théologiques et philosophiques,* 83: 117–134.
William of Ockham, 1967. S. F. Brown, G. Gál, A. Gambatese, and M. Meilach, eds., *Ordinatio. Prologus et distinction prima*. Saint Bonaventure, NY, The Franciscan Institute.
———, 1974. *Summa logicae*. G. Gál and S. F. Brown, eds., Saint-Bonaventure, NY, The Franciscan Institute.
Wöhler, H.-U., 1992. *Texte zum Universalienstreit,* 2 vols. Berlin, Akademie Verlag.
Wyclif, John, 1893. M. H. Dziewicki, ed., *Tractatus de logica,* c. 5. London, Trübner.
———, 1985. I. V. Mueller, ed., *De universalibus*. Oxford, The Clarendon Press.

4 Descartes on the Formal Reality, Objective Reality, and Material Falsity of Ideas

Realism through Constructivism?

Dermot Moran

From the standpoint of epistemology, René Descartes (1596–1650) is usually categorized as an indirect or representational realist in that he regards reality (*res,* things) as apprehended *through* our representations (*ideae,* ideas) and that what we immediately experience are our own conscious states. On the other hand, Descartes is also often conceived to be a direct realist in that 'ideas' are construed as being the actual 'things represented' (*res representata*) rather than the representings, and we have an immediate grasp of the essence of real entities through our ideas of them. It is not exactly clear where to locate Descartes' realism. Furthermore, how exactly does Descartes approach the issue of realism? How do 'ideas'—immediate conscious experiences—represent the real and how do they on occasion fail to do so? In this paper I shall explicate the nature of Descartes' realism by examining his complex and subtle account of the failure of ideas to represent reality, which he calls their 'material falsity': *falsitas materialis* (CSM 2:30/AT 7:43). I regard this discussion as illuminating a very rich and subtle account of mental and intentional content that would not again receive such a penetrating account until Edmund Husserl's (2001) discussion in the *Logical Investigations.* Descartes actually has quite a complex—and indeed coherent—account of the manner in which ideas present or display their content and also intend their objects. He identifies intentional and ideal elements or parts in conscious processes that need to be taken into account in any epistemological characterization of how consciousness gains a foothold in reality outside of it.

Descartes' brief discussion of 'materially false ideas' in his *Meditations* and *Replies* has attracted sustained attention from commentators in recent decades.[1] The general view is that Descartes' exposition of the notion of 'material falsity' is interesting and subtle but confused, and various efforts have been made to set him straight. Already in the seventeenth century, the brilliant Jansenist theologian Antoine Arnauld (1616–1698) in his Fourth Objections makes a number of telling criticisms of Descartes' account, and Descartes himself admitted to Mersenne in March 1641 that Arnaud "more than anyone else, has entered into the sense of what I wrote" (CSMK, 175/AT 3:331). Indeed, in a still classic account, Catherine Wilson (1978, 115) claims that Descartes, in response to Antoine Arnauld's probing questions, significantly alters his own position in the *Meditations*. She alleges that Descartes shifts his position from saying that materially false ideas mislead because they represent a "non-thing as a thing," to saying that they mislead because of their "obscurity." On the whole, however, Wilson characterizes Descartes' response to Arnauld as "a model of confusion confounded" (*ibid.*, 110). Likewise, Lilli Alanen (1994a, 206) has argued that the doctrine of material falsity contradicts Descartes' more usual account of truth in judgments, thereby engendering more confusion than clarity.[2] On the other hand, Norman Wells (1984, 26) argues that the confusion disappears if Descartes is read in the light of his Scholastic sources and especially Francisco Suarez's *Metaphysical Disputations*. The key, he argues, is Descartes' distinction between the truth and falsity of *things* and of *ideas*. For the Scholastics, beings are 'true' in and of themselves. Cecilia Wee (2006) in her recent book-length analysis separates what she takes to be Descartes' official account, according to which an idea is true if it genuinely comes from and accurately reflects its cause, from an alternative account, according to which an idea is true if it represents something as objective reality—though her discussion perhaps over-complicates the picture (see De Rosa 2008).

Before commencing our own discussion we need to clarify some ambiguities. Regrettably, there is widespread confusion about Descartes' use of Latin technical terms, such as *formaliter, materialiter, objective* and *esse objectivum* ('objective being'; for a good explication, see Wells 2006). Some of this confusion can be attributed to the varying usage found in Descartes' Scholastic sources (he claims to have remembered 'only the Coimbrans, Toletus, and Rubius', AT 3:185),[3] some to Descartes' own—quite typical among mathematicians—casual attitude towards language. But these terminological confusions and inconsistencies should not be taken as indicating that Descartes himself was confused. Rather I argue that Descartes is operating with a complex—but overall sound—conception of the phenomenological and conceptual content of our experience.

Descartes acknowledges that an idea can be "obscured and confused"—"because it contains some element of which we are ignorant" (CSM 2:105/AT 7:147). There are degrees of obscurity and clarity, confusion and distinctness (CSM 2:103/AT 7:145), and the obscurity and confusion of specific ideas can be produced as the result of quite different causes.

Furthermore, Descartes speaks both of ideas considered 'materially' (*materialiter*) and of the 'material falsity' (*falsitas materialis*) of ideas. These notions are quite distinct. An idea considered 'materially' is an idea considered as a *mode* of mind, *i.e.* as a temporal extended episode of thinking, a modification of a thinking substance, a moment of conscious life—as distinct from an idea considered in its representative capacity, as something that *represents* something else, or indeed the idea understood as the *thing* represented itself.[4] Some critics go further and claim—even more confusingly—that Descartes actually identifies *formaliter* with *materialiter.* Thus, in his classic study, Étienne Gilson (1930, 203) identifies the idea understood *materialiter* with the *formal being* of the idea. Descartes, however, when he talks of ideas *formaliter,* usually means ideas taken in their representative capacity, as representing something, or ideas as they are posited in acts of judgment. In this sense, John Yolton (1981, 79, for instance, identifies the notion of *formaliter* with the notion of the idea understood 'objectively' (*objective*); but this cannot be right—the formal and objective reality of ideas are quite distinct although related notions. Context, however, is usually sufficient to grasp the sense in which Descartes is employing a particular term. As is typical of a good communicator, he tends to reply to his critics employing their own technical terms.

Before focusing on the problem of material falsity, Descartes' overall position on ideas needs to be outlined briefly. (For background texts, see Perler and Haag 2010). As far as I can see, Descartes speaks of ideas in at least five different ways: as (a) *forms,* (b) *modes,* (c) *dispositions,* (d) *immediate* objects of the mind, and (e) *images.* He does not usually make explicit what sense he intends and therefore one has to apprehend his intended sense contextually. As a first move, Descartes distinguishes between ideas as entities or events in themselves, *i.e.* as 'modes or manners of my thinking' (*modes ou facons de ma pensée;* AT 9.1:29) and in their representative function. Thus, for instance, the *Preface to the Reader* (*Praefatio ad lectorem*) in the Second Latin Edition of the *Meditations* Descartes concedes that his initial account had been ambiguous with regard to the term 'idea':

> . . . there is an ambiguity here in the word 'idea'. Idea can be taken materially [*materialiter*], as an operation of the intellect [*pro operatione intellectus*] . . . Alternatively, it can be taken, objectively [*objective*], as the thing represented [*re . . . representata*] by that operation; and this thing, even if it is not regarded as existing outside the intellect, can still, in virtue of its essence, be more perfect than me. (CSM 2:7/AT 7:8)

The 'idea' taken on its own (*materialiter*) is to be understood as a temporal episode of 'thinking' and as possessing intentional content. The idea, in this sense, is a thinking episode in the broadest sense, a *cogitatio.* Metaphysically speaking it is a 'mode' (*modus*) of mind (which is a substance), *i.e.* the manner in which the mind itself manifests itself in its temporal flow of consciousness (think of a light bulb changing color). Descartes—and indeed subsequent thinkers including Hume—tends to think of an act of thinking as completely taking over the mind's attention at the moment it is being entertained; the mind is being 'minded' in a certain manner or mode. Descartes also thinks of an idea as a *form,* usually in terms of its causal power or its relation to its own cause. Ideas are forms in that they reflect their *formal cause.* Descartes offers a very elaborate discussion of formal *versus* efficient causes in reply to Arnaud in the Fourth Replies, and there he identifies formal cause with essence. Since God's essence and existence are not distinct, it is possible to say that God is formal cause of Himself and also efficient cause of Himself (see CSM 2:170/AT 7:243). The main principles Descartes defends are articulated in the Second Replies: 'whatever we perceive as being in the objects of our ideas exists objectively in the ideas themselves' and 'whatever exists in the objects of our ideas in a way which exactly corresponds to our perception of it is said to exist *formally* in those objects' (CSM 2:114/AT 7:161).

Descartes thinks of an idea not just as a 'mode' of mind, but also as having a certain representative 'content' for which he uses the Scholastic term 'objective reality' (*realitas objectiva*) which in turn points to, or in some sense represents some real or possible being (*realitas formalis*) beyond the idea itself. As Descartes puts it, "nature has taught" him to think of ideas as resembling the things they represent (CSM 2:26/AT 7:38). So, ideas do not just refer to actualities but in the first instance to possibilities, possible beings, essences *capable of* existing. The reality of some possible thing is expressed by its essence which is reflected in its 'objective reality'. As Descartes says in his Fifth Replies, "an idea represents the essence of a thing, and if anything is added or taken away from the essence, then the idea automatically becomes the idea of something else" (CSM 2:256/AT 7:371). Objective reality comes in degrees. There are degrees of 'objective being'—from the lowest (materially false or materially empty) to the highest—the infinite positive idea of God. Some 'real' entities also have, beyond their 'possible essence' an *actuality* or existence as a result of being caused. Existence is produced by the 'formal reality'; Descartes is clear about this in answering Arnauld, who wants existence to flow from the efficient cause alone (*cf.* CSM 2:149/AT 7:212). For Descartes certain ideas (*e.g.* God) have an objective reality so great that it can be accounted for only by them also having formal reality. Other ideas have objective reality but fail to have a formal cause and even mislead in presenting the kind of objective reality they possess. These are 'materially false' ideas.

For most of our ideas, there is a strong tendency to think of their objects as actually existing and also as being *like* the ideas of them. The assumption of actuality and the assumption of resemblance permeate our experience of our ideas. What Descartes is alluding to here is what Edmund Husserl calls the 'general thesis' of 'belief in being' (*Seinsglaube*), which he characterizes as the 'basic belief' (*Urdoxa*) that belongs to all perceptual acts. In the 'natural attitude', to invoke Husserl's conception which very much suits the naïve outlook Descartes' meditation is meant to challenge, there is a natural instinct to accept our experience as of the real and as adequately reflecting what reality is like. Descartes has exactly this understanding of the kinds of ideas that are experienced by an embodied sensory being. We have a natural if 'blind impulse': *ex caeco aliquo impulsu* (CSM 2:27/AT 7:40) to accept our experiences in a realist way.

Descartes often discusses the idea in itself in terms of what causes it or what it stands for in the outside world. Thus in a famous letter to Fr.

Giboeuf of 19 January 1642, Descartes says in relation to ideas being made inadequate by an act of abstraction from the intellect:

> I am certain that I can have no knowledge of what is outside me except by means of the ideas I have within me, and so I take great care not to relate my judgments immediately to things, and not to attribute to things anything positive which I do not first perceive in the ideas of them. But I think also that whatever is to be found in the ideas is necessarily also to be found in the things themselves. (CSMK, 201–2/AT 3:474)[5]

Ideas are what the mind primarily experiences. But we may also consider ideas in different ways.

In Descartes' convoluted language, when we consider an idea *materialiter,* we consider it *metaphysically* in terms of what it is, namely, as a 'mode' or 'operation of the mind' (*operatio intellectus,* AT 7:160; *opération de l'entendement,* AT 9.1:124). An 'operation' in Scholastic terms must be understood as the exercise of a *power.* A power, furthermore, must always be located in and dependent on a *substance.* No idea, then, can be 'greater' than the substance in which it inheres and the power of an idea ultimately depends on the substance in which it inheres. Substance underpins everything else.

On the other hand, when Descartes considers an idea *objective* he is focused on the 'thing' (*res*) or object presented in and through this operation of the mind (*pro re per istam operationem repraesentata*). This 'object', Descartes adds, need not actually exist outside the mind (*extra intellectum;* CSM 2:7/AT 7:8) but we are considering the thing in itself as given in the thought (this formulation is admittedly ambiguous). Thinking of an idea *objective* is thinking of that idea's 'objective reality', a concept introduced first—and controversially since it violates Descartes' own *epochē* concerning importing metaphysical doctrines—in the Third Meditation. At this point in the methodological self-scrutiny conducted in the *Meditations,* Descartes does not know whether there exists anything 'outside' the mind, and yet he knows he has ideas, because he *experiences* them immediately. He resolves to re-examine the flow of ideas to see if he can learn something more about them. At this point he introduces various taxonomies of idea (innate, adventitious, factitious; CSM 2:26/AT 7:37–38) based on how they appear to him and what their source appears to be—this is a phenomenological classification (*e.g.* some ideas seem subject to the will, others appear even if I will them not to).

But just to complicate things Descartes has another use of *materialiter.* He invokes this Scholastic term when he refers to ideas as capable of being 'materially false'. According to Descartes, ideas *scarcely* (*vix, à peine*) ever give material for error when they are considered *in themselves* (*i.e.* as modes of thought) and are not referred to anything else (CSM 2:26/AT 7:37). But they can be materially false as well as formally false. The distinction between formal and material falsity is explained in Meditation Three:

> For although as I have noted before, falsity in the strict sense, or formal falsity [*falsitatem proprie dictam, sive formalem*], can occur only in judgements, there is another kind of falsity, material falsity [*falsitas materialis*], which occurs in ideas, when they represent non-things as things [*cum non rem tanquam rem representant*]. For example, the ideas which I have of heat and cold contain so little clarity and distinctness that they do not enable me to tell whether cold is merely the absence of heat [*privatio caloris*] or heat the privation of cold, or whether both of them are real qualities, or neither is. And since there can be no ideas which are not as it were of things [*nullae ideae nisi tanquam rerum*], if it is true that cold is nothing but the absence of heat, the idea which represents it to me as something real and positive [*idea quae mihi illud tanquam reale quid & positivum repraesentat*] deserves to be called false; and the same goes for other ideas of this kind. (CSM 2:30/AT 7:43)

Formal falsity occurs where a judgment fails to correspond with reality, the *res*. Ideas on their own (as components of judgments) can be materially false if they supply my mind with material which in itself is misleading (*materia errandi*) and which will therefore mislead the unaided intellect or, on the other hand, if their obscurity means that they do not intentionally guide in the right direction. These are, as it were, *naturally* misleading ideas; we instinctively take up ideas in certain ways. As Descartes puts it, somewhat misleadingly, materially false ideas are ideas that 'represent a non-thing as a thing' (*non rem tanquam rem representant;* AT 7:43). The French translation of the *Meditations* adds a gloss on the phrase that there can be no ideas which are not 'as it were of things' which states that "since ideas, being like images, must in each case appear to us to represent something" (. . . *les idées étant commes des images, il n'y en peut avoir aucune qui ne nous semble representer quelque chose,* CSM 2:30 n. 2/AT 9.1:34–5). Ideas—as conscious experiences—must have some kind of phenomenologically apprehended content; there is something it is *like* to experience such an idea. Descartes also says that materially false ideas "arise from nothing [*a nihilo*]—that is, they are in me only because of a deficiency and lack of perfection in my nature" (CSM 2:30/AT 7:44). To say they arise from 'nothing' is not to violate the Aristotelian maxim *ex nihilo nihil fit,* as is clear from his discussion with Caterus in the First Replies. Descartes means that the phenomenological quality in the idea does not arise from something *outside* the mind. It is due to the mind's own sensuous character as an embodied mind. In his conversation with Burman, he is clear that the "body has an obstructive effect on the soul" (Cottingham 1976, 8/AT 5:150). Ideas—both as acts and as appearings of objects—have a distinct phenomenological character. Experiences have the character of appearing to be 'about' something (see García 1999, who claims Descartes is using a distinction between what an idea seems to represent and what it actually does). Cold, for instance, is a distinctly positive feeling—I can feel the *coldness* of my hand when I touch it to my face.

The experience of cold is not—as Arnaud wants to suggest—the experience of a *privation*. That would be to understand what cold is 'formally', *i.e.* to grasp its metaphysical nature, not materially. Cold, moreover, is not just a qualitative sensory feeling, it has an intentional arrow (as it were) that seems to point to some state of the world outside us. It tells us something (the air, the room, my hand) *is* cold. This, for Descartes, is perfectly natural; but precisely this natural intending must be resisted by the meditating scientist. Furthermore, this 'natural tendency' emerges from my embodiment and is ingrained since childhood and can be uncoupled or bracketed only by the meditating activity of the scientific mind.

On the one hand, Descartes is acknowledging in his usual manner that ideas are *always* representative or at least present themselves to us phenomenologically *as representative*. Ideas *seem* to represent something; they are 'as of'. Ideas in this sense are 'like images' (even if they are not all images) in that they convey something beyond themselves to the mind, namely that which they 'picture' or 'image' or, generally, *represent*. Now the question arises whether certain kinds of ideas which, when analyzed logically (or scientifically—the scientific meaning of 'cold' for instance) turn out to be 'non-things' or to arise from our 'defective' (*i.e.* embodied) nature, continue to exercise a representative function. Does, for instance, the idea of phlogiston continue to represent something even after it is shown to be an empty idea (in chemistry)? Inaccurate ideas—as the pagan's idea of God—can still refer in a genuine way to their object. I can *intend* or *mean* something even if I have an incorrect concept of what I intend. This referential function of an idea beyond itself is crucial for Descartes' argument for the existence of God, but it is also what can be misleading in the case of sensory ideas.

Descartes' overall discussion revisits a debate found in late-thirteenth- and fourteenth-century authors concerning the status of our thoughts of 'nothing' or privations, of impossible beings, possible but non-actual beings and so on. This discussion is taken up again in Arnaud's analysis of the idea of cold as a privation, and later, in the nineteenth century, by Bernard Bolzano (1781–1848) and Hermann Lotze (1817–1881), and thereafter by Franz Brentano and his followers (Meinong, Twardowski, Husserl), as well as by Frege and Russell. Indeed, Brentano's student Kasimiercz Twardowski (1866–1938) explicitly invokes Descartes' account of material falsity in his discussion of the possibility of Bolzanian 'objectless presentations'. Thus Twardowski comments:

> The doctrine of true and false presentations as it still appears in Descartes and his successors remains incomprehensible without the presupposition that to every presentation without exception there corresponds an object. Every presentation, according to Descartes, presents something as an object. Now, if this object exists, then the presentation is materially true; if it does not exist, then the idea is materially false. (Twardowski 1977, 23)

Twardowski continues:

> It is obviously Descartes' view that, irrespective of whether the object exists or not, it is always presented through the presentation. The question is only whether there corresponds to this intentional existence of the object in the presentation a true existence; and in presenting the really existing objects as well as the merely intentionally existing objects indiscriminately in the same way, the presentation easily leads to false judgments, since one can be as inclined to believe that the merely intentionally existing objects truly exist as that the really existing objects exist. (Twardowski 1977, 23)

We thus find in Descartes a confirmation of our view that to every presentation there corresponds an object (Twardowski 1977, 23–24).

Twardowski is enlisting Descartes in support of the Brentanian claim that every presentation presents an object; in other words experiences are essentially intentional. Twardowski thinks, however, that the *mode of relation* between the presentation (idea) and its object differs from the mode of relation of a judgment to its object because in the case of the judgment the 'existence or non-existence of the object' *matters*. Presentations, on the other hand, merely *present* or are entertained, and their formal truth or falsity is not immediately the issue; formal truth occurs only in judgments. Ideas on their own are merely entertainings, musings, appearings, and do not have judicative 'force'; they are not animated by the act of judging, the 'judgment stroke'. Certainly there are grounds for thinking that Descartes also holds that truth and falsity are generated in and by the judgment. There are many places where he says that strictly speaking the senses in and of themselves do not mislead but only the judgments that are hastily made on the basis of what is assumed to belong in the judging. The point is that, aside from formal positings and negatings, there are certain experienced aspects of our conscious episodes that have an intentional character, although their intentional direction needs to be treated with suspicion.

In his Fourth Replies to Antoine Arnauld's criticisms, Descartes admits having entertained a worry that his usage of the term 'materially false' might not be in conformity with what he calls the 'standard philosophical usage' (since he has never spent much time reading philosophical texts) but he is reassured by finding the term 'material' used in precisely the same sense in the very first author he went to consult, namely Suarez's *Metaphysical Disputations,* Part IX, Section 2, Number 4 (see CSM 2:164/AT 7:235). The term *materialiter* is then admittedly ambiguous for Descartes: it can mean the idea thought of as an operation or mode of the mind; or it can mean the manner in which this idea presents its intentional object in a way that it provides 'material' for error. The problem is this: In material falsity, what exactly in the idea provides the material for error? Is it the mental act or operation itself, the 'ideating'? Or is the failure a failure of content, of what elsewhere is called the 'objective reality' of the idea? Here one must be careful in supposing that the act and its content can be neatly distinguished. In fact, the thinking act is already contentful in a particular way. Thinking on its own (independent of its content) has a phenomenological character: there is something it is like to imagine as opposed to remember, will, judge and so on. But, to use Scholastic language, one can make a formal distinction between the kinds of content involved in an act of thinking.

I believe that Descartes' distinct uses of *materialiter* have confused almost all the commentators, because a material falsity is false because of our way of considering the idea *objective* not *materialiter*. Ideas considered *materialiter*, on the other hand, would seem to be all 'true', at least in the one sense that they really are mental occurrences appearing with an apparent phenomenological content that one's mind is entertaining, no matter whether they are perceptions, fantasies, or whatever. For this reason, in Meditation Three, Descartes says that ideas considered *in se* are all the same: that is there is no inequality among them, and moreover they cannot as such be false:

> Now as far as ideas are concerned, provided they are considered solely in themselves [*si sole in se spectentur*] and I do not refer them to anything else [*nec ad aliud quid illas referam*], they cannot strictly speaking be false. (CSM 2:26/AT 7:37)

That is: they have the experiential character they have, and that is that. And again:

> Of course if I considered just the ideas themselves simply as modes of my thought [*ut cogitationis meae quosdam modos considerarem*], without referring them to anything else [*nec ad quidquam aliud referrem*], they could scarcely give me any material for error [*materiam errandi*]. (CSM 2:26/AT 7:37)

Note here that Descartes does not say absolutely that ideas are *never* false but only that ideas *scarcely (vix) ever* give rise to falsity. This at least allows for the possibility that they may *sometimes* mislead—even within the manner in which they are being experienced and bracketing here their referential character. Normally, for an idea to mislead is for it to *misrepresent* in some way. Material falsity, then, most have concluded, must relate to the representative or objective side of ideas rather than to the *act* of thinking itself. I suggest instead that Descartes thinks literally of material falsity as a flaw in the sensory basis of the idea itself. Many of our sensorially experienced ideas have a 'natural' flaw in their nature: we are led to feel them as having a character which more careful scrutiny might lead us to question. It is not that they do not continue to have the phenomenological character of appearing to be 'real' and referring to real things outside the experience, but precisely that phenomenological character is what is suspect.

Consider a phenomenon such as referred pain. One really does have an experience or presentation of a distinctly located pain, as being *in the shoulder* for instance. But one can be told by the doctor (a doctor once told me that the first thing one was taught in medical school was never believe the patient) that this pain is actually 'referred pain'—it does not indicate damage in the shoulder but rather, let us say, in the vertebrae of the neck. The neck itself might separately be sore, and have its own phenomenologically experienced pain, but that manifests itself as a separate, unconnected event. The Cartesian account of the apparent shoulder pain is of course that it is a 'confused idea' that does not clearly and distinctly *represent its cause.* But is it a materially false idea? Is there something inherently misleading in the very concept of the very specific *pain* itself as felt in the shoulder? With the concept of material falsity Descartes is trying to express something more complex than simply the usual contrast between the phenomenological manifestation of the idea (with its quale and 'content') and its representational character. There must be something else involved in material falsity—that is, that the peculiar qualitative character of the experience is itself somehow 'misleading'. Let us follow this thought further.

As we saw, in his reply to Arnauld (which some consider a departure from the Third Meditation account), Descartes explains that material falsity means providing material for falsity, that strictly speaking the falsity is located in *judgment* (AT 7:233–5). Is Descartes' response to Arnauld merely a rhetorical concession to Arnauld's way of putting it, or is this account consistent with what he actually says in Meditation Three? In Meditation Three, it is unclear whether our ideas of sensible qualities "are ideas of real things or of non-things" (CSM 2:30/AT 7:44); and in the French version, this statement is expanded to include an explanation of 'non-things':

> I do not know . . . whether the ideas of those qualities of which we are aware are in fact the ideas of some real things, or whether they represent to me nothing but some *chimerical beings which cannot exist.* (AT 9.1:34; my trans. and italics)[6]

Certain experiences have a phenomenological character which is confused because the causal 'source' cannot be really determined. (Interestingly, Descartes here uses the late Latin term *author,* AT 7:44 ll. 9–10, rather than *auctor.*) In such cases, Descartes will conclude there is really no reason to seek a source for these ideas *beyond* myself. In other words, my embodied nature conveys experiences in a certain way which is due to my incarnation. I experience cold as something really out there, but on higher reflection I should rather regard it as a modality of my own incarnate being, for which 'I' am the source, and I should uncouple the intentional thrust of the experience. Many of my conscious experiences really can be related to *me* as a causal source. The ego (as a substance), for Descartes, has sufficient formal reality to be the formal or indeed eminent cause of anything which is 'below' it on the chain of being. According to Meditation Three, for instance, the notion of substance carries more objective reality, more content than that of accidents or modes. Furthermore, within the category of substance, the idea of an *animal* carries more objective reality than the idea of a *stone*. Let us look further at this conception of formal reality.

In the Third Meditation, to articulate the special relation between the idea of God and its cause, Descartes introduces what Suarez elsewhere calls a 'commonly found distinction' (*vulgaris distinctio*) between formal and objective reality. This distinction stems from Suarez and the Scotist tradition (Dalbiez 1929; Cronin 1966; Wells 1990, 2006). Indeed, the French text of the *Meditations* acknowledges this borrowing, adding a reference to "what the philosophers call actual or formal reality" (*cette réalité que les Philosophes appellent actuelle ou formelle;* CSM 2:28/AT 9:32). The Scotists used a fine-grained set of distinctions between objective reality, formal reality, ideal being, and so on; see Kobusch (1987, 79–104), who traces the origin of the discussion in Roger Bacon, Albertus, Siger of Brabant, Martin of Dacia and others. Francisco Suarez (1548– 1617) made a similar distinction between the objective and formal concepts, whereby the 'formal concept' refers to the form of the *act of thinking* itself and the 'objective concept' to the representative content of the thought. As Suarez writes in his *Metaphysical Disputations* (1597):

> The objective concept (*conceptus objectivus*) is said to be the thing (*res*) or *ratio* which is properly and immediately known or represented through the formal concept, as for example, when we conceive of a man, that act which we make in the mind in conceiving the man is called the formal concept [*conceptus formalis*], whereas the man known and represented by that act is called the objective concept [*conceptus objectivus*].[7]

Many commentators have thought that these sets of distinctions made essentially the same discriminations. If ideas are *forms* of some kind, then speaking of them as they actually are is to speak of them *formally*. However, this is precisely what Descartes calls speaking of an idea *materialiter*, as we have seen (and this is why Gilson identifies *formaliter* with *materialiter*); we are picking out the immaterial ontological nature of the act of thinking. In the 'Preface to the Reader' and also in the Third Meditation, Descartes uses *materialiter* to refer to ideas thought of *in themselves* without reference to their objects. In the Replies to the Fourth Set of Objections Descartes confirms this way of speaking when he says:

> Since ideas are forms of a kind [*formae quaedam*] and are not composed of any matter [*nex ex material ulla componantur*], when we think of them as representing something [*quatenus aliquid repraesentant*] we are taking them not materially [*materialiter*] but formally [*formaliter*]. If

> however, we take them, not as representing this or that, but simply as operations of the intellect [*operationes intellectus*], then it could be said that we were taking them materially, but in that case there would be no reference to the truth or falsity of their objects. (CSM 2:163/AT 7:232)

It is clear that referring to an idea as some kind of object, process or event in itself is to refer to it *materialiter.* In the Replies he says '*formaliter*' characterizes the representative character of an idea: we are thinking of what the idea is *of*, its intentional object.

With respect to Étienne Gilson's eminence in the field of mediaeval philosophy, we must still ask whether *formaliter* and *materialiter* here mean the same thing for Descartes? The majority of commentators have thought so, but Norman Wells (1990, 34 n. 3), in perhaps the most careful and erudite discussion of the topic, argues that the material/formal distinction here is quite different from the objective reality/formal reality distinction in Meditation Three. My own view is this: an idea considered as an activity which modifies the mind, or as a mode in which the mind manifests itself to itself consciously, is considered *materialiter.* When it is considered as *representative,* it is considered *formally* (what, confusingly, the 'Preface to the Reader' calls 'objective', CSM 2:7/AT 7:8). Descartes is confusing the issue because he is saying that ideas considered materially are forms of thinking, but to speak of them as 'forms' is not to speak formally. When some idea is considered representatively in terms of what those forms are forms *of* (*i.e.* what is objectified by the thinking) then the idea is considered 'formally' or 'objectively'; hence the formal/objective distinction applies only when we consider an idea in terms of the cause of its intentional object. Formally and materially are two ways of considering ideas which pick out the ambiguity of the term, *i.e.* as both act and object. But now we appear to be identifying *objective* (as found in the *Preface to the Reader,* AT 7:8) and *formaliter* as found in the Reply to Arnauld (AT 7:231). However Descartes' first proof for the existence of God in Meditation Three turns on whether we can distinguish between the objective and formal reality of an idea. Although considering an idea formally and objectively means thinking of it epistemologically in terms of the representative function or activity the idea possesses, nevertheless when we come to examine that representative function in detail we now need to distinguish the objective and formal components as referring to two *ontologically* distinct aspects of the representing process.[8]

Descartes himself offers an important formal definition of 'objective reality' in his Second Replies:

> Objective reality of an idea. By this I mean the being of the thing (*entitatem rei*) which is represented by an idea (*repraesentatae per ideam*), in so far as this exists in the idea (*quatenus est in idea*); . . . For what we perceive as being in the objects of our ideas (*in idearum objectis*) exists objectively (*objective*) in the ideas themselves. (CSM 2:113–4/AT 7:161)[9]

This is a complex—and initially confusing—definition, but at least he is clear that objective reality refers to the idea *in itself*—bracketed from its causal source. In the First Objections, Fr. Caterus therefore denies that 'objective reality' in the mind is anything real possessing formal reality and hence not anything requiring a causal explanation. Caterus denies that there is a 'real' intentional object between the external thing and the mind. Thinking of 'nothing', for him, is not apprehending an object which is not itself nothing; rather it is not thinking at all. Opposing him, Descartes adopts a Scotist stance whereby the 'objective reality' of an idea is something whose content requires a causal explanation. For Descartes, the thought of nothing is 'not nothing', though it is of course 'less perfect' than the thing itself (since it lacks formal reality or else any formal reality it has it owes to its causal source). The objective reality on this account contains the properties and the 'being' (*entitas*) of the thing represented in the manner in which they are so represented. Thus, for example, the objective reality of the idea of God contains the property of 'infinity' which is also a property of the infinite deity itself. Later Descartes will get into a tangle as to whether the objective reality is an exact mirror of the object represented (the formal reality) or is just a signpost pointing to it in a certain way. Clearly, as he recognized, the finite mind cannot *contain* within itself an idea that is itself actually infinite. The ego as a finite substance does not have enough formal reality to be the source of a genuinely infinite idea. But it can generate a *finite idea* of infinity (and this is Descartes' fallback position). A finite idea can have or somehow intend an infinite 'objective reality' and indeed, in the case of the idea of God, it does just that.

In her now classic study *Descartes*, Margaret Wilson (1978, 106) interprets material falsity as a falsity of ideas which "derives from their representative character." She distinguishes between the representational character of an idea and its objective reality. Wilson appears, then, to be in agreement with Twardowski (1977, 107)—whom she does not discuss—in holding that Descartes' idea is of a thing does not just concern existent things. The concept of a *res* is wider than the notion of an existent thing, as we saw earlier. A thing is a possible existent for Descartes. For instance, the Fifth Meditation argument moves from demonstrating that entities with mathematical properties are possible to the demonstration that they exist and also that they are actually as they are conceived to be. Wilson distinguishes an idea's 'of a thing' status and its representative character:

> Consequently, the claim that an idea or thought is 'of a thing (*res*)' must be distinguished *both* from the claim that it represents something that in fact exists, *and* from the claim that it has representational character. An idea may have representational character, yet fail, in the relevant sense, to represent *any thing*. (Wilson 1978, 108)

Wilson (*ibid.*) wants to distinguish (a) the claim that an idea is *of* a thing, (b) the claim that it *represents* something that in fact exists, and (c) the claim that it has 'representational character'. Wilson's tripartite distinction is not coherent, nor is it supported by the text of Meditation Three. Wilson thinks that something can have 'representational character' and yet fail to be 'of a thing'. But it seems to me that to represent is to have something presented through the idea.[10] Curiously, Wilson's discussion in her book does not require this tripartite classification. She does want to have ideas of a kind which are not sufficiently clear and distinct that we are able to tell whether they represent anything real (where 'real' means *possible*). Something is real if it does not contain a contradiction; everything conceived clearly and distinctly contains the notion of possible existence (AT 7:116, 166). Wilson does not give a very clear definition of the representational character of an idea, she simply uses it as a rendering of Descartes 'as if they were images of things':

> I will use the expression 'representational character' to designate that feature of ideas by which they are '*tanquam rerum imagines*'. (Wilson 1978, 102–3)

Wilson also interprets this Latin phrase in a very specific way:

> First, when Descartes says that ideas are '*tanquam rerum imagines*' I think he is saying more than that thoughts have 'objects', according to which they are classified (as ideas *of God, of heat,* or *cold,* etc.). He means also that ideas are received by the mind *as if exhibiting to it* various things—or as if making things *cognitively accessible*. . . . Second, when Descartes speaks of ideas being '*tanquam rerum imagines*', he does not mean that every idea involves a mental picture with visual properties: an idea, in other words, can purport to bring something into cognitive ken without purporting to represent it visually. . . . I will use the term 'representational character' to designate that feature by which they are '*tanquam rerum imagines*'. (Wilson 1978, 102–3)

Here Wilson acknowledges that all ideas seem to be *of* something; ideas have *aboutness,* as it were (*cf.* Shapiro 2012; Perler 1996). Yet she goes on to give a very confused picture. Of course, ideas do not necessarily have to have *visual* properties; we have representational ideas of smells and tones, for example, or of abstract entities. To smell an orange is not to have a *quale* of a certain olfactory character combined with a visual representation of an orange; this is a cartoon depiction of ideas. To my knowledge, Descartes does not use the term *imago* to mean pictorial image in this way. Indeed, he continues to employ the term 'idea' as *terminus technicus* to refer to 'forms of perception' in the divine mind in the traditional way. As he writes in the Third Replies:

> I used the word 'idea' because it was the standard philosophical term [*terminus technicus*] used to refer to the forms of perception belonging to the divine mind [*ad formas perceptionum mentis divinae significandas*],

> even though we recognize that God does not possess any corporeal imagination. (CSM 2:127/AT 7:181)

The term 'idea' does not include any reference to the imagination.

How should we understand the distinction between the *being* of a thing and its having representational character? Wilson tries to employ her tripartite distinction to sort out the puzzle of how there can be materially false ideas. Wilson (1978, 105) uses the representational character of ideas to explain the notion of 'objective reality'. Ideas exhibit to us different degrees of objective reality. She states:

> This account [of material falsity] forces us to recognize a distinction between the representative character of an idea and its objective reality. (Wilson 1978, 106)

It is not clear how she arrives at this conclusion and it makes no sense to me. It is entirely unclear how the representational character of an idea can be distinguished from its objective reality. Indeed there is no evidence that Descartes makes anything like such a distinction. There is, rather, the distinction between the phenomenological experience of having the idea, the conscious episode with the 'natural' intentional thrust, and the intentional object which is intended in and through this conscious process.

According to Wilson, material falsity in sensory ideas is to be explained by a gap between the objective reality of the idea and its representational character. Sensations for Descartes, according to Wilson (1978, 111), "lack objective reality, despite having representational character." It is again not clear what this means. Is the suggestion that sensory ideas apparently carry content but actually do not refer? Wilson (1978, 112) suggests that Descartes thinks of objective reality as something the idea 'wears on its face'. However, in the case of materially false ideas this does not hold and we cannot infer from the objective reality of the idea to its formal cause. Here Wilson seems to mean by 'representational character' the connection between the objective reality of an idea and its *formal cause*. The question is whether an idea has in some sense to *represent* what causes it. John Searle (1983, 61) has claimed, for instance, that a perception is uniquely characterized by representing its conditions of satisfaction, such that in seeing a yellow station-wagon, I know that that seeing precisely involves that it is caused by the presence of a yellow station wagon. Other philosophers of perception, *e.g.* Edmund Husserl, deny that perception necessarily involves grasping a causal connection. The question is, as Cecilia Wee puts it, whether an idea has to both represent its cause in some way, and also be an adequate representation of that cause. She writes:

> . . . an idea represents truly only if (1) the idea comes from the cause from which it purports to come, and (2) the idea accurately represents that cause. (Wee 2006, 39–40)

For example, consider a toy that says "made in China" on it. Some of our ideas—what Descartes calls our 'adventitious' ideas—come with 'made in the outside world' written on their faces, to use Wilson's expression. Other ideas could simply have written on them 'possible candidates for extra-mental existence'. In her discussion, Wee claims that the idea of the sun is not materially false, although it is inaccurate. It is still an idea *caused by the sun*. Now, the idea of the sun is problematic because Descartes thinks we operate with at least *two* different ideas of the sun. There is the commonsense idea and the astronomer's idea, which, as we shall see, is a constructed idea.

Descartes' argument in the Third Meditation will be that only in the case of God is this testimony (what is as it were written on the face of the idea, to use Wilson's locution) incontrovertible. Only 'God' is an idea which in and of itself is most true (AT 7:46), although in a certain sense not even chimeras are false in themselves (Meditation Three and *Letter to Clerselier* 23 April 1649, CSMK 376/AT 5:354).

Norman Wells (1984, 27–8) sees the problem as one of distinguishing between representing a true or false object and the notion of representing *falsely*. Wells is right to remind us that Descartes follows tradition in thinking of entities themselves as *true* in so far as they have being ('truth consists in being'; *Letter to Clerselier* 23 April 1649, CSMK 377/AT 5:356) and that falsity and deception are rooted in non-being, in defect. In other words, in this special ontological sense not just judgments but things may be said to be 'true'. Thus in Meditation Three the idea of God considered in itself is considered to be most true. What these true ideas 'represent', 'exhibit' (*exhibet*, AT 7:42 l. 30, 46 l. 14), or 'contain' (*contineat*, AT 7:46 l. 9) is something stable and immutable (a true essence, genuine *entia*). As Lilli Alanen (1994a, 208) remarks, the 'objective' truth of an idea is wedded to its objective reality and to the clarity and distinctness with which that objective reality is conceived. But the formal truth of the idea consists in its accurately representing its cause and conveying accurately the qualities of that cause as qualities contained in the objective reality of the idea.

From Suarez Descartes took the idea that an idea in and of itself cannot be false. In a sense whatever is given in *simplex apprehensio* must be true. Simple apprehension simply conforms to its object as Suarez says (*Disputationes Metaphysicae*):

> It is impossible that a concept of the mind be given which does not have something which represents the proper object; therefore, if it is compared to something else, it is not possible for it not to be a true conception of such an object since it necessarily is of it and represents it nor can it represent it naturally, if not intentionally and conforming to it, it is therefore true, since truth is nothing other than conformity of intellect to the thing. (*Disputationes Metaphysicae* 8,3,3; 25:284)[11]

To have an idea is to have something represented to one and this involves some kind of conformity with what is represented. This is the basis of both

Suarez's and Descartes' thinking on this subject. Ideas taken in themselves cannot strictly speaking be false. When I imagine a goat or a chimera I have a genuine object before my thought even if it does not exist; only in the case of God is an essential judgment also an existential judgement.

We cannot misapprehend at the level of *simplex apprehensio,* as Descartes emphasizes in reply to Gassendi (CSM 2:260/AT 7:378) and in his letter to Hyperaspistes August 1641: "falsehood is never apprehended as truth and those who deny that we have an idea of God do not really apprehend this even though perhaps they affirm it, believe it and argue for it" (CSMK 195/AT 3:432). Descartes here clearly is making use of the notion of apprehension as distinct from what we actually conceive or think.

What precisely is in the objective reality of an idea? The problem is whether the 'objective reality' of the idea is the total 'content' of the idea; *i.e.,* whether it actually possesses all the properties in the idea or whether the objective reality is a *subjective phenomenal appearance* which gives us just an aspect or profile, an *Abschattung* of the object, to use Husserl's terms. Is it the ideal essence of the object, or just those qualities we are currently able to think in the object? If I am thinking of 'gold', is the objective reality of that idea that it is yellow, heavy, precious, shiny metal, or is it the chemical composition, its place in the table of elements, and so on? Twardowski attempts to distinguish between properties of the object and properties of the content of my thought and with Descartes we must do likewise (and Malebranche makes this problem acute).

This question becomes crucial in regard to the argument for the existence of God. Because if the objective reality does not guarantee a firm toe-hold from which to argue to the need for a formal reality as the sole cause of that objective reality, then the whole notion of the objective reality of the idea becomes fraught with difficulty. It becomes unreliable in precisely the way in which Brentano's ambiguous understanding of content became unreliable as between the *psychical occurrence* and the *ideal* signification, to use Husserl's (1994, 388–395) terminology in his criticism of Twardowski.

Let us leave aside for the moment the question of the objective reality of the idea of God and concentrate on those ideas which Descartes acknowledges can be materially false, ideas of sense and ideas of privative and negative concepts (but not the ideas of infinite objects, which are positive for Descartes). Are sensory ideas intrinsically deceptive? Or are they merely deceptive because of some extrinsic context in which we normally find them embedded?

According to Descartes we assume two things when we have sensory ideas: that they come adventitiously from outside (AT 7:38, 9.1:30), and that they resemble qualities in the objects which cause them (AT 7:38). He is vehement in denying the second claim all through his life, from his early works to the *Principles.*

In his Conversation with Burman Descartes returns to the idea of the falsity of ideas (AT 5:152). He is challenged with the objection that there is no error within ideas considered in themselves but only in so far as they refer beyond themselves to something external (*relatio et applicatio ad res externas,* AT 5:152). In response, Descartes is recorded as saying that if he considers an idea (*e.g.* whiteness) in itself then even without referring the idea to anything else outside me, *ad nullam rem extra me referram* (*e.g.* a white thing), then even the idea of whiteness itself is somehow materially false and 'material for error' (*materia errandi*). I am making an error about the 'nature' (*natura*) of whiteness, as to what kind of quality it is. This occurs even if I am thinking of white *in abstracto* and am not referring it to anything else:

> Even if I do not refer my ideas to anything outside myself, there is still subject matter for error. . . . For example, I may say whiteness is a colour; and even if I do not refer this idea to anything outside myself [*ad nullas res extra me referram*]—even if I do not say or suppose that there is any white thing—I may still make a mistake in the abstract [*in abstracto*], with regard to whiteness itself and its nature or the idea I have of it. (Cottingham 1976, 11/AT 5:152)

Descartes here suggests that even the thought of 'white' on its own can be misleading if we think of it (as we naturally do) as a quality inherent in things. A materially false idea is one which naturally leads the mind to a certain assumption and that assumption unfortunately is incorrect (metaphysically and epistemologically).

In what sense does an idea contain all the content it ought to have? Descartes, as a mathematician, recognizes that one can frame an idea of a mathematical figure (*e.g.* a triangle) and consider many of its properties but there is still room to discover new properties of that figure without the idea becoming 'augmented', so that our grasp on the idea becomes clearer (see CSM 2:256/AT 7:371). Are these properties somehow already 'contained' in the objective reality of my idea? Descartes' answer is yes: we get clearer about the objective reality. He often speaks of what is immediately available in an idea; he does not want to say that in entertaining an idea, one is at the same time entertaining all the logical consequences of that idea. Clearly we need to distinguish between the 'aspect' of the objective reality of the idea that is available immediately to me, and the ideal objective content of the idea, such as would be available to the expert or perhaps as it is in the divine mind. Husserl in his review of Twardowski makes this precise distinction between the real psychological token, as it were, and the ideal sense that it instantiates. Malebranche and Arnaud develop Descartes' thoughts further in this area but it is clear that Descartes thinks that an idea, properly considered, can allow the essence of its object to show itself, *i.e.* the nature of that entity, as possible; this is at the heart of the exercise in the Fifth Meditation. But Descartes wants to distinguish between different ways of apprehending the idea. For instance, the sensory ideas of the geometrical shapes themselves do contain material for error; precisely because sensory experience clouds the intellect. It is not the actual drawn line that the geometer relies on to construct the idea and arrive at the essence, but an ideal line purified of extraneous and misleading sensory content. Descartes is aware that he is making a fine distinction here: sensory ideas of size and shape (according to the Fifth Meditation) can be relied on in so far as they convey accurate information about true size and shape (a right-angle not only looks like 90 degrees but can be measured to show it is 90 degrees). Descartes does not draw the inference that Berkeley does, namely, that sensory ideas of primary qualities are as misleading as ideas of secondary qualities. This is because Descartes somehow thinks that certain sensory ideas more clearly point beyond themselves to the nature (and not just the existence) of their source. Claudia Lorena García puts it as follows:

> Take, for example, our sensory ideas of sensible qualities: when Descartes classifies these as materially false and says that a materially false idea represents a non-thing as a thing, he is not merely saying that the putative qualities these ideas seem to represent *happen not to exist* in the physical world. He is saying something stronger: namely, that, unlike the shapes or figures which we sensorially perceive and which can actually exist in corporeal substances, those putative qualities we sense—such as colours, heat, or cold—cannot exist anywhere. (García 2001, 357)

Similarly, in the *Principles of Philosophy,* Descartes says that color and pain (*color et dolor*) "are judged to be real things existing outside our mind [*extra mentem nostrum existentes*], there is no way of understanding what sort of things they are" (CSM 1:217/AT 8.1:33), and concludes the passage:

> If he examines the nature of what is represented by the sensation of colour or pain—what is represented as existing in the coloured body or the painful part [*tanquam in corpore colorato vel in parte dolente existens, repraesentet*]—he will realize that he is wholly ignorant of it. (CSM 1:217/AT 8.1:33)

A Cartesian idea must be understood not just as a mental occurrence (of a certain temporal duration, qualitative feel, act quality, and so on), and not just as the object conveyed by the idea or thought, but also precisely as an act of intending, an intentional act with its own complex structure (as Husserl attempted to articulate). The occurrence in the mind is a temporal disturbance of the conscious flow; it has a certain peculiar character as an act (its 'act quality', to use the language of Husserl's *Logical Investigations*) and also a certain qualitative feel and experiential content (which belongs both to the act and the object-as-experienced-in-the-act). But beyond that there is an act whereby it intends or means something. I can intend this specific dog even if I am thinking the idea 'animal'. This specific referential feature of the act is not explicitly named by Descartes, but he invokes it all the time to characterize how finite humans possess a true idea of the infinite God, no matter what errant content their idea may have. This intending allows non-theologically expert believers—such as Mersenne's "the natives of Canada, the Hurons and other primitive peoples" (CSM 2:89/AT 7:124), invoked in the Second Objections—to have an idea of God that correctly is of God even if the phenomenological properties of that idea (God is a sacred tree or rock) are misleading.

Descartes wants to say that everyone can have a genuine idea of God. "What about the Hurons?," he was asked by the theologian Marin Mersenne, who think God is a tree or a stone? According to Descartes the Huron idea of God falsely represents God but still counts as a genuine idea of God. All humans, according to Descartes—including the Hurons—have an innate idea of God, though through inattention and through an indiscrimination between imagination and intellect the Hurons populate their idea poorly. The case of 'coldness' is not the same, because here it is a question of our corporeal incarnation cloaking our experience in a certain manner that is completely natural and, in daily life, unavoidable. We cannot shed our skin; so we will always have materially false experiences of cold. Only the scientist—from a 'transcendental' view, as it were—can have a genuinely scientific concept of 'energy' (or whatever) that cuts across our folk conceptions of heat and cold.

As Descartes clearly states in Axiom V of the Second Replies, the security of all our knowledge, both sensible and intellectual, depends on the acceptance and employment of the axiom that "the objective reality of our ideas requires a cause which contains this reality not merely objectively but formally or eminently" (CSM 2:116/AT 7:165). Descartes' notion of objective reality combines psychological or semantic content with a certain measure of ontological actuality or reality, a reality which requires a causal explanation. This content signifies, and may be causally linked to, but should not to be confused with, the idea's *object,* the thing (*res*) it actually represents.

Descartes stressed the distinction between act and content by insisting that they require separate causes: a cause of the *act* and a cause of the *content* of the thought.[12] In one sense all ideas are *of* the mind (they are mental entities, modes of thinking substance); in another sense, they are not, since they are about ('of' in a different sense) other things. Though, Descartes allows, we can recognize an object as much by its shape as by its color (*e.g. Principles* I §69, CSM 1:217–8/AT 8.1:33), nevertheless, our cognition of *what shape is* is more clear and certain than our cognition of what color is (Wilson 1993, 164–6, 173). This is the message of the account of the wax example in Meditation Two, but it is also the central message of the important discussion of the two ideas of the sun in Meditation Three. Although Descartes holds that the primary qualities manifested in our idea of that object are exhibited exactly as they are possessed in formal reality by that object itself (*e.g., Principles* I §70), this statement cannot be taken at face value, because Descartes is clearly aware that we sense size, motion, shape, and so on, in a way determined by our sense organs, bodily location relative to the object, and so on. Hence Descartes is not saying that the primary quality (*e.g.* size) as sensorially perceived exactly mirrors the property as it is in the object. We know *what* size and motion are and we further know that they must be in the body. But how precisely these primary qualities are in the body is determined by mathematical physics, the laws of perspective, and so on, and *not* by sense observation and its associated 'impressions'.[13]

Descartes' language is revealing, particularly his use of *tanquam* ('as it were') with regard to the source of our sensory idea of the sun:

> For example, there are two different ideas of the sun which I find within me. One of them, which is acquired as it were [*tanquam*] from the senses and which is a prime example of an idea which I reckon to come from an external source [*inter illas quas adventitias*], makes the sun appear very small. The other idea is based on astronomical reasoning, that is it is derived from certain notions which are innate in me [*ex notionibus quibusdam mihi innatis*] (or else it is constructed by me in some other way [*vel quocumque alio modo a me factam*]), and this idea shows the sun to be several times larger than the earth. Obviously both these ideas cannot resemble the sun which exists outside me; and reason persuades me that the idea which seems to have emanated most directly from the sun itself has in fact no resemblance to it at all. (CSM 2:27/AT 7:39)

Here Descartes is not even persuaded that the sensory idea of the sun actually comes from outside the intellect (which he normally characterizes as *extra intellectum* or *extra mentem*); rather the sensory idea is a product of certain implicit judgments made by us (CSM 1:170/AT 6:138) combined with a certain 'natural' tendency of our sensory experience. But he is certain that the astronomical idea is either constructed or derived from innate ideas within me by a process of reasoning—a process he had already described in the *Dioptrics* (CSM 1:172/AT 6:140). It is clear that Descartes is tending towards the view that all sensory ideas contain in them 'material for falsity' and ultimately may be considered to be materially false—not just heat, cold, heaviness, lightness, roughness, smoothness, but also the apparent sensory sense of distance, depth, and so on. Nothing in his express examples of 'materially false' ideas restricts the class of such ideas to those examples. Only ideas that lend themselves to *a priori* conceptual 'construction'—purely scientific or geometric ideas—will ultimately have nothing of material falsity in them. The exact opposite of a materially false idea, indeed, is the (genuine) idea of God, which contains only what is causally put there by its eminent source.

In conclusion, then, Descartes' analysis of the phenomenological content, conceptual content, referential character, and intended object of conscious experiences is far richer and more subtle than commentators in general have acknowledged. There is now (*cf.* Perler 1996) a growing recognition that

Descartes did have a conception of the intentionality of conscious experiences. He also carefully distinguished (though when pressed he unfortunately fell back on the Scholastic distinctions of his day) between the phenomenological character of the act and its object, the subjective (sometimes called 'psychological') and objective (sometimes called 'ideal') content of the experience, and the manner in which the idea represents both an object or thing independent of the experience and also carries a representation of the manner in which it is caused. This analysis in many ways foreshadows the complex distinctions made later by Husserl (2001) in his *Logical Investigations.* Of interest also is the manner in which Descartes identifies a kind of natural attitude with a general thesis of world-belief that needs to be systematically challenged by scientific ideas that are in some sense constructed geometrically out of clear and distinct innate ideas. The true idea of the sun is 'constructed' in this way. Descartes' discussion of the source of the content of the idea of a previously unimagined 'ingenious machine': *idea machinae alicujus valde ingeniose excogitatae* (CSM 2:97/AT 7:134)—a discussion quite neglected by commentators—provides a paradigm of Descartes' conception of the knowing process, *i.e.,* how an object is conceived or constituted 'through an idea' (*per ideam*). On several occasions—in the First Replies (CSM 2:75/AT 7:104), the *Synopsis* to the *Meditations* (CSM 2:10–11/AT 7:14), the Second Replies (CSM 2:97/AT 7:134), *Principles* I §§17 and 20 (CSM 1:198–9/AT 8a:11–12)—Descartes considers the case of the 'idea of a highly intricate machine' (*idea alicujus machinae valde artificiosae,* AT 8a:11), a 'very perfect machine' (*machina valde perfecta,* AT 7:14 l. 28) never before conceived by anyone. This example is invoked precisely to explain the concept of 'objective reality', or the 'objective perfection' or 'objective intricacy' of an idea. In a thought experiment, Descartes asks us to conceive of a machine of such intricacy that it bears no resemblance to any existing machine. What is the cause of such an idea? Descartes entertains a number of possibilities. The idea of a complex machine might be caused by an actual complex machine which I have seen at some time, "thus producing an idea resembling the original" (CSM 2:76/AT 7:104). This option is ruled out *ex hypothesi.* A second possibility is that "the cause might be the extensive knowledge of mechanics (*magna Mechanicae scientia*) in the intellect of the person concerned" (CSM 2:76/AT 7:104) or, thirdly, the cause might be a "subtle intelligence" (*magna ingenii subtilitas*) which enables someone "to invent the idea without any previous knowledge." The mind, since it is a substance, has the formal reality to generate the objective reality it requires for the idea from its own resources. Furthermore, Descartes wants clearly to separate the cause of the idea *as act of the mind* from the cause of the complexity of the idea's content:

> Nor will it suffice to say that the intellect itself is the cause of the idea, in so far as it is the cause of its own operations, for what is at issue is not this, but the cause of the objective intricacy [*artificium objectivum*]

> which is in the idea. For in order for the idea of the machine to contain such and such objective intricacy, it must derive it from some cause; and what applies to the objective intricacy belonging to this idea also applies to the objective reality belonging to the idea of God. (CSM 2:75/ AT 7:103–4)

Ideas—even of external objects—can be constituted in the mind out of its own resources. Even ideas of external things such as the sun which are occasioned in us through our sensory experiences need to be systematically replaced by the geometrically constituted clarified ideas of the sun as generated in astronomical reasoning. Descartes is, on this account, what one might call a 'constitutional realist'.

NOTES

1. See *inter alia* Alanen (1994a), (1994b), (2003); Beyssade (1992); Bolton (1986); Clemenson (2007); Field (1993); Kaufman (2000); Secada (2000); Shapiro (2012); Wee (2006); Wells (1984), (2006); and Wilson (1982), (1990), (1993). Descartes' works are cited according to the edition of Adam and Tannery (1996), designated as 'AT'. Except as noted, English translations are from Cottingham, Stoothoff and Murdoch, eds., (1985), designated 'CSM'; English translations of the letters are from Cottingham, Stoothoff, Murdoch and Kenny, eds. (1991), designated 'CSMK'. Citations are by volume:page numbers; occasionally line numbers are indicated by 'l.' or 'll.'.
2. See also Wells (1984, 25), who quotes Kenny, Cottingham and Robert McRae, as all thinking that Descartes' account of material falsity is confused.
3. Clemenson (2007, 70–91), shows the parallels between Descartes' analysis of ideas that misrepresent and the views of Rubio on obscure and confused sensations.
4. Descartes also discusses arguments materially and formally. He cites a syllogistic inference in the Second Replies which he says is 'materially true' (*i.e.*, the three statements in the inference are in fact all true) but that the formal inference is itself formally false and a 'sophism', see CSM 2:107/AT 7:151).
5. The French version reads: "Car étant assuré que je ne puis avoir aucune connaissance de ce qui est hors de moi que par l'entremise des idées que j'en ai eu en moi, je me garde bien de rapporter mes jugements immédiatement aux choses, et de leur rien attribuer de positif, que je ne l'aperçoive auparavant en leurs idées; mais je crois aussi que tout ce qui se trouve en ces idées, est nécessairement dans les choses."
6. The French text reads: "c'est-à-dire si les idées que je conçois de ces qualités, sont en effet les idées de quelques choses réelles, ou bien si elles ne me répresentent que des êtres chimériques, qui ne peuvent exister."
7. My translation. The Latin text reads: "Conceptus objectivus dicitur res, vel ratio, quae proprie et immediate per conceptum formalem cognoscitur seu representatur, ut verbi gratia, cum hominem concipimus, ille actus quem in mente efficimus ad concipiendum hominem vocatur conceptus formalis, homo autem cognitus, et representatus illo actu dicitur conceptus objectivus." See F. Suarez, *Disputationes Metaphysicae* 2.1.1 (Salamanca, 1597; rpt. Paris: L. Vivès 1856–67), 25:65. For commentary on this passage, see Cronin (1996, 79*ff*), and Wells (1993), esp. 522–528.

8. Wells (1990, 34 n.3) seeks to characterize the two uses of *objective* and *formaliter* with respect to ideas as an epistemological notion distinct from the ontological distinction between formal and objective reality in ideas. Wells (1990, 45 n. 51) likewise argues that Descartes' use of *formaliter* to mean the idea thought of as the act of representing is taken from Suarez, *Disputationes Metaphysicae* 25: I, 39 and 25,910.
9. 'Being of the thing' (*entitas rei*) clearly means here the thing understood as a possible essence, as a set of perfections understood by the mind. Descartes further claims that every genuine idea represents an essence.
10. Wells (1984, 2006) also correctly argues against Wilson that the notion of representing is none other than the idea of exhibiting. There is no need to separate as she does between being 'of a thing' and having 'representational character'.
11. My translation of a passage in Suarez quoted in Wells (1984, 29 n. 25).
12. This is recognized by Perler (1996, 78–99), who employs the notation of Chappell (1986, 177–98), whereby ideas as operations of the intellect ('$ideas_m$') are distinguished from ideas in the objective sense ('$ideas_o$'). I have not employed this notation as it tends to mask the more complex relation between act, content and object which Descartes employs.
13. In the Sixth Replies Descartes makes a very interesting distinction between 'three grades of sensory experience': the physical stimulation of the sense organs (in terms of the motion of particles), the immediate effects in the mind (*e.g.*, colors and smells) due to the "intermingling of mind and body," and the "judgments about things outside us that we are accustomed to make from our earliest years" (CSM 2:294–95/AT 7:437).

REFERENCES

Alanen, Lilli, 2003. *Descartes's Concept of Mind.* Cambridge, Mass., and London, Harvard University Press.

———, 1994a. "Une certaine fausseté matérielle: Descartes et Arnauld sur l'origine de l'erreur dans la perception sensorielle." In: J.-M. Beyssade and J.-L. Marion, *Descartes. Objecter et répondre* (Paris, PUF), 206–230.

———, 1994b. "Sensory Ideas, Objective Reality and Material Falsity." In: J. Cottingham, ed., *Reason, Will and Sensation: Studies in Descartes's Metaphysics.* (Oxford, The Clarendon Press), 229–250.

Ariew, Roger, and Marjorie Grene, 1995. "Ideas, In and Before Descartes." *Journal of the History of Ideas* 56: 87–106.

Beyssade, J. M., 1992. "Descartes on Material Falsity." In: P. D. Cummins and G. Zoeller, eds, *Minds, Ideas and Objects. Essays in the Theory of Representation in Modern Philosophy* (Atascadero, Cal., Ridgeview), 5–20.

Bolton, Martha, 1986. "Confused and Obscure Ideas of Sense." In: A. O. Rorty, ed., *Essays on Descartes' Meditations* (Berkeley, University of California Press), 389–403.

Chappell, Vere, 1986. "The Theory of Ideas." In: A. O. Rorty, ed., *Essays on Descartes' Meditations* (Berkeley, University of California Press), 177–198.

Clemenson, David, 2007. *Descartes' Theory of Ideas.* London, Continuum.

Costa, M., 1983. "What Cartesian Ideas Are Not." *Journal of the History of Philosophy* 21: 237–38.

Cottingham, John, 1976. *Descartes' Conversation with Burman.* Oxford, The Clarendon Press.

Cronin, T. J., 1966. *Objective Being in Descartes and Suárez.* Analecta Gregoriana 154; Rome, Gregorian University Press.

Dalbiez, Roland, 1929. « Les sources Scolastiques de la théorie cartésienne de l'être objectif à propos du 'Descartes' de M. Gilson. » *Revue d'Histoire de la Philosophie* 34: 64–72.

De Rosa, Raffaella, 2004. "Descartes on Sensory Misrepresentation: The Case of Materially False Ideas." *History of Philosophy Quarterly* 21.3: 261–280.

———, 2008. "Material Falsity and Error in Descartes's *Meditations*." (Review of Wee 2006) *Journal of the History of Philosophy* 46.4: 641–642.

Descartes, René, 1996. C. Adam and P. Tannery, eds., *Oeuvres de Descartes*, 11 vols. Paris, Vrin.

———, 1985. J. Cottingham, R. Stoothoff and D. Murdoch, eds. and trs., *The Philosophical Writings of Descartes*, vols. 1, 2. Cambridge, Cambridge University Press.

———, 1991. J. Cottingham, R. Stoothoff, D. Murdoch and A. Kenny, eds. and trs., *The Philosophical Writings of Descartes*, vol. 3: *The Correspondence*. Cambridge, Cambridge University Press.

Field, Richard W., 1993. "Descartes on the Material Falsity of Ideas." *Philosophical Review* 102: 309–334.

Flage, Daniel E., and C. Bonnen, 1999. *Descartes and Method: A Search for a Method in Meditations*. London and New York, Routledge.

García, Claudia Lorena, 1999. "Transparency and Falsity in Descartes' Theory of Ideas." *International Journal of Philosophical Studies* 7.3: 349–372.

Gilson, Étienne, 1930. *Études sur le rôle de la pensée médiévale dans la formation du système cartésien*. Paris, Vrin.

Hatfield, Gary, 2003. *Descartes and the Meditations*. London and New York, Routledge.

Husserl, Edmund, 1994. D. Willard, tr., *Early Writings in the Philosophy of Logic and Mathematics*. Collected Works 5. Dordrecht, Kluwer.

———, 2001. D. Moran, ed.; J. N. Findlay, tr., *Logical Investigations*, 2 vols. Introduction by Dermot Moran, Preface by Michael Dummett. London and New York, Routledge.

Kaufman, Dan, 2000. "Descartes on the Objective Reality of Materially False Ideas." *Pacific Philosophical Quarterly* 81.4: 385–408.

Kenny, Anthony J. P., 1968. *Descartes: A Study of his Philosophy*. New York, Random House.

Kobusch, Theo, 1987. *Sein als Sprache. Historische Grundlegung einer Ontologie der Sprache*. Leiden, Brill.

Normore, Calvin, 1986. "Meaning and Objective Being: Descartes and His Sources." In: A. O. Rorty, ed., *Essays on Descartes' Meditations* (Berkeley, University of California Press), 223–241.

Perler, Dominik, 1996. *Repräsentation bei Descartes*. Frankfurt am Main, Klostermann.

———, and Johannes Haag, 2010. *Ideen. Repräsentationalismus in der Frühen Neuzeit*, 2 vols. Berlin and New York, de Gruyter.

Searle, John R., 1983. *Intentionality. An Essay in the Philosophy of Mind*. Cambridge, Cambridge University Press.

Secada, J., 2000. *Cartesian Metaphysics: The Scholastic Origins of Modern Philosophy*. Cambridge, Cambridge University Press.

Shapiro, Lionel, 2012. "Objective Being and 'Ofness' in Descartes." *Philosophy and Phenomenological Research* 84.2: 378–418.

Simmons, Alison, 1999. "Are Cartesian Sensations Representational?" *Nous* 33: 347– 69.

Twardowski, Kasimir, 1982. *Zur Lehre vom Inhalt und Gegenstand der Vorstellungen. Eine Psychologische Untersuchung*. Wien, Hölder 1894; rpt. with an Introduction by Rudolf Haller, München: Philosophia Verlag, 1982.

Twardowski, Kasimir, 1977. R. Grossmann, tr., *On the Content and Object of Presentations*. The Hague, Martinus Nijhoff.

Vinci, Thomas C., 1994. *Cartesian Truth*. New York, Oxford University Press.

Voss, Stephen, ed., 1993. *Essays on the Philosophy and Science of René Descartes* Oxford, Oxford University Press.

Wee, Cecilia, 2006. *Material Falsity and Error in Descartes's Meditations*. London and New York, Routledge.

Wells, Norman J., 1984. "Material Falsity in Descartes, Arnauld and Suárez." *Journal of the History of Philosophy* 22: 25–50.

———, 1990. "Objective Reality of Ideas in Descartes, Caterus and Suárez." *Journal of the History of Philosophy* 28.1: 33–62.

———, 1998. "Descartes and Suárez on Secondary Qualities: A Tale of Two Readings." *The Review of Metaphysics* 51.3: 565–604.

———, 2006. *Descartes on Material Falsity.* New York, Peter Lang.

Wilson, Catherine, 2003. *Descartes's Meditations: An Introduction*. Cambridge, Cambridge University Press.

Wilson, Margaret D., 1978. *Descartes*. London, Routledge & Kegan Paul.

———, 1990. "Descartes on the Representationality of Sensation." In: J. A. Cover and M. Kulstad, eds., *Central Themes in Early Modern Philosophy* (Cambridge, Mass, Hackett Publishing Co.), 1–22.

———, 1993. "Descartes on the Perception of Primary Qualities." In: S. Voss, ed., *Essays on the Philosophy and Science of René Descartes* (Oxford, Oxford University Press), 162–76.

Yolton, John W., 1981. "Perceptual Cognition with Descartes." *Studia Cartesiana* 2: 63–83.

5 Quine's Conception of Objects

Beyond Realism and Anti-realism

Antti Keskinen

1 INTRODUCTION

If a philosopher professes to be a naturalist, it is natural to assume that she is also a realist, at least a realist about physical objects.[1] As perhaps the most influential twentieth-century naturalist, W.V.O. Quine, puts it, a naturalist leaves it to science to tell what there is and subscribes to an "unregenerate realism, the robust state of mind of the natural scientist who has never felt any qualms beyond the negotiable uncertainties internal to science" (1981a, 72). He describes his position as "robust realism," "unswerving belief in external things—people, nerve endings, sticks, stones" (1981c, 21). Hence, for Quine, physical objects like atoms, sticks, and stones are "ultimately real denizens of an ultimate real world" (1995b, 260). Since science, in a broad sense that also includes the social and the human sciences and parts of common sense, manifestly treats of material objects, there is no further philosophical question about their existence in general.

Thus, Quine explicitly adheres to realism about external objects, both macro- and microphysical ones. On the other hand, he has in many books and articles elaborated a view of objects as our constructions, posits or "reifications" of our theories of the world. He links the positing of objects to linguistic reference, and even describes objects, ontology, as a mere "spin-off" of our use of referential language (1986, 115). This part of Quine's philosophy makes it seem that he conceives of objects as dependent for their existence on us, more precisely on our holding specific kinds of theories. This in turn can be, and has been, understood as an anti-realist position about objects.

It seems that Quine's writings can be used to support a realist as well as an anti-realist reading. At a first glance, it might be thought that these do not really amount to two different readings at all: when Quine is speaking about ontology, he is a naturalistic realist, and it is only when he is engaged in epistemology that he speaks about objects being posits, constructed by us. I think this reaction does not by itself suffice to remove the unease one may feel about the coexistence of the realist and the anti-realist elements in Quine's philosophy. This is because for Quine, philosophy, in particular epistemology, is not a pursuit separate from the rest of science. Epistemology studies the relations between theories and their sensory data by studying denizens of the same world that other branches of science study. Thus, Quine's conclusion that any object talked about in any theory is a theoretical posit cannot be understood as some kind of extra-scientific, peculiarly philosophical point that would not concern the objects to which existence is granted in the science to which the Quinean naturalist looks for an answer to the question of what there is.

Roger F. Gibson has stressed the role of Quine's naturalism in the interpretation of those of Quine's epistemological views that seem to undermine his naturalistic acceptance of the parts of science on which his philosophy draws. In this paper, I argue that, in the case of Quine's conception of objects as theoretical posits, the appeal to naturalism does not resolve the apparent tension between the realistic and the anti-realistic elements in Quine's philosophy. I further argue that Quine's account of objects is not in tension with his robust realism, once we properly place that realism in the context of Quine's philosophy, especially his view that we cannot help but occupy the point of view of some theory or another. As regards the realism debate, which concerns realism as a metaphysical view, I argue that Quine cannot be classified as either a realist or an anti-realist.

Although Quine's view of the nature of objects does not amount to anti-realism, I think it nevertheless amounts to a substantial philosophical account. As an illustration of the consequences of Quine's conception of objects, I will consider the case of the idiom of *de re* propositional attitudes.[2] Quine thinks the idiom of propositional attitudes is practically indispensable in daily life and in some of the sciences, but he despairs of assigning any clear content to attitude ascriptions other than the *de dicto* kind, which he construes as an affirmation of a relation between a subject and a sentence. Quine's argument against *de re* attitude ascriptions has generated a slew of responses devoted to showing that Quine is mistaken and that *de re* ascriptions make sense. Here I examine the issue of *de re* ascriptions in connection with Quine's conception of objects, a connection he himself did not examine. I shall argue that his view of the nature of objects sets a strong necessary condition on the truth of an ascription of a *de re* attitude to another subject. I further argue that Quine's conception of objects entails that an ascription of a *de re* attitude to another person can be true only if that person's theory of the world has exactly the same empirical content as the ascriber's theory. Although this does not show that *de re* ascriptions make no sense, it constitutes a very substantial view about such ascriptions, one which appears highly counter-intuitive in view of everyday attitude-ascription discourse.

This paper is structured as follows. I begin with a discussion of the realistic and the anti-realistic elements in Quine's conception of objects and of why the coexistence of these elements in his philosophy may seem problematic (§2). In that connection, I clarify what I mean by 'realism' and 'anti-realism' concerning objects. §3 expounds the relevant details of Quine's conception of the nature of objects. After these preliminaries, I argue that there is no tension or problem in the coexistence of the so-called realistic and anti-realistic elements in Quine's conception of objects (§4). I also argue that Quine does not occupy either side of the metaphysical realism debate; in fact, I shall argue that the whole debate lacks sense from the Quinean point of view. I further argue that Quine's conception of objects has the implication mentioned above regarding the ascription of *de re* attitudes (§5). I conclude (§6) that Quine's account of objects is consistent, and transcends the realism–anti-realism debate about objects, but nevertheless may have implausible implications for the ascription of *de re* attitudes.

2 REALISM AND ANTI-REALISM ABOUT OBJECTS

The central thesis of Quine's naturalism is that there is no first philosophy: Science determines what there is and how we come to know what there is. It is not the task of philosophy to accomplish this independently of science. The Quinean philosopher operates from within our current cluster of theories about the world, and hence accepts many of the ontological commitments of common sense and of the special sciences. As Susan Haack (1993) has argued, it is not entirely clear what 'science' is supposed to mean in Quine's description of his naturalistic position. Sometimes Quine explicitly uses the term 'natural science', as when he (1992b, 9) says: "naturalism looks only to natural science . . . for an account of what there is and what what there is does." It is fairly clear, however, that Quine sees science as involving ontological commitment not only to theoretical entities of physics such as electrons, but also to many kinds of material objects of common sense, namely bodies. Quine's naturalistic commitment to science as the arbiter of what there is results in realism about objects in the external world, such as nerve endings, sticks, and stones. As mentioned, this realism about objects also underlies Quine's epistemology: Quine (1992a, 19) approaches the question about the relation of theories to sensory data as a question about "flesh-and-blood denizens of an antecedently acknowledged external world."

But Quine's unswerving belief in the existence of external things seems to be tempered with the conclusions of his epistemological investigations. From the perspective of Quine's (1981c, 20) epistemology, any object, not just the unobservable ones of mathematics and science, is nothing but a theoretical assumption, a posit of one theory or another. Quine connects the positing of objects to linguistic reference, more precisely to quantification in a theory expressed in the form of first-order logic.[3] The theoretical nature of objects arises from Quine's account of the relation between theory and sensory evidence. Central to this account is Quine's theory of observation sentences, observation categoricals, and the role of the quantificational apparatus of logic in the connections between theoretical sentences and the observation categoricals that serve as the test of theories. One of the aims of Quine's (1969, 83) study of the evidential relation is to clarify to what extent science is determined by sensory evidence and how much of it "transcends any available evidence." For example in *Word and Object* (1960, 5) he speaks of "getting man's net contribution" to science by "subtracting his cues from his world view"; in *The Roots of Reference* (1974, 3–4), he talks of seeing to what extent science is "man's free creation," "a put-up job." Quine's conclusion with regard to objects is that they are part of our contribution—they belong to the domain of our free creation. From the epistemological point of view, ontology is just a sort of by-product, a 'spin-off' of our use of quantifiers and predicates (1986, 115).

Robert J. Fogelin (2004, 38–39) calls Quine's epistemological conception of objects *anti-realist.* The realist position that Fogelin contrasts with Quine's anti-realism consists in the thesis that *independently of us, the world contains all sorts of objects,* while the anti-realist position consists in the view that *objects are posits that we introduce as part of our theoretical activities.* The anti-realist claims that objects are dependent upon us, and that this dependence consists in objects' being introduced through our theoretical activities. However, it is important to note that Quine's view of objects as dependent on us does not mean that theories would not be answerable to the external world. The purpose of Quine's theory of observation sentences and observation categoricals, discussed in the next section, is to show that there is a level of language at which a theory can be tested against the external world:

> The objectivity of our knowledge of the external world remains rooted in our contact with the external world, hence in our neural intake and the observation sentences that respond to it. . . . Man proposes; the world disposes, but only by holophrastic yes-or-no verdicts on the observation sentences that embody man's predictions. (Quine 1992a, 36)

So, the world does dispose of theories that conflict with it. It is the qualification concerning holophrastic yes-or-no verdicts on observation sentences that leaves room for what Fogelin sees as Quine's anti-realism about objects.

Some clarifications are in order concerning the realism–anti-realism debate with respect to which Quine's views. First, the realism and anti-realism at issue here should not be confused with the corresponding terms in Michael Dummett's (1993) discussions. By Dummett's semantic criterion, Quine would count as a realist since he accepts bivalence for sentences across the board. Second, the relevant notions of realism and anti-realism are connected to the question of whether objects are independent of, or dependent upon us. As already noted, Fogelin formulates the issue of realism and anti-realism about objects in terms of the independence and dependence of objects upon our theoretical activities. This formulation is in keeping with the terms of the contemporary realism debate. For example, Michael Devitt (2010, 33) defines realism as a metaphysical doctrine according to which the tokens of most commonsense and scientific physical types objectively exist independently of the mental. The kind of dependence at issue in this debate is dependence for "existence and nature on the cognitive activities and capacities of our minds" (Devitt 2010, 32). Against various kinds of anti-realists—or *idealists,* as Devitt also calls them—the realists insist that the physical world "is not *constituted by* our knowledge, by our epistemic values, by the synthesizing power of the mind, nor by our imposition of concepts, theories, or languages" (Devitt 2010, 33). Thus, the "cognitive activities and capacities of our minds," can mean very different things depending on the particular anti-realist view in question. The anti-realist position Fogelin attributes to Quine can be described as the view that objects are constituted by our imposition of theories, more precisely theories which utilize the referential idiom of quantifiers, variables, and predicates. Quine's view of objects as theory-dependent is based on his epistemological model of the evidence relation, which is discussed in more detail in the next section.

3 QUINE'S CONCEPTION OF OBJECTS AS THEORETICAL

In his most detailed epistemological studies, Quine investigates the relation between sensory evidence and theory from a genetic point of view, by looking at how a theory of the world could be acquired starting from neural impingements and certain innate psychological endowments like standards of perceptual similarity between episodes of neural stimulation. Quine conceives the genetic strategy as a way to investigate the gap between our "meager input," namely the stimulation of our receptor cells, and the "torrential output," namely, our theory about the world.[4] In this paper, I will not go into the details of Quine's genetic account as presented for example in *The Roots of Reference* (1974) and *From Stimulus to Science* (1995a). For my present purposes, it is sufficient to focus on some features of Quine's model of the evidence relation that arises out of his genetic speculations.

Observation sentences, in Quine's terminology, are occasion sentences on which members of the language community are disposed to give an agreeing verdict (assent or dissent) when jointly witnessing an occasion. They are sentences that "report intersubjectively observable occasions, observable outright" (Quine 1995a, 22), and they are the linguistic expressions most directly associated with ranges of neural intake. However, observation sentences do not by themselves serve to test theories empirically, nor do they by themselves provide theories with empirical content. The reason for this is that no occasion sentence can be derived merely logically from a theory, which is expressed in eternal sentences. Only as components of a special sort of eternal sentences, called free observation categoricals, can observation sentences fulfill those roles. *A free observation categorical* is a sentence of the form

(1) When *F*, *G*

where '*F*' and '*G*' are observation sentences.[5] On Quine's model of the relation between theory and sensory intake, a theory is tested by logically deriving an observation categorical from it and checking whether the prediction the categorical expresses holds. What Quine calls the *empirical content* of a theory is the set of observation categoricals the theory implies.[6] According to Quine's (1992a, 13–17) holistic model, the theory as a whole takes on empirical content in virtue of implying observation categoricals, and the sentences of the theory share in this content only as components of the theory, not in isolation.

In a theory couched in Quine's canonical notation,[7] the implied observation categoricals come in the form

(2) $\forall x(Fx \rightarrow Gx)$

which Quine calls *focal* in distinction to the free form (1), which can be logically represented as

(3) $F \rightarrow G$.

Quine (1993, 109–110) describes the observation sentence as "Janus-faced": it, so to speak, faces *inward* to a subject's neural intake in being keyed to a specific kind of stimulation of the subject's receptors, and *outward* to its subject matter, usually objects in the external world. In its outward orientation toward the subject matter, the sentence is conceived as structured, as consisting of component expressions which "recur in the theory to denote objects the very conception of which is pure theory" (Quine 1993, 110). From the point of view of neural intake, an observation sentence is an unstructured whole keyed to a range of sensory intake (1993, 109). In my view, the observation categorical can be described as in a similar way Janus-faced. Quine (1995a, 25–28; 1992a, 10–11) notes that there is a kind of difference in strength between the focal and the free aspects of an observation categorical, the focal being stronger. In its orientation toward the subject matter of theory, an observation categorical is focal, of the form (2). It contains predicate terms that recur in the theory and are true or false of *n*-tuples of objects. In its orientation toward neural intake, an observation categorical is free, of the form (3). It faces inward to the subject's neural intake, and is not refuted so long as each occasion on which the members of the linguistic community are disposed to assent to the antecedent *F* is an occasion on which they are disposed to assent to the consequent *G*. In a word, a categorical as free holds as long as each *F*-occasion is a *G*-occasion. In contrast, the categorical as focal requires for its truth that each *F* is a *G*, which is a stronger requirement.[8] Considered as focal, the observation categorical is not associated with neural intake in the same direct way that it is when considered as free, because the focal form contains no observation sentences as components (Quine 1992a, 10–11).

In his genetic account, Quine (1974, 99–101) speculates about how the focal observation categorical construction could be learned in what he describes as an "irreducible leap" in language acquisition. Before this leap, the free observation categoricals merely assert the "concomitance or close succession of separately specified phenomena" (Quine 1995a, 25–26), in the manner 'whenever this, that', with no ontological import, that is, with no reference to objects (Quine 1992a, 10). The main learning-theoretic difference between the free and the focal observation categorical constructions is that the focal, unlike the free, cannot be learned by first learning some of its instances on the basis of previous mastery of observation sentences. This learning-theoretic difference, of course, corresponds to an epistemological one. Considered as focal, an observation categorical is not composed of observational language. Its conditions of assent and dissent do not reduce to those of observation sentences. It is in this sense that all objects are theoretical constructions according to Quine's epistemology. The contribution of ontology—that is, of reference to objects by the apparatus of quantification, predicates, and variables—is a purely structural one: by adding logical structure, this referential idiom establishes logical connections between conjunctions of theoretical sentences and observation categoricals.[9]

Through the implication of observation categoricals, empirical content is holistically distributed to the sentences of the theory. Because of this holistic feature, any change in the fund of implied observation categoricals affects the content of every sentence of the theory, and consequently the content of all predicate terms appearing in these sentences. The predicate terms, in turn, are used in describing objects and in differentiating them from each other: for example, some objects are *planets,* others are not; some object *numbers the planets,* and no other object does. All objects being posits, there simply are no 'objects in themselves' in addition to objects-as-represented-in-a-theory. An object is like what the theory says it is like, and there is no room for the question of what the object is like independently of the theory. Because the content of the predicate terms used in saying what an object is like depends on the whole theory, an object depends not only for its existence, but also for its identity, on the whole theory.[10]

Quine (1981c, 20) sees his epistemological discussion of objects as an extension of an older point about unobservable theoretical objects to all objects. The point is that what matters for theory is structure, not objects. Quine bases his acceptance of this point on his view of the semantic primacy of the sentence over the term. At the observational level where theories can be intersubjectively tested, it is observation sentences, not terms, that are keyed to episodes of neural stimulation and that are agreed or disagreed upon by speakers. To briefly illustrate this point, consider two theories that both include the sentence "There are dogs" ($\exists xDx$), one of which tells us that dogs are mammals ($\forall x(Dx \rightarrow Mx)$) whereas the other says they are not mammals ($\forall x(Dx \rightarrow (\sim Mx))$). Now, from the point of view of epistemology, it is an empty question whether these theories share the subject matter of dogs, and disagree about these objects (namely, mammal or not), or whether these theories do not share a subject matter in the first place (namely, they are not both about dogs at all). The distinction between the objects a theory is about and what the theory says about them is not supported in Quine's epistemology. The only epistemological importance of the *term* 'dog' (the predicate '*Dx*') consists in its structural contribution, its role in establishing relations of logical implication between sentences. The theories can be intersubjectively evaluated only in virtue of the observation categoricals they imply, considered as free, namely, as compounds of observation sentences. In this way, theories are not evidentially related to objects that would constitute a shared subject matter for divergent views. All talk about objects—not just talk about unobservable objects like neutrinos[11]—is theoretical talk.

4 BEYOND REALISM AND ANTI-REALISM ABOUT OBJECTS

It thus appears that Quine subscribes to the view that objects depend for their existence and identity on our imposition of theories. Hence, by the standards of the contemporary realism debate as framed by Devitt, Quine would appear to be an anti-realist about objects, as Fogelin claims. This raises the question of how to understand Quine's explicit adherence to "robust realism" about physical objects in the external world, and whether there is a tension between his epistemological views and his realism. Can Quine be justifiably placed on either side of the realism debate? In this section, I address these questions and develop my own answers to them. I begin with a discussion of one rather well-known strategy of showing that Quine's epistemological views do not contradict his naturalistic acceptance of the science on which his epistemology is based.

Roger F. Gibson (1992, 21) argues that naturalism is the key to interpreting Quine correctly, especially with respect to Quine's view of the relation between science and epistemology. Quine (1969, 83) describes this relation as one of "reciprocal containment." Firstly, science is contained in empiricist epistemology. One of Quine's (1969, 75) two cardinal tenets of empiricism is that whatever evidence there is for science is sensory evidence, and epistemology studies the nature of this evidence, how and to what extent science rests on it. One conclusion of Quine's epistemological studies is that theories are not evidentially related to theory-independent objects, but rather to observation categoricals conceived as free, as compounds of observation sentences. That theories treat of objects at all is due to the referential constructions of language, constructions whose sole function from the epistemological point of view is to form logical links between theoretical sentences and the observation categoricals. This conclusion is meant to cover any theory, Quine's epistemology included.

Secondly, epistemology is contained in science. This direction of containment, according to Gibson, is important in avoiding misinterpretations. The containment of epistemology by science consists in the point that Quine's epistemology presupposes the existence of the external world of common sense and of science, which means that there is no exclusively epistemological perspective (Gibson 1994, 451; 1992, 24, 27). Gibson thinks this is the way in which Quine's naturalism provides the key to correct interpretation of his philosophy. Naturalism dictates that "Quine's epistemologizing takes place within an accepted theory, and his epistemological findings do not undermine or repudiate his acceptance of that theory" (Gibson 1992, 21). Quine's rejection of first philosophy extends to his own epistemology, which is not to be understood as a proposal for a piece of first philosophy that would somehow be able to show that the objects assumed in science or in common sense do not really exist. One possible misinterpretation one might seek to expose by drawing attention to this direction of containment is the reading of Quine as an anti-realist about objects. Thus, one might conclude that Quine's talk of objects as posits is to be understood as an epistemological account of theories, but not as an account of the nature of the objects in the external world. These objects are in no sense posited by us; they are independent of our theories and linguistic reference, since their existence is granted already in the naturalistic starting point.

However, such a line of thought should make us wonder what the distinction between objects in the external world and objects from the perspective of one or another theory could amount to. Quine (1960, 22, 274–275; 1981c, 22) is famous for holding the view that there is no "cosmic exile," that we have no choice but to occupy the standpoint of one theory or another. Could Quine see any sense in the claim that, although objects depend, epistemologically speaking, for their existence and identity on our theories and the empirical content holistically distributed to the sentences and predicates of these theories, this is only a point about our theories, not about the objects that exist in the world as described by the science that is presupposed in Quine's epistemology?[12] Consider the following statement about a theory *T*:

(4) *T* has been predictively successful, not one of the observation categoricals implied by it has been falsified. Nevertheless it is false because it is wrong about the objects it talks about.

Quine's view of objects as theory-dependent is quite compatible with the admission of a claim like (4) as a meaningful statement. Theories are of course fallible. *T* may come to conflict with experience, despite its remarkable predictive success. Someday an observation categorical it implies may turn out false. In that case, we may say that *T* was wrong about the objects it talked about or purported to talk about. *T* had to be revised—some of its sentences had to be repudiated, perhaps others added to it. From the point of view of the theory that results from the revision of *T*, call it *T*′, we may say, for example, that *T* assumed objects that do not exist, or that *T* was wrong about some objects that do exist. Or, we may want to say that *T* failed to concede existence to objects of some sort that we have come to recognize as existing after the replacement of *T* with *T*′.

However, according to Quine's account of theory revision, theories are not tested against theory-independent objects. A theory's conflict with sensory evidence consists in failed observational prediction, more precisely, the falsity of an observation categorical implied by the theory, which in turn means that the conflict consists in speakers' being disposed to assent to the antecedent observation sentence of the categorical, but to dissent from the consequent one, when witnessing an occasion. It is only as a free one, a compound of observation sentences, that an observation categorical connects its theory to the theory-independent world. And, as a free one, it involves no reference to objects. If we assume that the revised version of *T*, *T*′, happens to be a perfect predictor, there is no further meaningful standard we could invoke that *T*′ would still have to meet in order to be right about the objects it talks about (Quine 1981c, 22). We can still meaningfully say that *T*′ is a correct description of reality, that it is right about the objects it talks about. It is just that according to Quine, the very notions of 'reality' and of 'object' are always part of some theory or another. In "Structure and Nature," Quine puts this point as follows:

> The very notion of object, or of one and many, is indeed as parochially human as the parts of speech; to ask what reality is really like, however, apart from human categories, is self-stultifying. It is like asking how long the Nile really is, apart from parochial matters of miles or meters. Positivists were right in branding such metaphysics as meaningless. (Quine 1992b, 9; *cf*. 1993, 113)

It may be objected that in this passage Quine is talking only about the *notion* of object, and says that this notion is parochially human. It might be thought that Quine is making a point only about the term 'object' and its equivalents, and the use we make of these terms in our theories, not a point about the objects we talk about in our theories. However, I think Quine's way of speaking of "the notion of object" cannot be understood as limiting his point in this way. This is because Quine links the talk about the notion of object to the question, whether the idea of reality apart from human categories makes sense. He concludes that asking what reality is *really* like apart from our categorization is meaningless. The notion of object belongs within our means of categorization, and a distinction between reality as categorized by us and reality as independent of our categorization is declared meaningless metaphysics. Accordingly, I do not see that Quine's talk of the *notion* of object here would leave room for the idea that his point about parochiality would not concern objects.

However, Quine's view of objects as dependent upon our theories does not involve any claim to the effect that our parochially human reality of objects would be less than real:

> [W]e are now seeing ontology as more utterly a human option than we used to. . . . Must we then conclude that true reality is beyond our ken? No, that would be to forsake naturalism. Rather, the notion of reality is itself part of the apparatus; and sticks, stones, atoms, quarks, numbers, and classes all are utterly real denizens of an ultimate real world, except insofar as our science may prove false on further testing. (Quine 1995b, 260)

Since we refer to objects in the statements of our science, there are objects, and Quine's epistemological view about the theory-dependence of objects does not mean that the objects referred to would not be real. Quine's point about meaningless metaphysics, in the passage from "Structure and Nature" quoted above, concerns an urging of the question of whether a theory, no matter how predictively successful, gets the objects that really exist right. From Quine's perspective, this question is totally confused. According to Quine, the objectivity, or better, the intersubjectivity, our theories can have consists in their implication of observational predictions, in the form of observation categoricals. Hence Quine's conception of objects as theoretical posits does undermine the view according to which objectivity would derive from our theories measuring up to theory-independent objects. If the theory, or cluster of theories, that is presupposed by Quinean naturalistic epistemology contained this kind of view about objectivity, then Quine's epistemology would undermine the epistemologist's acceptance of that theory or cluster of theories. But assuming that proposing views about the nature of objectivity is left to the epistemologist, it seems that Quine's naturalism supports his robust realism about objects.

However, Quine also makes it clear that empiricism, especially his own conception of it, is a finding of science. Thus, his commitment to naturalism commits him to the cardinal tenet of empiricism that whatever evidence there is for science is sensory evidence, along with certain more specific views about the nature of this evidence.[13] These more specific views concern the nature of our meager channels of contact with the external world, namely the action potentials in receptor cells caused by the impact of light rays and molecules upon our sensory surfaces. According to Quine (1975, 68), science itself tells us that objects in the external world are not given to us in the sensory intake that is "our only source of information about the external world":

> Natural science tells us that our ongoing cognitive access to the world around us is limited to meager channels. There is the triggering of our sensory receptors by the impact of molecules and light rays. . . . Even the

> notion of a cat, let alone a class or number, is a human artifact, rooted in innate predisposition and cultural tradition. The very notion of an object at all, concrete or abstract, is a human contribution. . . . (Quine 1992b, 6)

Quine's empiricist position in epistemology is as much a consequence of his naturalism as is his acceptance of ontological commitments of science and of common sense. And Quine's empiricist epistemology produces the view of objects as theoretical. Consequently, I do not think that his naturalism could serve as the key to interpreting Quine's view on this matter, that it could alleviate the tension one may see between Quine's realism and his epistemological conception of objects. Since Quine's naturalism supports his empiricist epistemology, which in turn supports his view of objects as posits, I do not think his naturalism allows us to put any more stress on the containment of epistemology in science, and hence on the realistic side of Quine's philosophy, than on the converse containment. Because of the containment of science within empiricist epistemology, and due to Quine's view that we must conceive the world from the point of view of one theory or another, Quine's epistemological account of objects as theory-dependent cannot be seen *only* as an account of theories, but *not* an account of what objects in general are like. If Quine's epistemological conception of objects appears to threaten to undermine the acceptance of the theory (or cluster of theories) on which Quine's epistemology rests, then I do not see how Quine's naturalism could rescue the situation. Thus, Gibson's way of dissolving tensions within Quine's philosophy does not work with regard to the question about the compatibility of Quine's realism and his epistemological conception of objects.

Because of the reciprocity of the containment between epistemology and science, Quine's epistemological account of objects must be seen as applying to all theories. Hence, I think that Quine's account should be taken seriously as an account of the nature of objects in general. Does this mean that the apparent tension between Quine's robust realism about physical objects and his view of objects as posits is real and unresolvable? I do not think so. It is important to bear in mind that Quine's account of objects as posits does not mean that the objects assumed by our best science, whatever that encompasses, would somehow be less than real. The notion of reality is itself always part of some theory or another. The notion of reality "apart from human categories," as separate from the parochial point of view of one or another theory, is from Quine's perspective meaningless. The idea of an object somehow more real than a posit of a theory included in our best current science is nonsensical. The assumption that realism about objects must involve the view that such an idea is not nonsensical, gives rise to the appearance of the tension within Quine's philosophy. If we occupy, as we presumably do, the standpoint of a theory in which sticks and stones are referred to, then sticks and stones are "utterly real denizens of an ultimate real world." And sticks and stones have this status not because they would possess some special kind of reality apart from being posited in our theory, but because we have no choice but to occupy the standpoint of some theory or another and hence take on the ontological commitments of that theory.

Where does Quine stand in terms of the current realism debate? In this debate, realism is understood as a metaphysical view distinct from epistemological issues. Clearly, Quine cannot straightforwardly be associated with the realist camp because of his view that objects depend for their existence and identity on our theories. This kind of dependence Devitt (2010, 32–33) considers characteristic of anti-realism. Someone who seeks to save Quine from anti-realism might argue as follows. Although Quine's view of the nature of objects is completely general, as I have argued in the previous two paragraphs, this is strictly an epistemological view. Hence, Quine could also accommodate realism about objects as a metaphysical view. In fact, it could be argued that Quine proceeds in exactly the way Devitt (2010, 64–65) recommends, by starting with a realist metaphysics and basing a naturalistic epistemology on this metaphysics. This means that we "put metaphysics first," giving it priority over epistemology. We accept the existence of external objects, and their independence of our theories, prior to any detailed study of epistemology, which is just another theory about some of those theory-independent objects.

In a sense, Quine's naturalistic epistemology does "start with metaphysics," in that it is based on certain theories about us as denizens of the physical world. But this is not the sort of primacy of metaphysics over epistemology that would support the thesis that objects are independent of us in the sense that they are not theoretical posits. This is because Quine's epistemological view of objects as posits applies also to the theories presupposed in Quinean epistemology. We are again reminded of the reciprocity of the containment between science and epistemology. Hence, Quine cannot accommodate metaphysical realism as a view about the theory-independence of objects that would have a kind of priority over epistemology, a priority that would make such realism compatible with his epistemological view of the theory-dependence of objects.

For Quine, the notion of reality is always part of a theory, and the real objects in the real world are always objects of some theory or another, and hence, according to Quine's epistemology, are theoretical posits. In my view, Quine is indeed a realist about objects, as he insists he is, but not a metaphysical realist, since his realism does not involve the view that objects are independent of our imposition of theories. As regards anti-realism, his epistemological conception of objects entails nothing with respect to the question of whether there are objects which are real in some further sense than being posits of a theory included in our best current science. Such a question is rejected as meaningless by Quine. And neither does this conception entail that these objects would not be real in the admissible theory-internal sense of 'real'. As Quine reminds us in *Word and Object* (1960, 22), to call a posit a posit is not to patronize it, and we should not look down on the standpoint of a theory as make-believe, for we can never do better than occupy the standpoint of one theory or another.

Whatever one thinks of other aspects of Quine's epistemology, it may now seem that his epistemological theory of the nature of objects is not especially controversial. After all, if what I have said is correct, Quine's conception of objects as theory dependent does not amount to an anti-realist metaphysics; and on the other hand neither does his naturalistic realism amount to a realist metaphysics. Moreover, it may seem that Quine's inquiry into the epistemological standing of objects does not imply anything about any particular (class of) statements, for example that some statements would really be false *because* objects are theoretical posits. However, in my view this impression is not entirely accurate. In the following section, I consider the theme of *de re* attitude ascriptions from the point of view of Quine's conception of objects. Although Quine wrote on the propositional attitudes on several occasions over several decades and came to consider *de re* attitude ascriptions to be meaningless, he never approached the issue from the point of view of his epistemological conception of objects. While Quine's account of objects does not entail that *de re* ascriptions would be meaningless, I shall argue that it does entail that the truth of such an ascription requires that the ascriber's and the "attitudinist's" theories have, at the time such an ascription is made, exactly the same empirical content. This necessary condition on the truth of a *de re* ascription can be considered a substantial consequence of Quine's account, a consequence that is counter-intuitive and hence apt to raise doubts about Quine's conception of objects.

5 QUINE'S CONCEPTION OF OBJECTS AND ATTITUDES *DE RE*

Quine's discussions of the propositional attitudes like belief, desire, fear, wish, and so on, in several articles and books over several decades, have spawned a large amount of commentary. Most of this literature has, to my knowledge, concentrated on the problems pertaining to the ascriptions of *de re* attitudes, ascriptions of mental states that relate a subject to an object or objects. The space of the present paper does not allow for a detailed discussion of the various themes and proposals that have arisen from Quine's writings. My discussion is not meant to be a systematic contribution to the discussion about the attitudes nor a detailed historical exposition or interpretation of the development of Quine's view on the issue. Rather, I focus on what I take to be Quine's final view about the *de re* attitudes, with the sole purpose of illustrating one important consequence of Quine's conception of objects.

Quine's final view about the issue of the idioms of propositional attitude ascription consists in three key points: first, these idioms are indispensable in several contexts (1992a, 72–73); second, ascriptions of *de dicto* attitudes are free of any specific logico-semantic problems that would threaten their general intelligibility; and third, ascriptions of *de re* attitudes are devoid of any clear cognitive content. The second point is related to Quine's sentential account of *de dicto* ascriptions, namely, the account which construes an ascription understood *de dicto,* such as

(5) Ralph believes that Ortcutt is a spy,

as predicating a relation between the *attitudinist* (Ralph) and a sentence (a sequence of phonemes or characters):

(6) Believes (Ralph, 'Ortcutt is a spy').

This sentential construal rids *de dicto* attitudes of any problems regarding nonextensionality. Ascriptions are considered as normal two-place predications where the expression occurring in the second argument position—the designation of the sentence—does not contain as components any expressions occurring in the sentence designated. Quine (1976, 161) sees quotation as being on par with any other way of referring to an expression, for example spelling, in which case it is even more obvious that the sentence is only mentioned, not used. Hence, in Quine's view, although an ascription of a *de dicto* attitude like (5) contains in its complement clause an apparent reference to a particular object, the ascribed attitude is not really 'about' that object, such as the man Ortcutt in the case of (5).

Quine (1995a, 96–97) describes the use of quotation marks to refer to the complement clause of *de dicto* ascriptions as indicating a borderline between the ascriber's world and the world of the attitudinist as "fancied" by the ascriber. The quotation marks "mark an opaque interface between two ontologies, two worlds: that of the man in the attitude, however benighted, and that of our responsible ascriber of the attitude" (1992a, 69–70). The ascription of an attitude is always made from the point of view of the ascriber's ontology, and in the *de dicto* case no objects of this ontology are specified as the objects at which the attitudinist's purported attitude would be directed. For Quine, the *de dicto–de re* distinction is a distinction between dyadic and higher-degree predications. Thus, the *de re* sense of (5), that Ralph believes *of Ortcutt* that *he* is a spy, is construed by Quine, in contrast to the dyadic (6), as a triadic predication:

(7) Believes (Ralph, 'spy', Ortcutt).

In (7), the barrier between the worlds of the ascriber and the attitudinist is breached: The ascriber "interjects reality, by his own lights, into the attitudinist's world" (1995a, 96) by saying that the attitude (in this case, Ralph's belief) is directed at an object or objects, which belong to the ontology of the ascriber's world. Quine's talk of the ontologies of different subjects' "worlds" in this connection is, in my view, to be understood as figurative way of speaking about the subjects' respective theories of the world, their total belief systems. For Quine, it is theories that have ontologies; Quine is certainly not talking about possible worlds, since he rejects the notion of a possible world (1981d).

Quine's reason for claiming that *de re* ascriptions lack clear content is that he thinks it is hopelessly unclear when a *de re* attitude can be ascribed on the basis of an ascription of a corresponding *de dicto* attitude and the extra premise that the object of the purported *de re* attitude exists (from the point of view of the ascriber's theory, that is). Quine approaches this problem in terms of the inference step known as *exportation*. In the case of our example about Ralph and Ortcutt, exportation is the inference from (6) and '$\exists x(x = \text{Ortcutt})$' to (7). Quine thinks that the step of exportation should be allowed only if the attitudinist knows who or what the pertinent object (in this case Ortcutt) is. However, Quine points out that the notion of knowing who or what is thoroughly context dependent: there is no general, context-independent sense of knowing who or what someone or something is. For instance, in the case of people, in one context knowing who can require the ability to recognize that person by appearance while in another context such recognitional ability is irrelevant and one needs instead to be able to specify a social status (Quine 1981b, 121; 1995a, 97). Quine dismisses David Kaplan's solution to the problem of exportation on the grounds of the analogical context dependence of the notion of *believing* who or what someone or something is; Quine construes Kaplan's (1968) idea of a vivid name in terms of this notion.[14] Since Quine thinks the notion of knowing or believing who or what someone or something is does not make any sense "of itself," outside of specific contexts, he concludes that this notion does not yield the desired distinction between admissible and inadmissible cases of exportation.

Sean Crawford (2008, 82) claims that a construal of (5) as an ascription of an attitude that is not an attitude about any object, such as Quine's *de dicto* construal (6), is highly implausible, since (5) appears to say something not only about Ralph but about Ortcutt as well. Crawford argues that (5) is an ascription of a "relational" attitude, since it specifies an object and says that Ralph is belief-related to this object. In this, (5) contrasts with "notional" ascriptions that do not relate the attitudinist to any particular object or objects, such as "Ralph believes there are spies," in the sense of mere belief in the existence of spies.[15] Crawford (2008, 85, n. 17) thinks that in the case of a relational attitude ascription like (5), there is no problem of exportation since the relevant terms (in this case 'Ortcutt') are already exported and the ascribed attitude is really *de re*. Crawford goes on to develop his own account of *de re* ascriptions, on the basis of an earlier proposal by Brian Loar (1972), according to which the truth of a *de re* ascription does not require that the attitudinist have a "strong" epistemic connection to the object, for example that she knows what the object is in the sense of being "acquainted" with it or being able to recognize it.

In what follows, I will try to show that Quine's conception of objects as theoretical sets a peculiar necessary condition on another person's mental state being relational, that is, relating that person to a specific object or objects. Quine never discusses the theme of *de re* attitudes from the point of view of his conception of the nature of objects; he thinks the problem of exportation clinches the case against the meaningfulness of *de re* (that is, relational) ascriptions. The point I want to make about the implication of Quine's account of objects with regard to relational attitude ascriptions is independent of the problem of exportation and also of Quine's construal of attitude ascriptions as predications, exemplified by (6) and (7).

Consider Quine's famous example about Ralph and Ortcutt. Ralph believes that the man he has seen on a local beach one summer afternoon is a spy, and that the man in a brown hat he often sees walking down the street where he lives is no spy but rather a pillar of the community. Assume that Martha, on the other hand, believes that these fellows Ralph has been observing are actually one and the same person, Bernard J. Ortcutt. To continue the story (*per* Quine 1974, 75–78), assume further that Martha discusses with Ralph and convinces him of the pertinent identity claims, and that their discussion leaves them with agreeing dispositions to assent, dissent, and abstain with regard to sentences of the form 'Ortcutt is *F*', where '*F*' is a predicate term, possibly a complex one. At first sight, we might say that in this situation, Ralph's and Martha's theories agree about Ortcutt and that hence there is a common object in their respective ontologies. Thus it might be thought that coming from Martha's mouth, the ascription (5) should be understood as relational and as such considered true.

However, given Quine's holism, the notion that Ralph's and Martha's theories are in agreement about Ortcutt turns out to be very strong. As explained in §3, on Quine's holistic model the sentences of a theory share in the empirical content of the theory only as the logically structured system which is the theory itself. *Any* difference between the empirical contents of the theories held by Ralph and Martha will entail that the sentences 'Ortcutt is *F*' have different contents in Ralph's and in Martha's mouths. In view of what was said in §3 about Quine's epistemology not supporting the distinction between an object as subject matter and what is said about this subject matter, if Ralph's and Martha's theories diverge in empirical content, these theories do not have any common objects in their respective ontologies. Hence the "agreement about Ortcutt" turns out to be a mere surface impression. This divergence in empirical content amounts to the falsity of Martha's purported ascription of a relational attitude to Ralph. Martha's ascription of a relational attitude to Ralph can be true only if her theory has exactly the same empirical content as Ralph's theory.

What I have just said does not entail that Quine's epistemological conception of objects would rule out ascriptions of relational attitudes as meaningless, only that this conception of objects entails that such ascriptions are true *only if* the ascriber's and the attitudinist's theories have exactly the same empirical content. If the theories differ in empirical content, their ontologies are also completely disjointed; in terms of Quine's metaphor of worlds, the ascriber's and the attitudinist's worlds have no objects in common.

At the end of §4, I noted that the necessary condition that Quine's view of objects sets on the truth of *de re* ascriptions is both a substantial consequence of Quine's view, and also counter-intuitive. For this consequence of Quine's view of objects to seem substantial or counter-intuitive in the first place, one needs to be convinced that relational attitude ascriptions are made at all, or made commonly enough for the matter to have any importance. In the context of the present paper, I cannot undertake the task of persuading the reader of this; here I can only say that I tend to agree with Crawford that ascriptions like (5) appear to say something not only about the attitudinist but also about the object referred to by means of the singular term in the complement clause ('Ortcutt' in the case of (5)), and that in my experience ascriptions of this sort occur rather commonly in everyday speech. I think the consequence of Quine's account of the nature of objects at issue here is substantial since it basically says that because of the theoretical nature of objects, the truth of a *de re* or relational ascription depends upon a factor that is never considered in the practice of making and evaluating such ascriptions, namely, upon the empirical contents of the ascriber's and the attitudinist's total belief systems. Thus, from the point of view of common practice, this is a surprising and novel point that may also appear counter-intuitive; at least it violates my intuitions. Furthermore, I would expect that in most cases different subjects' theories differ somewhat in the fund of observation categoricals they imply; differences in personal histories may suffice to bring this about. If this is indeed the case and such differences in subjects' theories are more of a rule than an exception, then Quine's conception of objects entails that false relational ascriptions are more of a rule than an exception.

If you happen to believe, for whatever reason, that most ascriptions of a relational attitude to another subject are indeed false, then likely you will not regard this consequence of Quine's view of objects as problematic. Since we cannot rule out in principle the case where two subjects' theories have exactly the same empirical content, it would be too much to say that according to Quine's epistemology *all* relational attitude ascriptions are false. However, Quine's conception of objects does appear to entail that all such ascriptions of a specific sort are, without exception, false, namely: cases in which *A* ascribes a relational belief to *B* such that *A* considers that belief to be false. If the belief is to be false, then it cannot concern the objects in *A*'s world (that is, in the ontology of *A*'s theory) that *A* portrays the belief as being about. Hence, the ascription itself is after all false: *B* does not believe (however falsely) *of that (those) object(s)* that such-and-such is the case. If we accept Quine's account the nature of objects, then we must also concede that a true ascription of a false relational belief never occurs, and is indeed impossible on epistemological grounds. Again, if you happen to think that all such ascriptions are in fact false, this result may not violate your intuitions. What you may still find implausible is the idea that the reason why these and most other relational ascriptions are false lies in the nature of objects instead of, say, the epistemic limitations to our access to the mental states of others.

6 CONCLUSIONS

Quine's naturalism entails a realistic take on the existence claims of science, and the rejection of any extra-scientific tribunal that could rule on what there is—hence Quine's robust realism about physical objects. Moreover, science is not only the arbiter of what there is, but of how we come to know what there is—hence Quine's view of his empiricism as a consequence of naturalism. In his empiricist epistemology, Quine comes to the view that objects are posited by us in the course of our theoretical activities. Fogelin thinks this epistemological view merits classifying Quine as an anti-realist about objects. He conceives of the realism/anti-realism distinction in terms of independence/dependence with respect to our theoretical activities. This conception of the distinction comports with the current realism debate as framed by Devitt.

I have argued that Quine's naturalism does not remove the apparent tension between Quine's realism and his epistemological conception of objects. The containment between science and epistemology is genuinely reciprocal. Naturalism cannot be used to give primacy to either direction of containment. Quine's epistemological conception of objects should in my view be taken seriously as a general account of the nature of objects—because of the science-in-epistemology containment and Quine's view that there is no cosmic exile, this conception cannot be understood as being only about our theories and not about the objects that we discuss in our theories. I argued that there is no tension between Quine's "robust realism" and his epistemological conception of objects, because Quine holds that the notion of reality is itself always part of a theory. The notion of reality divorced from the point of view of one or another theory is, Quine contends, meaningless. Hence the conception that objects can be real in some further sense than being posits of a theory included in our best current science is nonsensical. The appearance of the tension within Quine's philosophy arises only if it is assumed that realism about objects must involve the view that such a conception does make sense.

I have argued that Quine's position cannot be classified as either realist or anti-realist. On one hand, Quine cannot accommodate realism as a metaphysical view that has primacy over epistemology; hence his realism is not contradicted by his epistemological view of objects as theory-dependent. On the other hand, Quine's epistemological conception of objects does not entail that the objects talked about in our best current science would be less than real, in the admissible theory-internal sense of 'real'. With regard to the reality of objects in some sense other than being posits of our science, Quine's epistemology entails nothing, since he rejects any such sense of reality as meaningless. In my view, we should take Quine at his word and consider him a realist about objects, but not a metaphysical realist. According to his naturalism, there are objects in the external world, and science tells us what objects there are and what these objects are like. But the real objects in the real world are always objects of some theory or another, and hence posits of that theory.

Although Quine's conception of objects does not entail a commitment to the controversial metaphysical doctrine of anti-realism, it does have at least one quite substantial—and perhaps also controversial—consequence. As argued in §5, Quine's conception of objects entails that an ascription of a relational attitude to another subject is true only if the ascriber's and the attitudinist's theories have exactly the same empirical content. From the point of view of the common practice of making and evaluating relational attitude ascriptions, this necessary condition seems highly unintuitive; the total sets of the empirically testable consequences of subjects' belief systems are not considered relevant within this practice. It may also be the case that two different subjects' theories very rarely have exactly the same empirical content; if so, Quine's account of the nature of objects entails that most ascriptions of a relational attitude to another subject are in fact false. I do not wish to go so far as to claim that this consequence with respect to *de re* or relational ascriptions constitutes a *reductio* of Quine's conception of objects. Much depends here on the weight given to the actual practice of attitude ascriptions, and one's theoretical interpretation and intuitions about this practice. At the very least, I think that a theory about the nature of objects that implies such a specific stance on relational attitude ascriptions should be examined with a critical eye.

NOTES

1. Parts of the present paper have been previously published as Keskinen (2012). I wish to thank *Theoria, Stiftelsen Theoria,* and Blackwell Publishing for permission to use them here. I also thank the editor, Ken Westphal, for comments which helped improve §5.
2. I think Quine's conception of objects bears on other issues besides the *de re* attitudes; see my (2010).
3. Quine's way of tying the assuming of objects to language has to do with his empiricism. He complains that the assuming of objects is a mental act which is "notoriously difficult to pin down." Quine (1981c, 2) concludes that the only way to study the assuming of objects is to study linguistic reference. His preference for logically regimented theories derives from his view of the ontological unclarity of natural language (1981c, 9–10).
4. For Quine's view of the epistemological question his genetic investigation is meant to answer, see *e.g.* Quine (1974), 2–3, (1969), 82–83.
5. The term 'observation categorical' is introduced only in Quine's later writings; in his (1974) it was called simply 'categorical'. The qualifications 'free' and 'focal', which Quine uses, *e.g.*, in (1995a, 27) and (1992a, 10–11), correspond to the somewhat earlier 'primitive' and 'objectual' in his (1990b, 9).

6. More precisely, the set of *synthetic* observation categoricals it implies; Quine (1992a), 16–17.
7. Quine's canonical notation consists of standard first-order syntax without primitive singular terms.
8. See, *e.g.*, Quine's examples of the raven (1995a, 27) and the willow (1992a, 10–11); for a critical note on the raven example, see his (2008a, 453).
9. See, *e.g.*, Quine (1992a), 31–36, (2008b), 360–361.
10. According to Quine, some sentences of a theory may play no role in the implication of any observation categorical. This kind of sentences would be empty of empirical content, though it is no part of Quine's (1995a, 48–49) program to repudiate them on this score. As Panu Raatikainen (2003) points out, Quine does not impose his notion of empirical content as a positivist criterion of meaningfulness. Hence, to be more precise, according to Quine's epistemology an object depends for its existence and identity on the whole empirically meaningful part of the theory.
11. See Quine's (1960, 16) example of the term 'neutrino', which is analogous to the present one of 'dog'.
12. Gregory (2008, §5.5) contains an illuminating discussion of this question. My account here repeats some points he makes; unlike his discussion, I connect these points to the realism–anti-realism debate and to the apparent tension between Quine's realism and his epistemological conception of objects.
13. Which should not properly be called 'evidence', as Quine notes. For Quine (1992a, 2), 'evidence' is not a technical term of epistemology. In particular, Quine does not want to claim that we would be generally aware of our neural intake, as equating of neural intake with evidence might suggest.
14. This, I believe, oversimplifies Kaplan's position, but I cannot discuss this issue here. For details see Kaplan (1968), 197–203.
15. An attitude is relational if and only if it is *de re;* but no such equivalence holds between notional and *de dicto*. Crawford (2008, 80–81) argues that Quine's and others' discussion of the issue has at times been plagued by conflating the *de re–de dicto* distinction with the *relational–notional* distinction. The terminology of *relational–notional* comes from (Quine 1956).

REFERENCES

Crawford, S., 2008. "Quantifiers and Propositional Attitudes: Quine Revisited." *Synthese* 160: 75–96.

Devitt, M., 2010. *Putting Metaphysics First: Essays on Metaphysics and Epistemology.* New York, Oxford University Press.

Dummett, M., 1993. "Realism and Anti-realism." In: *idem., The Seas of Language* (New York, Oxford University Press), 462–479.

Fogelin, R. J., 2004. "Aspects of Quine's Naturalized Epistemology." In: R. F. Gibson, ed., *The Cambridge Companion to Quine* (Cambridge, Cambridge University Press), 19–47.

Gibson, R. F., 1992. "The Key to Interpreting Quine." *The Southern Journal of Philosophy* 30.4: 17–30.

———, 1994. "Quine and Davidson: Two Naturalized Epistemologists." *Inquiry* 37.4: 449–465.

Gregory, P. A., 2008. *Quine's Naturalism: Language, Theory, and the Knowing Subject.* London, Continuum.

Haack, S., 1993. "The Two Faces of Quine's Naturalism." *Synthese* 94: 335–356.

Kaplan, D., 1968. "Quantifying In." *Synthese* 19.1–2: 178–214.

Keskinen, A., 2010. *Quine's Critique of Modal Logic and his Conception of Objects.* Dissertation, Tampere, Tampere University Press.

———, 2012. "Quine on Objects: Realism or Anti-realism?" *Theoria* 78: 128–145.

Loar, B, 1972. "Reference and Propositional Attitudes." *The Philosophical Review* 81.1: 43–62.

Quine, W.V.O., 1956. "Quantifiers and Propositional Attitudes." *The Journal of Philosophy* 53.5: 177–187.

———, 1960. *Word and Object.* Cambridge, Mass., MIT Press.

———, 1969. "Epistemology Naturalized." In: *idem.*, *Ontological Relativity and Other Essays* (New York, Columbia University Press), 69–91.

———, 1974. *The Roots of Reference.* La Salle, IL, Open Court.

———, 1975. "The Nature of Natural Knowledge." In: S. Guttenplan, ed., *Mind and Language* (Oxford, Oxford University Press), 67–83.

———, 1976. "Three Grades of Modal Involvement." In: *idem.*, *The Ways of Paradox and Other Essays, revised and enlarged edition* (Cambridge and London, Harvard University Press), 158–177.

———, 1981a. "Five Milestones of Empiricism." In: *idem.*, *Theories and Things* (Cambridge and London, Belknap/Harvard; first published 1981), 67–73.

———, 1981b. "Intensions Revisited." In: *idem.*, *Theories and Things* (Cambridge and London, Belknap Harvard; first published 1981), 113–124.

———, 1981c. "Things and their Place in Theories." In: *idem.*, *Theories and Things* (Cambridge and London, Belknap Harvard; first published 1981), 1–24.

———, 1981d. "Worlds Away." In: *idem.*, *Theories and Things* (Cambridge and London, Belknap Harvard; first published 1981), 124–129.

———, 1986. "Reply to Dagfinn Føllesdal." In: L. E. Hahn and P. A. Schilpp, eds., *The Philosophy of W. V. Quine* (La Salle, Ill., Open Court), 114–115.

———, 1990b. "Three Indeterminacies." In: R. B. Barrett and R. F. Gibson, eds., *Perspectives on Quine* (Cambridge, Basil Blackwell), 1–17.

———, 1992a. *Pursuit of Truth.* Cambridge and London, Harvard University Press.

———, 1992b. "Structure and Nature." *The Journal of Philosophy* 89.1: 5–9.

———, 1993. "In Praise of Observation Sentences." *The Journal of Philosophy* 90.3: 107–116.

———, 1995a. *From Stimulus to Science.* Cambridge and London, Cambridge University Press.

———, 1995b. "Naturalism; or, Living Within One's Means." *Dialectica* 49. 2–4: 251–261.

———, 2008a. "Assuming Objects." In: *idem.*, *Confessions of a Confirmed Extensionalist and Other Essays* (Cambridge and London, Harvard University Press), 449–461.

———, 2008b. "The Elusiveness of Reference." In: *idem.*, *Confessions of a Confirmed Extensionalist and Other Essays* (Cambridge and London, Harvard University Press), 352–364.

Raatikainen, P., 2003. "Is Quine a verificationist?" *The Southern Journal of Philosophy,* 41.3: 399–410.

6 Did Sherlock Holmes Inhale Pipe Smoke through a Hole in His Forehead?

Peter Swirski

> There is a dangerous critical heresy, typical of our time, according to which we can do anything we like with a work of literature, reading into it whatever our most uncontrolled impulses dictate to us. This is not true. Literary works encourage freedom of interpretation, because they offer us a discourse that has many layers of reading and place before us the ambiguities of language and of real life. But in order to play this game, which allows every generation to read literary works in a different way, we must be moved by a profound respect for what I have called elsewhere the intention of the text.
>
> —Umberto Eco, *On Literature*

1 THE PROBLEM OF FICTIONAL TRUTH

Who would doubt that, when smoking his pipe, Sherlock Holmes inhales through the mouth and not through a hole in his forehead? Or that Peter Walsh, who in *Mrs. Dalloway* trembles uncontrollably on encountering his old flame, does so out of excitement and not because he suddenly develops Parkinson's? To any competent reader these things are as clear as the late Mother Teresa's conscience, even though one would look in vain in the respective stories for any words to that effect. On the other hand, no reader would ever assume that Sherlock Holmes or Peter Walsh are humanoid aliens from space even though nothing in the text explicitly contradicts this outlandish interpretation.

Figuring out what happens in the story—which is to say, what is true in it—is the first step in contact with any piece of writing, be it literary fiction or a philosophical counterfactual. After all, we must know what is going on in the narrative in order to proceed with interpretation. An account of fictional truth is thus a prerequisite for any and all kinds of cognitive, symbolic, or sociocultural analyses readers and critics typically engage in. What strategies readers (should) use in retrieving the implicit contents of fiction, and what is the role of authorial intentions in this process, are the questions I will try to answer in this chapter.

The problem of figuring out the implicit story content is known in analytic philosophy as the problem of fictional truth (or the problem of truth in fiction). It is a matter of competence that allows readers to 'flesh out' the implied contents of what they read. After all, when making sense of stories—be they fully fledged literary narratives or thought-experimental sketches so beloved by philosophers—we never limit ourselves to what is printed on the page. Sometimes, we even justifiably disregard what *is* printed there.

Prima facie, this seems very strange. How does a reader know how to fill in what, in *The Act of Reading,* Wolfgang Iser correctly identified as textual gaps?[1] Whence does a reader get the authority to occasionally overrule the author in this process? And how is it that this process of inferring the missing content is so automatic and, at least for the most part, successful in the first place?

To illustrate the issues at stake, let me briefly turn to Edgar Allan Poe's quirky story, "The Thousand-and-Second Tale of Scheherazade," which adroitly forefronts some aspects of these considerations. Poe is keenly aware of what the interpretation of fictional truth entails—so much so, that he writes the entire story to play havoc with the different types of backgrounds needed to flesh out its implicit contents. It would be an exaggeration to claim that "Scheherazade" delineates a solution the problem of fictional truth, but Poe's astute manipulation of the conventions of realism and fantasy points us in the right direction.

Mockingly embedded in a narrative framework of the Arabian Nights, "Scheherazade" is a story (within a story within a story) of Sinbad's marvelous expedition through the natural and scientific wonders of nineteenth-century America. Befitting a tale which inverts what is true and what is fantastic, Poe cautions in the motto: "Truth is stranger than fiction" (1151).[2] This is only the first of several cautions against using reality as the interpretive background to make sense of the story. Playing with the conventions of fantasy and realism, Poe proposes that believability and not factual soundness may be the determining factor when it comes to fiction.

Indeed, what is true and fantastic in each layer of Poe's story turns out to be *relevant to that fictional layer.* Thus, in the middle layer, Scheherazade entertains her husband with tales of wonder, from the giant sky-blue cow to a blue rat to a pink clockwork horse with green wings wound up with an indigo key. Peculiar and incredible as all these are, however, the Sultan accepts them as real and true. Clearly the criteria that determine plausibility for the fairy tale monarch are at odds with those that we would employ for that purpose. In fact, they are symmetrically reversed: what is fantastic for us, creatures of the real world, is real to denizens of fairy tales, and *vice versa.*

Throughout the story, Poe deliberately estranges real-life marvels of nature and triumphs of technology into fairy tale miracles. To amplify the effect, instead of a straightforward and unambiguous narration, he carefully selects facts and artifacts with singular properties, and employs a number of rhetorical tropes to increase the impression of futuristic novelty. In one hyperbole, for example, he describes ash ejected by a volcanic explosion to be so dense that, at one hundred and fifty miles away, it is "impossible to see the whitest object" (1161), however closely held to the eye. Then, with no warning, the next passage adopts the viewpoint of a Lilliputian, describing "monstrous animals with horns resembling scythes" that inhabit "vast caverns in the soil" (1162). It takes the unsuspecting reader a while to realize the relative size-shift in the description, and recognize in the rapacious giant a tiny lion-ant.

Poe's oblique allusions to facts of science have lost none of their grip today. At one point Sindbad describes the local atmosphere as being "so dense as to sustain iron or steel, just as our own does feathers" (1161). Poe refers in this roundabout way to powdered steel which, indeed, can float in atmospheric air. Without knowledge of this fact, readers who contemplate this seemingly outrageous lie in light of something like the Reality Principle, are bound to deny its truth. The narrative ruse jolts readers into a new perspective on the familiar, as estrangement is supposed to.

By means of such clever transpositions, Poe alerts the reader to some important interpretive principles of fictional truth. For one, he shows that the background knowledge *relevant to a given fiction* is indispensable to the understanding of its contents. Second, he suggests that such background knowledge may take precedence over nominal facts of science.

2 STORYWORLDS AND POSSIBLE WORLDS

It is time to refine Poe's scattered insights in a systematic account of fictional truth. To begin with, much like in nonfiction, the grasp of story content must obey the same basic causal constraints that obtain in reality. Notwithstanding the fact, highlighted by Baudelaire and Borges, that writers do create their own predecessors, any claims about the influence of Sherlock Holmes (first edition in 1892) on Poe's Dupin (first story in 1841) are patently anachronistic and unacceptable.

Naturally, this imposes only the broadest restriction on admissible story entailments. Our account must specify where to look for the right kind of background to flesh out any variety of story. After all, depending on the choice of background, inferences about the contents of various fictions may differ dramatically. Needless to say, they also differ between fiction and nonfiction. Given that all nonfictions are perforce embedded in reality, readers flesh out their contents using something like a Reality Principle: everyday empirical and psychosocial assumptions about the real world. We do not need to be told, for example, that people inhale pipe smoke through the mouth or that lovers miss each other when apart. In nonfiction all of those facts are a given.

Not so in fiction. Propositions can, after all, be true in the story but not true *simpliciter* (*i.e.*, not true in our world, or simply not true). The background relevant to understanding a work of fiction may also differ—more or less dramatically—from the background relevant for nonfiction. The whole point of "Scheherazade" is that the natural and scientific wonders described by Poe really exist. But it is equally the case that we cannot employ the Reality Principle to interpret what happens elsewhere in the story.

For example, it is true in Poe's fiction and true in our world that one can engineer "a deep darkness out of two brilliant lights" (1168) using light-wave superposition. But, at the same time, while it is true in "Scheherazade" that the Queen recounts Sinbad's adventures, it is not true that either character ever existed in reality. Truth is one thing, and fictional truth another. They can, but need not correspond to each other.

Distinct though they are, there has been no shortage of attempts to relate truth and fictional truth to each other. Typically it has been assumed that fictional 'worlds'—*i.e.* heuristic projections of the semantic space in which stories unfold—map onto possible worlds, those coherently complete and completely coherent projections of states as understood in possible-worlds semantics.[3] If fictional propositions are not true in our world, perhaps they are true in some possible world associated with the 'world' of a given fiction?

Although not implausible at first glance, the correspondence between storyworlds and possible worlds fails for several reasons:

1) In contradistinction to storyworlds, possible worlds are always in principle determinate with respect to truth. In a possible world the answer to a question may not be known, but by definition it exists. Not so in fiction. Take the number of crewmen on Sinbad's ship. Although it must be determinate, neither the text nor any form of background knowledge can flesh out the exact value of "a vast number" (1157).

2) Because of (1), possible worlds and storyworlds do not map onto one another. Fictions, in other worlds, do not affix themselves to a unique possible world. In fact, any storyworld is consistent with an infinity of possible worlds. The entire contents of "Scheherazade," for instance, are compatible with the possible worlds in which Sinbad owned a flying carpet at the age of five and those in which he did not, in which Martians invaded the Earth the day after Orson Welles's broadcast and did not, in which Frank Zappa released *Sheikh Yerbouti* in 1979 and did not, and so on *ad infinitum*.

3) Like our own reality, possible worlds are always and in principle consistent. Nothing logically impossible can be true in them. Not so in fiction. To take one illustration, Stanislaw Lem invokes a constellation of time-travel paradoxes in "The Seventh Voyage" (see below), and chronomotion entails all manner of logical inconsistencies.

4) Logic aside, it is psychologically and epistemically impossible that readers do, or even could, reconstruct complete possible worlds from

works of fiction. Because so much in fiction remains indeterminate, no readers could hope to reconstruct an entire and complete possible world from a story. Jorge Louis Borges's story, "Tlön, Uqbar, Orbis Tertius," in which a corpus of encyclopedists labor to invent another world down to the last detail, vividly narrating the infinitude of detail needed to accomplish such a supertask. Mapping out a possible world in totality is a psychological and empirical dead end.

3 NOT THE TEXT ALONE

We must rescind the idea that fictional truth can be analyzed in terms of what is true in one possible world. But what about truth in various possible worlds, all consistent with a single fiction? Philosopher David Lewis, who did more than anyone else to spearhead research into this topic, approached fictional truth precisely in terms of such a counterfactual setup.

Lewis begins by proposing that what is true in fiction can be treated as if it was uttered in the world of the story as known fact. In other words, he wants readers to think of propositions relating to a story as prefixed by the operator: "It is fictional in the work *X* that. . . ." For example, it is fictional in Mario Puzo's and Francis Ford Coppola's *The Godfather* that Don Corleone is a mafia boss.

This is, of course, nothing new. When interpreting what is true in the story, we implicitly apply this fictional modifier. Note, however, that inferences about a story are legitimate only if preceded by the same modifier. Nothing follows from a mix of modified (prefixed) statements and unmodified ones. From the fact that Sherlock Holmes lived in London (fictionally true), and that Conan Doyle lived in London at the same time (true in reality), it does not follow that the detective could run into his creator.

To get to what is true in the storyworld, Lewis considers a plurality of possible worlds (which he calls *S*-worlds) that can be associated with it. Though different from one another, they all have a common core consisting of all that is made explicit in the text. In other words, the intersection of all possible worlds associated with any story corresponds to what is explicitly stated in the text. Why is this important? Because as a consequence it makes no sense to equate fictional truth with what is common in all possible worlds since that would make it identical with the text.

This is a crucial point, worth reiterating for the sake of literary critics and philosophers who are still inclined to think that what we can legitimately say about the contents of a story is determined entirely by the text. This is emphatically not so. Let us come back for a moment to Poe's "Scheherazade." Is it true in the story that Sindbad is a keen observer of natural and scientific wonders on his voyage? The proposition is so obvious as to be almost truistic, yet there is nothing in the text that explicitly states so. But if the text determines fictional truth, it ought *not* to be true in the story.

Clearly, we can understand (retrieve) much more from what we read than is stated on the page. This knowledge is, of course, symmetrical. Just as we can infer a number of extratextual propositions to be true, we can judge others to be false—even when there is nothing in the text to contradict them. Is Poe's Sindbad a contemporary black comedian and occasional Hollywood movie actor? Ridiculous as the proposition is, there is nothing in the text of "Scheherazade" to say that he is not. Is the proposition true, however? Decidedly not.

The same reasoning applies, of course, to what we can legitimately say about Hector, Ahab, Sherlock Holmes, and any other fictional character, event, or environment. This is because we know much more about them and about the stories we read than what is explicitly enunciated in the text. The upshot is clear: we must abandon the idea that fictional truth is limited to what is on the page, *i.e.*, to the text. Clearly, fictional truth is determined by the text fleshed out in accordance with an appropriate background. Which background is appropriate is, of course, the mother of all questions.

4 THE REALITY PRINCIPLE VS. THE MUTUAL BELIEF PRINCIPLE

In "Truth in Fiction" Lewis (1983) submits two alternatives for selecting the relevant background to retrieve implicit story contents. Though they look rather forbidding, I reproduce them to give you a flavor of his approach to the problem:

> Analysis 1: A sentence of the form "In the fiction *f*, Φ" is non-vacuously true iff some world where *f* is told as known fact and Φ is true differs less from our actual world, on balance, than does any world where *f* is told as known fact and Φ is not true. It is vacuously true iff there are no possible worlds where *f* is told as known fact. (Lewis 1983, 270)

> Analysis 2: A sentence of the form "In the fiction *f*, Φ" is non-vacuously true iff, whenever *w* is one of the collective belief worlds of the community of origin of *f*, then some world where *f* is told as known fact and Φ is true differs less from the world *w*, on balance than does any world where *f* is told as known fact and Φ is not true. It is vacuously true iff there are no possible worlds where *f* is told as known fact. (Lewis 1983, 273)

In *Mimesis As Make-Believe*, Kendall Walton (1990) christened Analysis 1 the "Reality Principle." The name fits, for Analysis 1 argues that what is true in the story is *as close as the text allows to the real world*. Whether it is the world of the author or the reader remains unclear (Lewis speaks only of *our* world), but we can finesse this as we go along. In other words, what is true in the story is the text augmented by what was known or believed in the real world. On first blush it might look like a sound rule for generating inferences about fictional truth. It works, after all, for all historical, realistic or naturalistic novels, and for many philosophical counterfactuals that deal with the real world, where things not explicitly stated can be imported whole-hog from the world as we know it.

Perhaps motivated by such considerations, a number of thinkers have succumbed to the illusion that the Reality Principle is the master key to all fiction. The most prominent among them is Umberto Eco, who endorsed the Reality Principle in *Six Walks in the Fictional Woods* (1994, 83) with these words: "everything that the text doesn't name or describe explicitly as different from what exists in the real world must be understood as corresponding to the laws and conditions of the real world."

Unfortunately, this is not so for two elementary reasons. First, Analysis 1 generates inappropriate inferences outside realistic fiction. In *Aladdin*, magic carpets levitate by dint of magic and not magnetic levitation, yet the Reality Principle would have us believe otherwise. Second, the Reality Principle is too indiscriminate, for it generates an infinity of irrelevant truths-in-fiction. When we think about what happens in a story, we never invoke *all* the inferences warranted by the Reality Principle.

Is it true in Poe's "Scheherazade" that Frank Zappa released *Sheikh Yerbouti* in 1979? Nothing in the text disputes it, and the proposition is as close as the text allows to the real world (of the reader, in this case). Yet it is starkly anachronistic and aesthetically irrelevant. Limiting the Reality Principle to the real world of the author rules out the Zappa example, but succumbs to others, contemporary to or predating Poe. Is it true in Poe's "Scheherazade" that Hannibal crossed the Alps with thirty seven war elephants in tow? No way. The proposition is true, and was true in Poe's world, but it is interpretively unintended and, as such, irrelevant.

Given these evident and irreparable drawbacks, let us turn to Lewis's Analysis 2, the Mutual Belief Principle. Here the appropriate background for fleshing out textual gaps is limited to the *conventional beliefs in the author's community*. The immediate advantage of the Mutual Belief Principle is that it rules out the Zappa example. Nothing about Zappa was believed in Poe's community, hence such anachronistic inferences can be correctly ruled out. Similarly, Analysis 2 is sensitive to various beliefs that may not be part of our reality but which were the intellectual coin of the writer's times. Geocentrism, theory of humours, or phrenology are not true but were believed at one point or another and, as such, may be indispensable to the understanding of certain stories.

However, while clearing some of the hurdles that tripped the Reality Principle, the Mutual Belief Principle runs into its own brick wall. First of all, it does not clear the irrelevance of Hannibal's elephants. Second, the background needed to make sense of certain fictions may be idiosyncratic rather than a part of a network of communal beliefs. For example, in Poe's vignette, "The Oval Portrait," we are to assume that the painter's transfer of his wife's beauty onto the canvas drains her life spirit until, upon the completion of the work, she dies. The doctrine of mimetic art depleting life from people was neither mutually believed in Poe's times, nor is it part of the modern reader's world knowledge, yet it is necessary to make any sense of this story.[4]

In general, what is distinctive about Lewis's account, *i.e.*, his treatment of truth in fiction along the lines of truth in possible worlds, is also the source of his problems. Among others, it forces him to assume that "Truth in a given fiction is closed under implication" (1983, 264). This means roughly that all possible implications are true in a given story. Yet, as we already noted in the discussion of possible worlds, one of their most unlikely features is the need for readers to make all inferences that could be made.

In psychological terms it is highly dubious, to say the least, that anything resembling such a total inferential process could ever take place. Moreover, even if we could, we would never want to generate all implications of a story anyway, since most would be aesthetically irrelevant (again, recall Hannibal's war elephants or Zappa's *Sheikh Yerbouti*). Moreover, Lewis's theses are helpless vis-à-vis inconsistent fictions which, by default, cannot be mapped onto any (logically) possible world.

All the same, galvanizing research into fictional truth, Lewis laid down the rules of engagement. What is fictionally true is by default what is in true in the (fair) text. But, given that we can assert much more than what is printed on the page, fictional truth is delimited not only by the text but also by the appropriate inferential background.

5 PLUS THE RELEVANT BACKGROUND

Before I say more about the inferential background, we need to say a few more words about the fair text. As I already mentioned, most critics and philosophers—including Lewis himself—automatically assume that the text is inviolable: what is on the page must be true in a given fiction. Thus if Mario Puzo says that Don Corleone had three sons and one daughter, this is what is true in *The Godfather.* Most of the time, just like with the Newtonian mechanics, this is indeed a good approximation of how things work. But just as Newtonian mechanics can sometimes give inaccurate results, so does the assumption that what is in the text is necessarily true in a given fiction.

Since more people are familiar with the film than the novel, let us take a closer look at the cinematic text of Coppola's *The Godfather.* The film is set for the most part in the 1940s. However, in the scene when Don Corleone shops for fruit just before he is gunned down, boxes of Sunkist oranges that are visible in the upper part of the frame clearly display graphics from the 1970s. Is it true, then, that in *The Godfather* during the 1940s time-travel devices were available to local grocers? No. The clear anachronism is an error, and even though it features in the (cinematic) text, it must not guide any inferences about the story.

When Sonny is gunned down at the toll-booth, the windshield of his automobile is utterly shattered by machine-gun fire, but almost intact a camera cut later. Similarly, when the police captain McCluskey readies to strike Michael at the entrance to the hospital, the latter's coat is wide open at his chest but, in the very next shot, it is closed and straight when the punch lands. What gives? Did Michael ask for time-out to neaten up? Was Sonny's windshield partially replaced by a repair crew who happened to pass by? No. Once again, even though they are in the text, both of these unintended textual elements are unintended lapses, and are not fictionally true.

As I documented at length in *Literature, Analytically Speaking* (2010), the common notion of 'text' is thus only a convenient shorthand for 'intentionally configured text'. The restoration of the text in accordance with authorial intentions will in almost every case (barring typos and other unintended errors), correct suchlike lapses. Here is one more example from *The Godfather*. In the scene when Michael Corleone flies to Vegas, while he is getting out of the automobile in a hotel driveway, viewers can clearly see a hirsute, bearded, unmistakably 1970s hippy inside the hotel lobby. Confessing on the film DVD to acute embarrassment about the oversight, the director Coppola effectively excises that part of the text. Not part of his executive intentions, this textual element does not thus validate the inference that, like Rip Van Winkle, the story suddenly skips twenty years in chronology.

Bearing all this in mind, let us return to our formula for determining fictional truth. By now, it should be familiar: what is true in any fiction is the joint product of the intentionally configured text *plus* the relevant background:

$$\Phi = \mathrm{T} + \beta$$

While it may not be immediately apparent how to pick the right background for each and every story, it is clear that to retrieve what is fictionally true—and thereby to execute the opening gambit in any literary or philosophical interpretation—we must look beyond Lewis's analyses. What we need is a rule that will derive only true and only relevant inferences about the story.

6 CIRCULAR, INSUFFICIENT, AND UNNECESSARY

Before I present my own recommendation, let me review one more proposal that attempts to do just that. In *The Nature of Fiction* (1990), Gregory Currie grounds fictional truth in the persona of the fictional author, defined as that fictional character whom we take to be telling us the story as known fact. Fictional truth is thus offered as the text plus inferences from the belief system of the fictional author. Unfortunately, the linchpin of this entire approach—that is, the concept of the fictional author—is demonstrably circular, insufficient, and unnecessary (not to mention several lesser drawbacks).

For instance, according to Currie (1990, 76), readers cull the belief system of the fictional author from the text: "as we read we learn more about his beliefs." In other words, we interpret the story to get to the fictional author's beliefs, so that we can use his beliefs to interpret the story, in a vicious circle that offers no point to jump-start the interpretation of fictional truth. Moreover, Currie's postulates fail in the face of 'mindless' fictions—stories without an apparent narrator—from which it is impossible to reconstruct the belief system of the fictional author and, consequently, what is fictionally true.[5]

Between textual under-determination and finite cognitive resources, it is also impossible that readers could reconstruct a complete belief system of the fictional author, no matter how interminable the work. Neither can Currie's treatment of inconsistent fictions handle stories like Lem's which I describe below. Finally, as a consequence of his recourse to the fictional author, he pays inadequate attention to genre as one of the relevant background beliefs that reflexively regulate interpretation. Fictions, after all, 'belong' to genres only because authors reflexively intend for readers to recognize that genre conventions are being employed in order to be recognized as such.

Considerations of genre can, after all, deeply affect interpretation of what happens in the story. The canons of verisimilitude and coherence in Henry James's "Turn of the Screw"—or, in the modern reworking of James, in Thomas Disch's *The M.D.: A Horror Story*—dramatically alter the perception of what is true in their respective fictions. Whether the reader chooses to follow a realistic or supernatural interpretation, the ambiguities in the plot cast a long shadow over the entire story. The point is, knowledge of genres belongs to the creative tactics of the real author, from which the heuristic construct such as the fictional author is entirely cut off.

Curiously, although in other aspects of interpretation Currie is a realist about intentional attitudes, he embraces hypothetical intentionalism when it comes to story contents (in this he mirrors some other aestheticians).[6] On the other hand, he cannot simply dispense with the author altogether. To understand why, consider the interpretive puzzle of the Flash Stockman, first brought up by Lewis.

7 FLASH STOCKMAN

In this traditional folk song, the narrator brags:

> I'm a stockman to me trade and they call me Ugly Dave,
> I'm old and grey and I only got one eye.
> In the yard I'm good, of course, but just put me on a horse
> And I'll go where lots of young 'uns daren't try. . . .
> Just watch me use the whip, I can give the dawdlers gyp,
> I can make the bloody echoes roar and ring.

> With a branding-iron, well, I'm a perfect flamin' swell,
> In fact I'm duke of every blasted thing. . . .

There is no doubt (*i.e.,* it is true) in the poem that the narrator is a braggart who is not to be taken at his word. Even though the stockman may believe every boast he makes, things are just the opposite of his elaborate verbal hyperboles. But the latter point of view is clearly not the narrator's, leading Currie—who wants no truck with the real author—to fall back on the fictional author, with all the problems that follow.[7]

One last case in which Currie's approach clearly breaks down is when the beliefs of the real author's community, the beliefs of the fictional author, and the fictional events of the story are all temporally distinct. Suppose I, a contemporary author, want to portray a pre-Christian Viking community but from the point of view of an enlightened late-mediaeval cleric (thus automatically ruling out both the Reality and the Mutual Belief Principles). As long as the beliefs of the mediaeval narrator coincide with those of his Viking subjects, we could indeed rely on the belief system of the fictional author to make appropriate inferences about the story. But as soon as those beliefs begin to diverge—as they certainly would in this case—Currie's proposal yields anachronistic inferences that could not satisfy any competent reader.

Casting about for the inferential background to make competent inferences about the stories we read, Currie contemplates a proposal operationally similar to Lewis's Analysis 2: the background for a correct interpretation of what is true in a given story is the mutual beliefs in the author's community. On this view, to retrieve what is true in fiction, readers should follow what was conventionally (in another remix, only predominantly) believed in the real author's times, and deviate only when the text explicitly contradicts these beliefs.

Note that, although superficially alike, Lewis's and Currie's proposals are categorically distinct. While the former involves possible worlds, the latter relies on a fallible, incomplete, and potentially inconsistent *system of beliefs*. This is in itself an important improvement, inasmuch as it replaces the dubious demands of inferential completeness, which plagued Lewis's account, with a much more realistic view of actual readers' actual reading strategies. In one move, replacing a set of possible worlds by a set of beliefs solves the problem of generating irrelevant truths about fiction.

Still, apart from the drawbacks discussed heretofore, there are other reasons to reject this refurbished Mutual Belief Principle. Here is why. W. W. Jacobs's short story, "Monkey's Paw" (1902), puts a novel spin on the folk tradition of magical objects granting wishes that may be indistinguishable from curses. An aging couple comes into possession of a magic amulet said to grant three wishes. Immediately they wish for two hundred pounds, which they indeed get the next day, but in the form of compensation for their son who dies in a work accident. Bereaved, they wish for his return, and in the dead of night someone starts breaking into their house. Petrified with terror, the old man uses the third wish to send the intruder away.

What is remarkable about Jacobs's story is that, *prima facie,* there is not a single supernatural event in it. People do get killed while on the job, and strangers do knock on doors at night, only to desist when no one answers. The Mutual Belief Principle (which in this case coincides with the Reality Principle), which follows the conventional beliefs of the author's community, would thus have us conclude that there are no ghosts, no resurrections, no magic. But such an interpretation contradicts the purpose of the story which derives its power and coherence precisely from the implication that the monkey's paw does after all work.

In the end, Currie is well aware that his version of the Mutual Belief Principle does not make the grade. Admitting that he has "no rules to substitute for this one," he (1990, 80) closes the matter thus: "I take it there would be considerable agreement in practice about how such inferences as this ought to proceed." This misses the point. No one has ever doubted that readers are capable of making appropriate inferences about the fictions they read. The whole purpose of the analysis was to fashion a workable model of this process.

8 REAL INTENTIONALISM

Currie's abstention from possible worlds resolves the problems that dogged Lewis. It makes good sense to treat a story *not* as an intersection of logically complete and necessarily consistent worlds, but as a set of propositions which are not only incomplete, but sometimes openly contradictory. The source of Currie's own problems, however, is his source of relevant propositions: the heuristic device of fictional author.

I submit that a shift from the hypothetical author to the real author, and from hypothetical intentions to real intentions, resolves all outstanding problems. More formally, I submit that what is true in fiction follows from:

(T) What is enunciated or directly warranted by the fair text, if need be reconstituted according to the author's reflexive intentions.

(β) A reflexive recognition of the *real* author's successfully executed reflexive intentions *relevant to a given fiction.*

Being a realist about intentions entails that, much as in other aspects of everyday life, one can sometimes fall short of the intended goal. A theory of fictional truth based on real intentions allows, therefore, for failed intentions in cases when the evidence, textual or otherwise, shows that the author tried to express a proposition *P,* but the results mean *not-P.* Real intentionalism does not, of course, mean that we interpret the meaning as *P.* But it makes it possible to distinguish between *P* and *not-P,* which is crucial when assessing writerly skill and artistry.

Since the point is crucial, let me highlight the advantages of real intentionalism with regard to retrieving fictional truth:

1) The circular concept of the fictional author yields to a determinate reality of the real writer with real goals and designs, while the focus on the author's executive intentions—*i.e.,* those that actually went into the work's execution—rules out true but irrelevant inferences.
2) Since we are now dealing with a real-life person, there is no need to generate complete belief systems. The relevant background beliefs are now limited to those associated with the author's executive reflexive intentions for the fiction in question.
3) A significant part of the real creator and the real readers' Mutual Contextual Beliefs is the knowledge of genres and other tacit conventions, used reflexively by the former to signal the relevant context for interpretation.
4) In cases of explicit incoherence, we can interpret the entailments that follow from contradictions only to the extent warranted by the author's reflexive intentions.

The last point is, in some ways, the litmus test for any account of fictional truth. The challenge for any account of fictional truth is how to interpret logically inconsistent fictions, such as time-travel stories. After all, in classical logic anything follows from internal incoherence (since *ex contradictio quodlibet*). Naturally, the issue here is logical incoherence, such as in time travel, rather than inconsistency, which may arise out of simple forgetfulness. An example of the latter may be the placement of Watson's war wound in the Sherlock Holmes stories, sometimes said to be in the shoulder, other times in the leg.

Real intentionalism, grounded in the real-life fallible author, provides a sensible and, it must be pointed out, universally practiced editorial solution to such inadvertent inconsistencies. All else being equal, if most of the evidence suggests that the wound is in one place rather than the other, we should go with the greater number, one mistake being more likely than many. If, on the other hand, the numbers are roughly equal and the matter cannot be settled by an appeal to authorial intentions, there is no determinate answer. It is true, then, that the wound is in *either* place, but not true that it is in one or the other.[8]

9 THE SEVENTH VOYAGE

The more thorny problem arises in stories like Lem's "The Seventh Voyage," where time-travel paradoxes could hardly be more explicit. In the story Ion Tichy embarks on a cosmicomic voyage, only to be thwarted by his rocket which breaks down in a region of space plagued by gravitational vortices. Fittingly enough for a caper in which our space-age Crusoe fries a steak on the overheating atomic pile, Tichy becomes chrono-duplicated in all stages of his life. With the spaceship full of appearing and disappearing Tichys, the protagonist is forced to interact in all kinds of loopy ways with his own multiple selves, trying to sort out the causal and temporal entanglements of his (their?) chrono-multiplication.

Although logical incoherence in a possible world entails the latter's complete collapse, this clearly need not be so in fiction. Lem deals with inconsistencies in the following way: he narrates individual scenes in a logically consistent way, but avoids their self-contradictory entailments. For example, although the Tuesday Tichy knows by virtue of being his own older chrono-twin that the Monday Tichy will not help him repair the rocket, he still tries to persuade the other to change his mind, as if it were indeed possible.

The above pattern is common to most of time-travel fiction, from H.G. Wells's prototypical *Time Machine*, to contemporary blockbusters like *The Terminator.* Although at the deeper level all such stories are irredeemably incoherent, the audience is implicitly asked to suspend judgment with regard to what follows from such incoherence. This is psychologically plausible since readers do not—and could not—trace all implications of all textual propositions anyway. After all, even when we try our best, we fall prey to deductive corruption, memory lapses, or mental partitioning.

Thus from explicit inconsistencies it does not follow that any arbitrary proposition is true in a given fiction. If Lem's story is anything to go by, in fact, the correct interpretive strategy is to follow standard logic within the text but only to the extent warranted by the author's reflexive intentions. When the contradictory narrative 'bubbles' overlap, one must acknowledge the global incoherence of the story, without necessarily transferring it to the individual propositions—all the more so, that narrative illogic can be employed with a serious purpose in mind.

Chronomotion can, after all, be harnessed to artistic goals which can outweigh the price levied by the logic department. Not to look too far, time travel can model the sociocultural or psychological effects of a future shock—or conversely the effects of repeatedly undergoing a controlled experience, as in Alain Resnais's (1976) film *Je t'aime, je t'aime.* Such cognitive and artistic goals are, in fact, a good guide to the extent of inferential entailments relevant to the story. I thus fully agree with David Davies (1996, 53) that "'truth *in* fiction' and 'truth *through* fiction' are *not* as distinct as some would have us believe."[9]

What does all this tell us about the logical calisthenics in "The Seventh Voyage"? It is true in the story that time travel is possible, but not true that any arbitrary proposition follows. It is true that the past is fixed and determined, and that the characters are aware of that fact. Indeed some of them invoke it as an argument not to repair the rocket (if I fix it today, why does the Tomorrow Me ask me to fix it today?). But this fact is not the focus in the understanding of the characters' behavior and beliefs. Instead, these are approached by the author as standardly consistent. The hilarity and finesse of Lem's spoof stems, in fact, from the murderous logic with which he tracks some, but not all, of the implications of this illogical setup.

10 THEORY OF MIND

We have come a long way toward the answer to the problem of fictional truth. At the end of the day, the solution is clear and compelling: we should follow the author's intentions in trying to retrieve what the author is trying to say. There is only one thing left to explain: Where does the remarkable facility for reading intentions come from?

The short version of the answer is, of course: from evolution. As human beings, we love to share information about what we think and feel. But we succeed in doing so only thanks to our shared cognitive presets for *intention* and *attention*. You and I could never communicate via this chapter, for example, if it were not for this prosocial preset and for the Theory of Mind ('ToM') module that enacts it. Together with the ability to appreciate each other as intentional agents, our ability to form a joint-attentional frame is the basis for our chapter-mediated 'conversation'—and for all human communication, period.

Wittgenstein's *Philosophical Investigations* was perhaps the first major work to direct attention to the way in which common contextual knowledge is absolutely necessary for people to exchange information successfully and efficiently. Over the decades Wittgenstein was joined by a growing number of philosophers and psychologists, including Jerome Bruner, Herbert Clark, and Michael Tomasello. Today, their joint conclusions are as well supported empirically as one could wish for: joint attention, 'we' intentionality, and shared cultural background are indispensable elements of all human communication.

Philosophy and literature come together in this enterprise. The most widely discussed example of why common contextual knowledge is indispensable for communication comes originally from Quine, but it may be more familiar in literature from Eco's novel *Baudolino* (2000). In the story, a visitor to foreign lands witnesses a creature streaking by, only to hear a native exclaim 'Gavagai'. What does it mean? Does the indigene simply name the moving creature? Or call attention to the movement? Or tell you to get out of the way? Ask you to catch it? Run after it? Delight in its fleet escape? It could be any of these or a million other things. Unless he is already attuned to the native's mentally (that is, intentionally), the visitor will never understand what the exclamation means.

In our everyday world, every normal person relies on the ability to project mental states onto other beings and on having mental states attributed to self as part of making sense of one another's behavior. So instinctive, so universal, and so accurate is this process that it must be a domain-specific adaptation hardwired into the brain rather than any sort of incultured skill. Studies of apes, children, twins, and patients with chronic or acquired neural damage reveal the essential elements of its structure, as remarkable as they are commonplace.

The central aspect of ToM is our astonishing and astonishingly reliable ability to read people's minds—or, more precisely, to read minds off people's behavioral and linguistic cues. This ability is part and parcel of our everyday psychology by means of which we account for other folk's actions in terms of beliefs and desires. Each of us has his own little interaction engine that could, dedicated to detecting intentions in actions. More than a mere facilitator of human communication, ToM is its actual *enabler.* It enables us to perceive others, be they human, animal, or even machine, as intention-driven agents and as a result to frame what they do or say in terms of goals and purposes.

The evidence from human ontogeny and primatology is compelling in this respect. Underscoring its fundamental role in our species, the capacity for mind-reading crystallizes in all normally developing infants as young as nine to twelve months old. As soon as they begin to point referentially, say, to their favorite toy, they keenly attend to the participating adult's grasp of what is going on, usually by means of eye contact. Strikingly, if the adult responds by showering them with attention but ignoring the toy, they persist in pointing to it. Do not attend to me, you big oaf, attend to what I want you to attend to!

Psychologist Michael Tomasello summarizes these findings as follows in *Origins of Human Communication*:

> from almost as soon as they begin pointing infants understand the most important aspects of how human-style, cooperative communication works—they communicate on a mental level, within the context of common conceptual ground, and for cooperative motives—thus displaying something approaching the complete social-cognitive infrastructure of mature cooperative communication. (Tomasello 2010, 134).

However, while infants acquire the ability to engage others mentally already in this prelinguistic phase, apes never develop it at all. They certainly do not lack the perceptional and motor skills to follow, for instance, the direction of a pointing finger. But they just do not 'get it'. They see the finger and they follow its extension with their gaze, but no light bulb lights up in the "Eureka" moment when they ought to realize that it is meant to *mean* something. In contrast, in all ordinary one-year-olds this joint-intentional light bulb lights up without fail and brightly. Significantly, all apes are quite able to perceive other agents as intentionally motivated. Experiments demonstrate that they ably distinguish good-faith attempts thwarted by acts of God from situations in which somebody stiffs them for no reason.

If there is a tasty morsel in a bucket that just keeps slipping out of the zookeeper's hands, the ape will wait patiently. But it will raise a ruckus if the keeper begins to hand it over and stops arbitrarily. Unable or unwilling? Our evolutionary cousins cope with this critical distinction very well. Manifestly, both apes and human infants recognize that agents have goals. Both grasp that agents strive to bring the state of the world closer to their internal (volitional) representations of it. Both recognize agents as rational, in the sense of making allowances for their motives. Unable? Okay. Unwilling? Time to hoot and display.

11 MEANING

But this is where all parallels end—and where human-level communication kicks in. Whereas the synchronization of joint attentions does not emerge in apes at all, in little people it emerges before the first year of age, never to fade away again. Just think about it. Our normal everyday conversations are chock full of dead-ends, such as: "Wanna go out?" "My mom is sick." You do not get logically or semantically or in any other way from an invitation to a date to a parent's illness—and back. Except that in shared cultural space it is tacitly understood that ill parents need attending to.

The operative word is 'tacitly'. There is absolutely nothing in this mini-dialogue to suggest how to flesh it out. You have to bring this knowledge from the outside, from the common background that enables this and an infinity of other social exchanges: verbal, gestural, or even wholly implicit. Where is the missing part of this exchange? In the context, obviously. But such context is never all-encompassing. Indeed, for practical reasons, it could not be. Instead, in a clear pointer to the retrieval of fictional truth, it is selected with a view to *what is reciprocally assumed to be relevant to a given conversation.* The context is what each participant sees as relevant and knows that the other sees as relevant and knows that the other knows that the first one knows, and so on.

The guiding principle behind this communicative turning the tables on each other was first described by Paul Grice in a 1957 paper called "Meaning," and refined over the next three decades in subsequent publications. Taking stock of what is going on when people communicate with each other, Grice homed in on intentions. He called them *reflexive* intentions, for the simple reason that they are intended to be recognized as having been intended to be recognized. A communication is successful if the meaning intended by the speaker (or writer) is identified by the listener (or reader) by recognizing the reflexive intention to express it.

Mind-reading is once again at the heart of the process, as it is in every exchange, from every water cooler gossip to every baseball coach's pantomime. In all cases, participants have to parse for intentions to generate relevant inferences on the way to generating relevant replies. After all, generating only logically correct inferences is not enough. They may be correct but irrelevant—accurate but inadequate. Much like chess-playing computers, we constantly need to prune our search space so that, instead of chasing all possible conversational gambits, we home in on the relevant ones (allowing that at times it may be relevant to be irrelevant, say, out of boredom, antipathy, or just whimsy).

In sum, communication could not have taken off without the social-cognitive and social-intentional adaptations which acted as a ratchet that progressively developed into language as we know it. Put differently, the communication zeppelin must be of the social-cognitive and social-motivational kind to get off the ground at all. Human communication is enabled not just by cognitive adaptations for understanding individual intentionality, but by social-cognitive and social-motivational adaptations for shared intentionality. The Theory of Mind module generates joint goals and joint attention which generate the common conceptual ground on which communication flourishes.

Another related facet is that human communication is hyper-cooperative. One of its unique features is that the mere expression of desire for a certain informational state of affairs often suffices to induce that state. This is due to what philosophers of action refer to as shared or 'we' intentionality (Bratman 1992; Searle 1995). It is a perception of common goals and a common frame of reference and—no less crucially—mutual awareness of this shared attention frame. It is the joint awareness of the joint awareness that enables all kinds of collaborative activities to emerge, including notably communication.

Human-like communication is possible only because of the underlying shared bedrock of common intentional and attentional ground. That, in turn, is enabled by our species-unique cooperative motivation for communicating. More formally, human communicative acts are grounded in joint attention to the common ground and enacted by prosocial motives such as exchanging information and sharing emotions and dispositions. Communicators and recipients have not only a vested interest in cooperative communicating, but a natural inclination for doing so, including adjustment to the other's mental picture of the world and of the exchange, and message repair.

Given the very structure of the communicative process—its interdependence, reciprocity, recursiveness, and reliance on explicit and tacit common ground—the structure of shared intentionality, itself deeply implicated in ToM, is the only one that makes sense of the whole setup. Take even such a basic part of everyday conversation as irony. It entails stepping into each other's shoes to ensure that the message gets across. You monitor your friend's attention to see if she is getting it, your friend monitors yours to make sure she is getting what you want her to get, and *vice versa*.

This is as far away from the traditional understanding whereby the emergence of language enabled communication and thus cooperation.

If anything, it was the other way round. Facility of perceiving others as intentional agents and engaging them in joint attention frames kick-started human-level communication—of which reading fiction, whether literary or philosophical, is such an everyday instance.

NOTES

1. For a critique of Iser's approach, see Swirski (1994). The first part of my analysis draws on Chapter Six of Swirski (2010).
2. All page references to Poe are from the Mabbott edition.
3. Among literary critics, see Pavel (1985); for possible world semantics, see Cresswell (1994), Chihara (1998), and Stalnaker (2003).
4. Although the early Hebrews forbade fashioning of artistic images of living things, they had nothing to say about painting killing the subject, nor were such beliefs held either by Poe or by his contemporaries. Stecker (1994, 198) endorses the Mutual Belief Principle.
5. It seems to me that the argument also extends to self-reflexive fictions; see Byrne (1993), Le Poidevin (1995), and David Davies (1996).
6. For example, Tolhurst (1977), Nathan (1992), Levinson (1992) and (1996), Chapter Ten.
7. *Cf.* Lewis's (1983, 280) confession in the closing paragraph to "Truth in Fiction": "But there is a real problem nearby, and I have no solution to offer. Why doesn't the iteration collapse? When the singer pretends to be Ugly Dave pretending to tell the truth about himself, how does this differ from pretending to be Ugly Dave *really* telling the truth about himself? It must be the former, not the latter; else we should conclude that there is no inner fiction and that what is true in the outer fiction—now the only fiction—is that Ugly Dave is duke of everything and tells us so. That would be to miss the point entirely. We must distinguish pretending to pretend from really pretending. Intuitively it seems that we can make this distinction, but how is it to be analyzed?" See also Lamarque (1987).
8. This checks with Currie's (1990, 87) argument.
9. See also Davies (2004).

REFERENCES

Bach, Kent, and Robert M. Harnish, 1979. *Linguistic Communication and Speech Acts*. Cambridge, Mass., MIT Press.

Baron-Cohen, Simon, 2005. "The Empathizing System: A Revision of the 1994 Model of the Mindreading System." In B. J. Ellis and D. F. Bjorklund, eds., *Origins of the Social Mind: Evolutionary Psychology and Child Development* (New York, Guilford), 468–492.

———, Helen Tager-Flusberg, and Donald J. Cohen, eds., 2000. *Understanding Other Minds: Perspectives from Developmental and Cognitive Neuroscience*, 2nd ed. Oxford, Oxford University Press.

Behne, T., M. Carpenter, J. Call, and M. Tomasello, 2005a. "Unwilling *versus* Unable? Infants' Understanding of Intentional Action." *Developmental Psychology* 41: 328–337.

———, 2005b. "One-year Olds Comprehend the Communicative Intentions Behind Gestures in a Hiding Game." *Developmental Science* 8: 492–499.

Bratman, Michael, 1992. "Shared Co-operative Activity." *Philosophical Review* 101: 327–341.
Bruner, Jerome, 1983. *Child's Talk*. New York, Norton.
Byrne, Alex, 1993. "Truth in Fiction: the Story Continued." *Australasian Journal of Philosophy* 71: 24–35.
Call, J, 2004. "Inferences about the Location of Food in the Great Apes." *Journal of Comparative Psychology* 118: 232–241.
Carrol, Nöel, 1992. "Art, Intention, and Conversation." In: G. Iseminger, ed., *Interpretation and Intention* (Philadelphia, Temple University Press), 97–131.
Chihara, Charles S., 1998. *The Worlds of Possibility: Modal Realism and the Semantics of Modal Logic*. Oxford, The Clarendon Press; New York, Oxford University Press.
Clark, Herbert, 1996. *Using Language*. Cambridge, Cambridge University Press.
Cohen, Philip R., Jerry Morgan, and Martha E. Pollack, eds., 1990. *Intentions in Communication*. Cambridge, Mass., MIT Press.
Cresswell, M. J., 1994. *Language in the World*. New York, Cambridge University Press.
Currie, Gregory, 1990. *The Nature of Fiction*. Cambridge, Cambridge University Press.
———, 1991. "Interpreting Fiction." In: R. Freedman and L. Reinhardt, eds., *Literary Theory and Philosophy* (London, Macmillan), 96–112.
———, 2004. *Arts and Minds*. Oxford, The Clarendon Press.
Davies, David, 1996. "Fictional Truth and Fictional Authors." *British Journal of Aesthetics* 36: 43–55.
———, 2004a. *Art As Performance*. Oxford, Blackwell.
———, 2004b. *Definitions of Art*. Ithaca, NY, Cornell University Press.
Dutton, Denis, 1987. "Why Intentionalism Won't Go Away." In: A. J. Cascardi, ed., *Literature and the Question of Philosophy* (Baltimore, Johns Hopkins University Press), 192–209.
Eco, Umberto, 1994. *Six Walks in the Fictional Woods*. Cambridge, Mass., Harvard University Press.
———, 2000. *Baudolino*. Trans. William Weaver. Orlando, FL, Mariner.
———, 2004. *On Literature*. Orlando, FL, Harcourt.
Focquaert, Farah, and Steven M. Platek, 2007. "Social Cognition and the Evolution of Self-Awareness." In: S. M. Platek, J. P. Keenan, and T. K. Shackelford, eds., *Evolutionary Cognitive Neuroscience* (Cambridge, Mass., MIT Press), 457–497.
Gomez, Juan-Carlos, 1990. "The Emergence of Intentional Communication as a Problem-Solving Strategy in the Gorilla." In: S. T. Parker and K. R. Gibson, eds., *'Language' and Intelligence in Monkeys and Apes* (Cambridge, Cambridge University Press), 333–355.
Gomez, Juan-Carlos, 2004. *Apes, Monkeys, Children, and the Growth of Mind*. Cambridge, Mass., Harvard University Press.
Grice, H. Paul, 1957. "Meaning." *Philosophical Review* 66: 377–88.
———, 1968. "Utterer's Meaning, Sentence Meaning and Word Meaning." *Foundations of Language* 4: 225–242.
———, 1969. "Utterer's Meaning and Intentions." *Philosophical Review* 78: 147–177.
———, 1971. "Intention and Uncertainty." *Proceedings of the British Academy* 57: 263–279.
———, 1982. "Meaning Revisited." In: N. V. Smith, ed., *Mutual Knowledge* (New York, Academic), 223–243.
Iseminger, Gary, ed., 1992. *Intention and Interpretation*. Philadelphia, Temple University Press.
Iser, Wolfgang, 1978. *The Act of Reading*. London, Routledge and Kegan Paul.

Lamarque, Peter, 1987. "The Puzzle of the Flash Stockman: A Reply to David Lewis." *Analysis* 47: 93–95.
———, and Stein Haugom Olsen, 1994. *Truth, Fiction, and Literature.* Oxford, Oxford University Press.
Lem, Stanislaw, 1977. "The Seventh Voyage." In: *idem., The Star Diaries.* Trans. M. Kandel (New York, Avon), 1–22; http://english.lem.pl/home/bookshelf/the-seventh-voyage
Le Poidevin, Robin, 1995. "Worlds within Worlds? The Paradoxes of Embedded Fiction." *British Journal of Aesthetics* 35: 227–238.
Leslie, A, 1987. "Pretense and Representation: The Origins of 'Theory of Mind'." *Psychological Review* 94: 412–426.
Levinson, Jerrold, 1996. "Intention and Interpretation in Literature." In: *idem., The Pleasures of Aesthetics* (Ithaca, Cornell University Press), 175–213.
Lewis, David K, 1969. *Convention: A Philosophical Study.* Cambridge, Mass., Harvard University Press.
———, 1983. "Truth In Fiction." In: *idem., Philosophical Papers, Vol. 1* (Oxford, Oxford University Press), 261–280.
———, 1986. *Counterfactuals.* Oxford, Blackwell.
Liszkowski, Ulf, 2005. "Human Twelve-Month-Olds Point Co-operatively to Share Interest and Provide Information for a Communicative Partner." *Gesture* 5: 135–154.
———, 2006. "Infant Pointing at 12 Months: Communicative Goals, Motives, and Social-Cognitive Abilities." In: N. J. Enfield and S. C. Levinson, eds., *Roots of Human Sociality: Culture, Cognition and Interaction* (Oxford, Berg), 153–178.
Livingston, Paisley, 1993. "What's the Story?" *SubStance* 71/72: 98–112.
———, 2005. *Art and Intention.* Oxford, Oxford University Press.
Poe, Edgar Allan, 1988. T. O. Mabbott, ed., *Collected Works of Edgar Allan Poe,* 3 Vols. Cambridge, Mass., Harvard University Press.
Premack, David, and Guy Woodruff, 1978. "Does the Chimpanzee Have a 'Theory of Mind'?" *Behavioral and Brain Sciences* 4: 515–526.
Quine, W. V. O., 1960. *Word and Object.* Cambridge, Mass., MIT Press.
Rizzolatti, Giacomo, and Corrado Sinigaglia, 2008. *Mirrors in the Brain.* Oxford, Oxford University Press.
Saxe, Rebecca, and Simon Baron-Cohen, eds., 2007. *Theory of Mind.* London, Psychology Press.
Searle, John, 1995. *The Construction of Social Reality.* New York, Free Press.
Sperber, Dan, and Deirdre Wilson, 1995. *Relevance: Communication and Cognition.* Oxford, Blackwell.
Stalnaker, Robert C., 2003. *Ways a World Might Be: Metaphysical and Anti-metaphysical Essays.* Oxford, The Clarendon Press.
Stecker, Robert, 1993. "The Role of Intention and Convention in Interpreting Artworks." *The Southern Journal of Philosophy* 31: 471–489.
———, 1994. "Art Interpretation." *The Journal of Aesthetics and Art Criticism* 52: 193–206.
———, 1997. *Artworks: Definition, Meaning, Value.* University Park, PA, Pennsylvania State University Press.
———, 2003. *Interpretation and Construction: Art, Speech, and the Law.* Malden, Mass., Blackwell.
Storey, Robert, 1996. *Mimesis and the Human Animal: On the Biogenetic Foundations of Literary Representation.* Evanston, IL, Northwestern University Press.
Swirski, Peter, 1994. "Iser's Theory of Aesthetic Response: A Critique." *Reader* 32: 1–15.
———, 2007. *Of Literature and Knowledge: Explorations in Narrative Thought Experiments, Evolution, and Game Theory.* London and New York, Routledge.

———, 2010. *Literature, Analytically Speaking: Explorations in the Theory of Interpretation, Analytic Aesthetics, and Evolution.* (Cognitive Approaches to Literature and Culture Series.) Austin, University of Texas Press.

Tolhurst, William E., 1977. "What a Text Is and How it Means." *British Journal of Aesthetics* 19: 3–14.

Tomasello, Michael, 1999. "The Human Adaptation for Culture." *Annual Review of Anthropology* 28: 509–29.

———, 2005. "Understanding and Sharing Intentions: The Origins of Cultural Cognition." *Behavioral and Brain Sciences* 28: 675–735.

———, 2010. *Origins of Human Communication.* Boston, MIT Press.

Walton, Kendall, 1990. *Mimesis As Make-Believe: On the Foundations of the Representational Arts.* Cambridge, MA, Harvard University Press.

Wittgenstein, Ludwig, 1953. *Philosophical Investigations.* Oxford, Basil Blackwell.

Zunshine, Lisa, 2006. *Why We Read Fiction: Theory of Mind and the Novel.* Columbus, OH, Ohio State University Press.

Part II

Scientific Realism

7 Realism

Metaphysical, Scientific, and Semantic

Panu Raatikainen

1 INTRODUCTION

Today realism comes in many philosophical guises. In this essay, I shall review and compare three somewhat different, influential forms of realism. My own sympathies in all three cases lie on the realist side. However, I do not intend to defend realism conclusively and in detail here; that would surely be too large a task for one essay. Here I briefly characterize what sorts of arguments can be given in their support. My main aim is to clarify these positions and their mutual relations; later chapters examine these and related forms of realism in detail.

2 METAPHYSICAL REALISM

2.1 Putnam's Metaphysical Realism

Let us begin with what I call metaphysical realism. A potential source of confusion needs to be sorted out first. In contemporary philosophical literature, 'metaphysical realism' often refers to the specific view that the later Putnam took as his target while defending his own 'internal realism'. The view that Putnam calls *metaphysical realism* is a complex view. According to Putnam, it is committed to the following three doctrines:

1) There is exactly one true and complete description of the way the world is.
2) It requires 'a ready-made world'; the world itself must have a built-in structure.
3) Truth involves some sort of correspondence relation between words or thought-signs and external things and sets of things. (Putnam 1981, 49–50)

The last doctrine is really a semantic thesis and not metaphysical at all (see below). The first two are expressions of an extreme conceptual absolutism, and of an extreme reductionist variant of scientific realism, which is what I think Putnam primarily had in mind. This is a view very few contemporary scientific realists advocate.

2.2 Realism, Idealism, and Phenomenalism

Be that as it may, this is emphatically not what I mean by 'metaphysical realism' here. Rather, I simply want to focus on realism construed as a metaphysical doctrine. The question of the nature and plausibility of realism, in this sense, arises with respect to a large number of subject matters: commonsense physical objects, universals, non-existing possible objects, numbers and other mathematical objects, concepts, meanings, moral values, *etc.* It is obviously possible to be selectively realist or non-realist about various topics: For example, one could be a realist about the everyday world of macroscopic objects and their properties, but a non-realist about, say, mathematical objects or moral values.

What I wish to focus on here is the most general and uncontroversial issue: common-sense realism, or realism about the external world, that is, realism about the middle-sized commonsense macroscopic physical objects. 'External' here means external with respect to the knowing mind. This is the view that trees and stones, tables and chairs, dogs and cats, exist—even when we, knowing subjects, do not think about or perceive them. That is, they exist independently of the mental.

There are two important qualifications: First, realism obviously does not deny the possibility of error and even hallucination; the thesis is not that all middle-sized physical objects which are believed to exist must exist, but only that in order to exist, an object need not necessarily be (actually or potentially) thought or perceived. Second, realism does not deny the obvious fact that there are all sorts of mundane causal interactions between mental and physical reality; and clearly artifacts such as tables and chairs even owe their existence to the creativity of the human mind. The thesis is simply that the existence of an object at any time does not conceptually depend on being (actually or potentially) thought about or perceived. Realism, in this sense, is typically contrasted to idealism and phenomenalism.

By 'idealism' I mean the view that asserts only the existence of that which is actually thought about or perceived. One must be careful here, though: the idealist does not usually deny that stones, tables and other such things exist. She merely adds that they are all constituted of ideas, thoughts, sensations, or sense data, or more generally, that their existence is not independent of mental states. Historically, at least, the key motivation for turning towards idealism has been the desire to avoid radical skepticism, by closing the gap between reality and the knowing mind. Whether it really helps is quite another issue, as will become evident below.

Such idealism faces a serious problem: According to this view, a new object comes into existence whenever a subject perceives it, and ceases to exist when it is no longer perceived. However, when the subject perceives the object again, it makes no sense to ask whether it is the same object as before. Idealism also leads to the counterintuitive conclusion that a rock in the forest, or a beer can in the refrigerator with the door closed, does not exist. However, a certain kind of continuity and temporal identity is fundamental for our very concept of an object. In other words, adopting this kind of idealism would render all our deeply rooted beliefs about objects mistaken, and almost everything we think we know about the world would be in error. That is, radical skepticism would follow.

For such reasons, many philosophers wishing to remain skeptical about realism have preferred phenomenalism instead. According to this view, something really exists if and only if it *could be* perceived. According to phenomenalism, the rock in the forest, or the beer in the fridge, exists even if nobody actually sees it, as long as it would be at least possible to see it. Already Berkeley (1710, §3; 1713, 251) briefly reflected on phenomenalism, although more often he stuck to idealism. Nonetheless, he did rely on all ordinary objects being actually thought about by God, in order to guarantee their continued existence. John Stuart Mill (1867) first clearly formulated and defended phenomenalism. Phenomenalism became very popular in the first half of the 20th Century, partly with the rise of Logical Positivism. The influential proponents of phenomenalism have included A. J. Ayer (1940) and C. I. Lewis (1946).

2.3 Problems of Idealism and Phenomenalism

Phenomenalism accords with the common-sense belief about the persistence of physical objects better than pure idealism. Still, according to both idealism and phenomenalism, there could be no existence without minds; the existence of matter depends upon the existence of perceivers. We can then ask, following Armstrong (1961), what would happen if there were no conscious minds in the world. This scenario is not contradictory and is not known to be false purely *a priori*, but only empirically; it is only a contingent fact that this is not the case. Certainly the universe could have developed in a way that no creatures of a 'higher consciousness' would have emerged. It follows from idealism that in this case neither trilobites nor stars would have existed. Also, in the actual world, stars and trilobites did not then exist until there was a conscious mind that perceived or at least could have perceived them. And yet, we know perfectly well that stars and trilobites existed long before us, and also that stars and trilobites would have existed even if humankind or something similar had never evolved. Both idealism and phenomenalism thus entail that we have been wrong about all these issues, and radical skepticism threatens again.

Or imagine, following Stout (1938–39), someone in a room which is supported by foundations which no one perceives. What the person perceives—*i.e.* the room—is actual, but the foundations either do not exist (idealism), or are at best mere unfulfilled possibilities (phenomenalism). But how could something actual be supported by unfilled possibilities (or even something non-existent)? It seems that this would lead us to reject many of our ordinary causal and other explanations.

Idealism and phenomenalism may also be accused of being circular. They submit that a given material object 'consists of' certain (actual or potential) thoughts or sensations. But which sensations (or thoughts)? Peter Strawson (1959) and Roderick Chisholm (1976, 138–44) note that providing an answer seems impossible without referring to that very material object (*e.g.*, "the sense experience of *this* tomato"); the idealistic account of material objects, therefore, moves in a circle. The problem is that objects and things are typically identified by reference to their location in time and space, which in turn is determined by referring to other external material entities. They cannot, however, be identified simply by sensory experiences alone, because a person might have qualitatively identical sensations of two distinct objects in two completely different places.

Further, it may be asked what then is the foundation of the existence of the mind itself on which the existence of physical objects is supposed to depend? For idealism claims that all existence is based upon being thought about or perceived by some conscious mind. The existence of the mind itself must therefore be based either on that mind itself, or on the fact that some other mind is aware of it. But the idea that the existence of an object is based upon the object itself seems odd. If, on the other hand, the existence of the mind is based upon the existence of another mind, on what is the existence of the latter based? This idea leads to an infinite regression, and the existence of minds would be left without any support—except, perhaps, Berkeley's or Malbranche's all-perceiving God.

An argument that is widely taken as decisive against phenomenalism is the argument from perceptual relativity, due to Chisholm (1948). The gist of the argument is to point out how perceptual experiences of unchanging physical objects vary with changes in the conditions and perceivers: "Whether a material thing will ever present, say, a red appearance or sense-datum depends partly upon the thing and partly upon the conditions under which it is observed." The lighting may be abnormal, the observer may be color-blind, and so on. Therefore, a statement concerning a physical object, *e.g.*, 'This table is red', is *not* as such equivalent with any definitive statement about sensations, or sense-data, but at best a statement about sensations or sense-data is equivalent with the statement together with a statement about observation-conditions. Consequently, both observation-conditions and things-perceived should be definable in terms of what might appear. Due to the relativity of perception, however, the task of the phenomenalist seems to be "similar to that of an economist who hoped to define both supply and demand in terms of possible prices."

These and other problems which have accumulated have made idealism and phenomenalism increasingly unpopular among philosophers. At the same time, though, it is unlikely that realism can ever be absolutely proved correct by deductive logical reasoning. D.C. Williams (1966) has argued, however, that an inductive argument can be presented on the behalf of the "probability" of realism. Hilary Putnam (1975b) has developed this argument further. They do not, though, mean the familiar textbook cases of simple inductive reasoning, such as, "All ravens observed so far are black; therefore (probably) all ravens observed in the future will be black." Rather, what Williams and Putnam have in mind is the so-called inference to the best explanation, or, in Peirce's terminology, "abduction": the theory which explains and systematizes the phenomena we observe much better than any of the alternatives, is the theory according to which there are permanent material objects, existing independently of any mental states. Williams notes that this theory is confirmed by our daily observations in innumerable ways.

Putnam adds that the situation is exceptional in that no exactly formulated alternative theory that could be taken seriously has ever been devised; this same point has also been made by Popper. Putnam further notes that the situation is all the more peculiar in that our language already assumes realism, and all the attempted alternative theories must be formulated in a language (*e.g.*, a "language of appearing") which already presupposed the "thing language" (*e.g.*, "looks like a stone," which already presupposes the concept of 'stone'), a point made already by Sellars (1956).

This is closely connected to Wittgenstein's critique of "private language" and Sellars' refutation of the myth of the "Given." Both argue that forming a language that would be essentially private, and would *only* make reference to one's own sensations, would be impossible. This is because language is essentially social and public; and all talk about sensations and thoughts depends both conceptually and epistemologically on the "thing language" in which reference is made to physical objects. Thus, idealism would make all meaningful language impossible.

Philosophy is unlikely to ever achieve absolutely conclusive conclusions. The above analysis, however, gives good reasons to conclude that idealism faces a number of difficult problems. And because nothing actually speaks against it at all, there is excellent reason to believe that ordinary physical objects are real and that they exist independently of the mental.

3 SCIENTIFIC REALISM

3.1 Objects of Scientific Enquiry

Natural science is replete with expressions that appear to talk about all sorts of exotic entities unobservable by human senses, such as electrons, neutrinos, and genes. According to scientific realism, scientific theories, with all their talk about the unobservable, are meaningful and serious attempts to describe and explain mind-independent reality. It could perhaps be claimed that scientific realism, not radical empiricism, is the natural attitude of scientists themselves. Einstein and Planck, for example, endorsed realism even in the heyday of Positivism. And indeed, one can see the classical debate between Galileo Galilei and the Catholic Church as an instance of the disagreement between realism and instrumentalism. The Church was well prepared to admit that the heliocentric theory was better in practice, in simplifying calculations and predicting observations; the heresy lay in any claim for its truth. Mere calculative simplicity was not enough for Galilei. He insisted that the world actually is as the heliocentric theory says. In other words, Galilei did not defend just the theory, but also its realistic interpretation (*cf.* Popper 1963, Kitcher 2001). In contemporary philosophy of science, realism was first defended against Logical Positivism, or Logical Empiricism, starting in 1950s, by Smart (1956, 1963), Popper (1956, 1963), Sellars (1961), and Maxwell (1962). Later, realism has competed mainly with relativism and constructivism.

In all its forms, Logical Empiricism gave a privileged status to empirical language or observational concepts. Early Logical Positivism even hoped to show that all scientific concepts could be explicitly defined in observational terms: it was committed to 'descriptivism' or 'reductive empiricism'. This task was soon realized to be impossible, as Carnap (1936–37), a leading theorist of positivism, admitted already in 1936. This led Logical Empiricists to accept a more sophisticated version of empiricism, which had been advocated earlier, specifically, called 'instrumentalism'. According to this view, it is not possible to eliminate theoretical concepts from science, or define them in terms of observational concepts, but these theoretical concepts do not refer to anything real; they are only practically useful fictions which enable one to systematize observations and predict new observations on the basis of old ones. Contemporary scientific realism emerged largely in opposition to this view.

Beginning in the 1960s, certain radical views about science have also enjoyed popularity. Kuhn (1962) and Feyerabend (1962, 1963, 1965) argued that an old theory and the new one which replaces it, are "incommensurable"; they concluded it makes no sense to say that science can progress and that the new theory is closer to the truth than its predecessor. Some remarks by Kuhn about "different worlds" have also inspired a constructivist view, according to which each comprehensive scientific theory creates its own reality. Such views have been popular especially in science studies. Latour and Woolgar (1979), in their famous book, *Laboratory Life: The Social Construction of Scientific Facts*, for example, argued that not only scientific facts, but also the theoretical entities postulated in science, are social constructions. In general, radical constructivism either denies that there is any reality which is independent of language, theories, or conceptual schemes (pure idealism), or states that the independent reality is beyond the reach of our knowledge (as in Kantian transcendental idealism). In either case, whatever reality is knowable is created through the imposition of concepts. After the collapse of Logical Empiricism, such views have been the main opponents of scientific realism.

3.2 Varieties of Scientific Realism

It is possible and indeed useful to distinguish a weaker and stronger version of scientific realism.

1. *Minimal Scientific Realism* is the modest claim that, *pace* instrumentalism, scientific theories and their existence postulates should be taken at face value, that is, all talk about non-observable entities in science should be interpreted literally and realistically, and is not to be given some anti-realistic reinterpretation. Here are a couple of representative expressions of such a minimal view:

> The theoretical claims of scientific theories are to be read literally, and so read are definitely true or false. (Leplin 1984b, 2)

> Theoretical terms of scientific theories (*i.e.* nonobservational terms) should be thought of as putatively referring expression; that is, scientific theories should be interpreted 'realistically'. (Boyd 1983, 45)

Unfortunately, these formulations involve the concepts of 'reference' and 'truth', which seems to make them in part semantic theses. However, upon closer scrutiny, only the trivial 'disquotational' properties of truth and reference are needed, and merely deflationary notions may be invoked; that is, these formulations do not, after all, commit scientific realism to any particular substantial view of truth and reference. (This same observation applies also to the stronger formulations below.) Here is a more kosher formulation, also by Leplin:

> Scientific theories make genuine existential claims. (Leplin 1984b, 2)

Note that this formulation is silent about the success of scientific postulation; it is perfectly compatible with a thoroughgoing skepticism, *i.e.*, the view that science generally fails here and that the postulated entities do not typically exist. Most realists certainly want to endorse more, but this formulation is enough to contradict instrumentalism, for example, and it is useful to distinguish it from the other, stronger theses.

2. *Standard Scientific Realism* claims, very roughly, the following:

> One ought to believe in the existence of the unobservable entities posited by our most successful scientific theories.

However, the lessons of the actual history of science already make it necessary to revise this formulation: namely, it has turned out that at least some theoretical existence-postulates of science fail. For example, in chemistry, it was once assumed that there is a special burning substance, 'phlogiston', which leaves the burning material during combustion, but it is now agreed that nothing of that sort exists. Or, in astronomy, it was once postulated that there is another planet—it was christened 'Vulcan'—between Mercury and the Sun causing the observed deviations in Mercury's orbit; but it turned out that no such thing exists. And there are further cases. Therefore, if we want a thesis which is not clearly false, it is better to weaken it somewhat:

> Most unobservable entities posited by our scientific theories exist independently of the mental.

Here is a representative statement from Devitt:

> The central idea of scientific realism is that science really is committed and is, for the most part, right in its commitments. (Devitt 2004, 767)

A couple further qualifications are needed. First, in their early stages, sciences typically make stabs in the dark, and often get things badly wrong, and their postulated theoretical entities may well fail to exist. Therefore, it is better to restrict the realist thesis to mature sciences. Second, at the frontiers of scientific research, scientists often tentatively put forward speculative hypotheses with their postulates, but do not even themselves firmly believe in them; the realist does not obviously need to commit herself to anything more. The realist thesis is not plausible if it is not restricted to well-established theories. As our refined formulation, let us take the following:

> Most unobservable entities posited by the well-established theories of mature sciences exist independently of the mental.

Standard scientific realism is thus a form of metaphysical realism, applied to the non-observable entities postulated in science. It does not involve any semantic theses, such as the correspondence theory of truth.

3. *Strong Scientific Realism,* or *Convergent Realism,* is a stronger view often espoused by scientific realists, in terms such as these:

> The best current scientific theories are at least approximately true. (Leplin 1984b, 1)

Naturally, the same qualifications as above are in order: the thesis is plausible only with respect to the well-established theories of mature sciences. The difference between this stronger view and standard scientific realism is close to the distinction between *entity realism* and *theory realism,* put forward by Cartwright (1983) and Hacking (1983).

3.3 Arguments for Scientific Realism

What kind of arguments can be presented to support scientific realism? To begin with, it was vital for Logical Empiricism in all its forms to have a clear-cut and absolute distinction between the observable and the theoretical. But this dichotomy turned out to be quite unclear and problematic.

It is reasonable to doubt the claim that all that exists is restricted to what is observable by human senses. This claim has no particularly convincing arguments in its favor, and it is problematically species-chauvinistic and arbitrary: dogs, for example, can smell and hear things that are not observable to humans, not to mention what is observable to bats with their echolocation. It would be odd to insist that what dogs or bats can observe does not exist. The borders of the observable begin to look hopelessly blurry.

Hanson (1958) captured the criticism of the empiricist idea of pure observation with the now popular slogan that "observation is theory-laden." According to him, it is not the case that first there is some pure observation, which is then interpreted, but rather that observations are always immediately interpreted. Observation always involves a conceptual structuring. That is, in a sense, nothing is observable. Soon also Thomas Kuhn and Paul Feyerabend began to emphasize the theory-ladenness of observation as part of their critique of Logical Empiricism (Kuhn 1962; Feyerabend 1962, 1963, 1965). Hilary Putnam (1962) and Peter Achinstein (1965) argued convincingly that the whole distinction between the observable and the theoretical was confused and untenable.

We have seen above that there are very good reasons to accept metaphysical realism about mundane observable objects—such as rocks, flowers, and cats; its denial is extremely implausible. Then, because of the relativity and blurriness of the boundary between the observable and the theoretical, there is no principled reason to limit reality to the former. This gives some reason to believe at least in a minimal version of scientific realism.

However, the most popular argument in favor of scientific realism is the so-called *miracle argument* or the success argument; different versions of this argument have been put forward, *e.g.*, by Grover Maxwell (1962), J.J.C. Smart (1963), Hilary Putnam (1975a, 1978), and in particular, Richard Boyd (1981, 1983); also see Carrier (1991, 1993), Psillos (1999).

The point of departure of the argument is the apparent empirical and practical success of science. It is assumed that all parties of the debate agree on this point. Mature science is informed at all levels by theoretical presuppositions about "the deep structure of reality." The methods of science regularly result in correct predictions, and a wide range of working practical applications (such as satellite channels, DVD players, or the GPS navigation system). This dramatic success of science, at a purely instrumental level, would be inexplicable, 'a miracle', unless these presuppositions were, in fact, at least approximately true of an independently existing reality, and the theoretical objects postulated by them really existed. It's the best explanation for all the pragmatic success of science. This is a very convincing argument on behalf of a moderate realism. It is an instance of the so-called inference to the best explanation, and thus an abductive argument. Consequently, realism in this sense is an empirical hypothesis, even if a very general one.

This argument has, I think, considerable force. On the other hand, however, it operates on a rather general and abstract level, and may have a somewhat sublime flavor for some. Therefore, I often illustrate the issue with the following story:

Consider radiation and radioactivity. It is something one cannot see, taste or smell; that is, it is not observable. Nevertheless, both scientists and laymen nowadays strongly believe that there is such a thing. The investigation of radiation was prepared by Röntgen's pioneering work with x-rays. This inspired Henri Bequerel to conduct experiments with uranium compounds; in 1896 he observed a phenomenon that led to the discovery of radiation. In the following year, Marie and Pierre Curie discovered two new radioactive elements, polonium and radium. At the time there was hardly any idea of the health effects of radiation, and the researchers did not realize that they should have protected themselves properly against them. Consequently, many pioneers of radiation research, including Röntgen, Bequerel, and Marie and Pierre Curie, had to experience the health effects of radiation personally—they suffered suppurant burns. In 1904 the first person known to have been killed by x-ray exposure, a research assistant of Edison, died. Marie Curie died in 1934, overtaken by pernicious anemia clearly caused by years of radiation exposure.

Now it is extremely difficult to make any sense of these historical events if one does not accept some form of scientific realism. Were these deaths and other health effects caused by "mere fictions"? Did these pioneer scientists "socially construct" something that caused them burns or even killed some of them? Certainly such effects were not part of the ideas they formed *circa* 1896.

Actually, the first radiation accidents took place even before radiation was discovered. Already in 1895 Emil Grubbe studied cathode rays. He injured his hands with wounds which resembled serious burns but which his doctors could not explain. Moreover, already centuries earlier, in Central Europe, a mysterious, lethal 'mine disease', now identified with lung cancer, was strikingly frequent among miners, whereas otherwise it was quite rare. This was apparently caused by radiation from radon in mines.

So it seems that contemporary science provides us with a detailed notion of radioactivity—a notion well-supported with observational evidence. It can also explain the mechanism through which radiation causes its health effects, *e.g.* burns or cancer. But the view that scientists' talk of radiation does not refer to anything real—that the concept of radiation is just a practically useful fiction which enables one to systematize observations; or that radiation was only socially constructed by the researcher that discovered it—in addition to being quite implausible, also implies that useful fictions can have causal effects, or that social constructions may have causal effects, even before they began to exist (*i.e.*, prior to their formulation)! That is, a social construction may work causally backwards in time! In sum, views which deny the mind-independent reality of the theoretical entities of science lead to totally absurd accounts of such rather familiar stories, and to a totally preposterous conception of causes and effects. Cases like this should, at least, give pause to anyone attracted to non-realist views in this context.

3.4 Objections to Scientific Realism

Let us next briefly review the most popular objections to scientific realism.

1. *The Argument from Underdetermination.* One common objection to scientific realism is based on the alleged radical underdetermination of theory by observation: the thesis that for any theory, it is possible to find an indefinite number of empirically equivalent theories, that is, theories which are compatible with all the observations. The standard response is to grant that *deductive* underdetermination (*i.e.* that the theory cannot be logically derived from the evidence; or, that two incompatible theories may have exactly the same deductive observational consequences) is unavoidable, but to insist that we also must take into account the *inductive* relationships between observations and theory—and that this allows one to discard any radical underdetermination (see, *e.g.*, Laudan 1990, 1998; Ladyman 2002; Psillos 2005).[1]

2. *Pessimistic Meta-Induction.* The inductive argument against scientific realism based on the history of science, the so-called pessimistic meta-induction, proceeds, roughly, as follows: "Many past theories in science have turned out to be to a large extent false, and their theoretical terms non-referring. Therefore, it is not justified to expect that the theoretical entities postulated by present theories exist either, or that present theories are (approximately) true" (see, *e.g.*, Laudan 1981). However, the conclusion of this argument can be resisted. Is it not at least equally plausible to assume that our methods and theories have improved through trial and error, and that as science proceeds, the erroneous existence assumptions have become, in the long term, less frequent? (see, *e.g.*, Devitt 1991).

3. *The Circularity Accusation.* It has also been claimed (Laudan 1981: Fine 1986) that "the miracle argument" of realists is circular because it uses inference to the best explanation. Arguably this criticism, however, misses its target, because the dispute is about realism and not the reliability of this type of reasoning. Further, inference to the best explanation is applied regularly in science also to observable phenomena. The critics of scientific realism do not usually contest the reliability of this type of inference in that area. The problem cannot concern the type of reasoning. There is no reason to limit radical skepticism to science only, but from the skeptical perspective, observable everyday reality is equally at risk. The conclusion would be that there could be no justified knowledge about anything besides the current states of one's own consciousness. Such an extreme position is difficult to prove false, but there is also no good reason to believe it. Not even the opponents of scientific realism usually want to go that far. Their skepticism is thus arbitrarily selective.

In sum, the standard arguments against scientific realism are not a serious threat, and scientific realism is on firm footing. There is some room for controversy about the best and the most defensible formulation of scientific realism. But in some form or other, it is a very plausible view.

4 SEMANTIC REALISM

4.1 Dummett's Recasting of the Contrast between Realism and Anti-realism

The way that philosophers have understood realism underwent considerable changes during the 1970s, largely due to Michael Dummett. He was also influential in Putnam's conversion to what he called "internal realism" in contrast to "metaphysical realism" (as he defined it; above §2.1). Dummett (1978) argued that at least in some cases, it is not fruitful to focus on the existence of entities of a certain type: First, he pointed out that phenomenalism (or even idealism), as we have seen, need not say that macroscopic physical objects, such as rocks and trees, cats and dogs, tables and chairs, do not exist. Rather, the claim typically is that such objects are reducible to (or are constructions out of) sense data, or other such entities.[2] Second, he noted that in some cases, such as with realism about the past (or about the future), the question is not at all about the referential character of terms. Moreover, in the case of realism in mathematics, focusing on the reference of terms would be, according to Dummett, at least highly misleading, because the fundamental issue there is the objectivity of mathematical statements, as also with statements about the past.

For such reasons, Dummett prefers to characterize the dispute between realism and anti-realism in a way that concerns, not a class of entities, nor a class of terms and their reference, but rather some class of statements. These may be statements about the physical world, about mental events, mathematical statements, the theoretical statements of science, statements in the past tense, *etc.*—"the disputed class," as Dummett calls it.

'Realism'—in the Dummettian sense—is the belief that statements of the disputed class possess an objective truth value, independently of our means of knowing it. Realism so understood thus considers truth as at least potentially verification-transcendent, as some kind of correspondence to reality. Dummett and his followers often prefer a slightly different characterization and say that realism is committed to the Principle of Bivalence, that is, to the view that every meaningful statement is determinately either true or false, regardless of whether we are in a position to verify or falsify it. Anti-realism, in contrast, submits that the meanings of these statements are tied directly to what we count as evidence for them, so that a statement, if true at all, can be true only in virtue of something which we could know—and similarly for falsity. As a consequence, the Principle of Bivalence may well fail. Famously, Dummett defends such anti-realism with respect to, *e.g.*, mathematical statements and statements about the past.

The realism dispute in this context thus concerns the notion of truth appropriate for statements of the disputed class; it is also a dispute concerning the kind of meaning which these statements have. Consequently, it is sometimes appropriately called *semantic realism,* in order to distinguish it from some other philosophical views regarding realism. According to semantic realism, the meaning (or a central component of the meaning) of a (declarative, indexical-free) sentence can be taken to be the conditions in which the sentence is true, where truth is, as before, understood to be at least potentially verification-transcendent. Actually, in his earlier papers, Dummett suggested that truth, taken as something verification-transcendent, cannot serve as the central notion of the theory of meaning and must be replaced by verifiability. Later, however, he suggested that truth itself must be equated with verifiability. Either way, the question is whether a sentence can have verification-transcendent truth-conditions.

What, exactly, is the relation of the Dummettian semantic realism to the metaphysical construal of realism? The question is elusive. To begin with, Dummett often seems to suggest that the metaphysical views of realism and idealism are merely two unclear metaphors (*e.g.* Dummett 1978, *xxv*) which cannot be rationally argued for or against, but must be replaced with this semantic setting (*cf.* Devitt 1983, Miller 2006):

> How [are] we to decide this dispute over the ontological status of mathematical objects[?] As I have remarked, we have here two metaphors: the platonist compares the mathematician with the astronomer, the geographer or the explorer, the intuitionist compares him with the sculptor or the imaginative writer; and neither comparison seems very apt. The disagreement evidently relates to the amount of freedom that the mathematician has. Put this way, however, both seem partly right and partly wrong: the mathematician has great freedom in devising the concepts he introduces and in delineating the structure he chooses to study, but he cannot prove just whatever he decides it would be attractive to prove. How are we to make the disagreement into a definite one, and how can we then resolve it? (Dummett 1978, *xxv*)

This interpretation of Dummett—ascribing such a "metaphor thesis" to him—is the prevailing one (*e.g.*, both Devitt and Miller endorse it), and there certainly is textual evidence for it. Nevertheless, there is some room, it seems to me, for a different reading. Namely, Dummett typically applies the metaphor thesis more specifically to *mathematical* objects, and this leaves open the possibility that the metaphysical question might be perfectly meaningful in the case of, for example, ordinary middle-sized physical objects. And on still other occasions, Dummett rather seems to admit that (even in the case of mathematical objects) the metaphysical realism question and the semantic realism question are simply two independent and equally meaningful questions:

> To characterize a type of realism as a thesis about (putative) objects of some kind focused attention, I thought, on the wrong issue. For example, a neo-Fregean platonist about mathematical objects, such as Wright or Hale, could still deny that they have any properties other than those we are capable of recognizing, whereas, conversely, a Dedekindian who maintained that mathematical objects are free creations of the human mind might nevertheless insist that, once created, they have properties independently of our capacity to recognize them. It appeared to me evident, and still appears to me evident, that, interesting as the questions about the nature of mathematical objects, and the ground for their existence, may be, the significant difference lies between those who consider all mathematical statements whose meaning is determinate to possess a definite truth-value independently of our capacity to discover it, and those who think that their truth or falsity consists in our ability to recognize it. Hence, from my standpoint, the Dedekindian would be a species of realist, and the neo-Fregean a species of constructivist. Put more generally, what reality consists in is not determined just by what objects there are, but by what propositions hold for good: the world is the totality of facts, not of things. This was the reason for the concentration on acceptance or rejection of the principle of bivalence. (Dummett 1993, 465)

Now is Dummett's semantic approach a good substitute for the more traditional, metaphysical issue of realism? For example, is the commitment to the correspondence theory of truth a good indicator for someone's being a realist? The answer is arguably negative: On the one hand, some influential scientific realists such as Peirce, Sellars, and Bhaskar have favored an epistemic view of truth, and viewed truth as some kind of idealized verifiability, perhaps the later Putnam could be included here as well. Some philosophers, such as Field and Horwich, in turn combine scientific realism with the deflationist theory of truth rather than with the correspondence theory.

On the other hand, it is possible to advocate the correspondence theory of truth, but hold that the relevant correspondence is, for example, to sense data rather than to any objective mind-independent entities, as the spirit of realism would require. For example Schlick (1932, 1934, 1935), the leader of the Vienna Circle, held such a view, and combined the correspondence theory of truth with a kind of phenomenalism; he explicitly stated that realism is just as meaningless a metaphysical doctrine as idealism. Consequently, the correspondence theory of truth is neither a necessary component nor a sufficient criterion of (metaphysical) realism.

Neither is the commitment to the Principle of Bivalence a good criterion of realism. We might be forced to give it up, for example, because of complications in quantum physics—a possibility suggested by Quine and Putnam—and still remain scientific realists. Also, some considerations in the philosophy of logic—the need to avoid the paradoxes of truth, for example—might similarly lead one to abandon the Principle of Bivalence, though without altering one's realistic commitments in metaphysics. Hartry Field (2008) might again provide a good example. And again, someone like Schlick can combine the Principle of Bivalence with a non-realist view, *e.g.* phenomenalism.

Therefore, although the questions of the theory of truth and the theory of meaning are certainly philosophically very interesting, they cannot replace the more traditional metaphysical question of realism.

4.2 Dummett's Argument for Anti-realism

Let us briefly review Dummett's master argument for anti-realism, the so-called "manifestation argument" (*cf.* Shieh 1998, Miller 2003): We understand the sentences of the disputed class. Assume then that the meaning (or a central component of the meaning) of such a sentence were its recognition-transcendent truth-conditions. Then, so the argument goes, we would know the truth-conditions of the sentence.

To suppose, in general, that knowledge of meaning consisted of verbalizable knowledge would involve an infinite regress: it would be impossible for anyone to learn a language if they were not already equipped with a fairly extensive language. Hence, Dummett concludes, the knowledge which constitutes the understanding of language must be, in the end, implicit knowledge.

Implicit knowledge cannot, Dummett continues, meaningfully be ascribed to someone unless it is possible to specify what counts as the manifestation of that knowledge: there must be an observable difference between the behavior or the capacities of someone who is said to have that knowledge and someone who is said to lack it; hence it follows that grasping the meaning of a statement must, in general, consist of a capacity to use that statement in a certain way.

Consequently, our knowledge of the verification-transcendent truth-conditions of the relevant sentences should be manifested in our use of those sentences, *i.e.,* in our exercise of the practical abilities which constitute our understanding them. Dummett, however, submits that such knowledge is never manifested in the exercise of the practical abilities. Consequently, we do not possess knowledge of verification-transcendent truth-conditions. Therefore, the assumption must be discharged, and the sentences of the disputed class do not have verification-transcendent truth-conditions, hence semantic realism about the subject matter of disputed class must be rejected.

4.3 Some Problems for Anti-realism

Essential to Dummett's argument for anti-realism is the idea that understanding amounts to the knowledge of meaning, in a very literal sense. Namely, Dummett takes it for granted that if the meaning (or a central component of the meaning) of a declarative sentence is the conditions in which the sentence is true, then understanding the sentence is a matter of knowing its truth-conditions. Miller (2006, 994) aptly calls this key premise in Dummett's argument "the epistemic conception of understanding." Although Dummett and many others take it more or less as a tautology—Wright (1993, 18) states that it is "the immediate consequence of a series of platitudes"—it is far from uncontroversial.

No theory of meaning is constitutive of scientific realism. However, as soon as one accepts even modest scientific realism, the door is opened for semantic externalism, the view of meanings made famous by Putnam (1975c) with his slogan, "Meanings just ain't in the head!" If radical empiricism is given up, and if it is granted that there can be differences between two substances without there being some observable difference—that not everything that looks like the same substance really must be the same substance—then it also seems possible that there be a case in which everything appears the same but nevertheless two speakers speak about different things. And this is all that semantic externalism needs. Externalism, however, entails that there is a definite sense in which we do *not* actually know the meanings of many of our words. Semantic externalism therefore conflicts with the key premise of Dummett's argument.[3] Consequently, inasmuch as the arguments for semantic externalism have force, Dummett's master argument against semantic realism is undermined.

Dummett submits that we should accept an epistemically constrained notion of truth, and identify truth with verifiability. Is this notion really as clear and plausible as Dummett and others assume? I have argued elsewhere in detail (Raatikainen 2005) that the kind of notion of truth Dummett asks for has never been adequately explained even in the most well-defined case, the case of mathematical statements, where verifiability amounts to provability. All attempts to explicate this idea arguably turn out to be either clearly implausible, or circular.

The verificationist notion of truth is even less plausible if we turn our attention to physical reality and empirical knowledge of it. Different epistemic utilities, *e.g.*, simplicity, explanatory power, or support from observations, are effective in the acceptance of scientific theories and hypotheses (*cf.* Levi 1967, Kuhn 1977, McMullin 1983, Laudan 1984). Different dimensions of the goodness of an explanation include nonsensitivity, cognitive salience, precision, factual accuracy, and degree of integration (Ylikoski and Kuorikoski 2010). These various utilities and explanatory virtues, however, often pull in opposite directions, and the choice of which one to maximize may be pragmatic if not arbitrary. Therefore, it is doubtful that 'ideally justified' converges toward a determinate, unique theory. There may well exist indefinitely many mutually incompatible theories which deserve equally to be called 'ideally justified'. But that cannot be identified with truth, for whatever truth is, it is not contradictory.[4]

It is important to note the difference between the traditional verificationism and Dummett's approach: Dummett does not declare a statement meaningless if it is not verifiable. However, his view is that a competent speaker must know its verification-conditions, that is, she must know what *would* count as a verification of the statement. Dummett's whole approach requires that the theory of meaning must necessarily be strictly atomistic, or molecular. It must be possible to ascribe determinate and unique verification-conditions to each and every meaningful statement. Therefore, it is striking how little attention Dummett and his followers pay to the widely accepted idea of confirmation holism, or the Duhem-Quine thesis.[5] Namely, it is arguably a sound observation in the philosophy of science that isolated sentences of the more theoretical sort often have no connection to observations in themselves, but only when taken together in relatively large bundles, as a theory; and even then often only with some further auxiliary assumptions; moreover, different choices of auxiliary hypotheses lead to different inferential connections with observations. This suffices to show that not all apparently meaningful scientific statements have verification-conditions of their own (*cf.* Devitt 1983). Dummett grants that a statement can often be verified in a number of different ways. However, he adds that we must distinguish between direct (or 'canonical') and indirect means of verification, and that the former plays an essential role in the knowledge of meaning. In the light of confirmation holism, however, it seems very difficult to make any clear sense of the talk about some specific direct means of verifying a theoretical statement. Consequently, the whole Dummettian positive picture of the meaning of a sentence as its conditions of verification becomes deeply problematic.

5 CONCLUSION

We have discussed various 'realisms' in philosophy. They are conceptually independent, although one view can provide some degree of support for another. The best arguments in their favor are very different. It is quite implausible to deny general metaphysical realism, according to which there is a mind-independent reality—that ordinary material objects exist whether or not we cognize them. Furthermore, it is very natural and arguably justifiable to extend this idea to the unobservable entities postulated by science, and thus accept some form of scientific realism. Whether semantic realism *á la* Dummett is tenable is an entirely separate issue, and cannot replace the more traditional metaphysical question. Dummett's argument against semantic realism can be resisted, and the semantic anti-realism advocated by Dummett and his followers can be effectively criticized.

NOTES

1. The response is popular, and plausible, but there does not seem to have been an exact demonstration. However, in (Raatikainen 2012) I present a concrete and exact example, a pair of theories which are deductively underdetermined,

but once inductive relations are taken into account, observations clearly support one and not the other.

2. Dummett seems to miss the relevance of the *independence* dimension in the characterizations of metaphysical realism; mere existence is not sufficient. Consequently, this is not a genuine problem for metaphysical realism. Both idealism and phenomenalism deny the *mind-independent* existence of physical objects, and that is sufficient to distinguish them from realism.
3. I discuss this conflict in some detail in (Raatikainen 2010). I am, of course, not at all the first one to suggest this kind of objection. Perhaps the first to use explicitly externalist views against the Dummettian arguments was Millar (1977); see also Currie & Eggenberger (1983), Devitt (1983), Miller (2006).
4. For a useful review of several problems for epistemic theories of truth, see David (2004).
5. I don't intend to suggest that Dummett was not aware of the Duhem-Quine thesis; he certainly was. Dummett's discussions of this theme simply are not very helpful. On the one hand, he comments on some of Quine's observations more or less approvingly; on the other hand, he does not tire of emphasizing how disastrous a holistic view of meaning would be. Yet he says very little that would help to show how his view would avoid the problems mentioned here.

REFERENCES

Achinstein, Peter, 1965. "The Problem of Theoretical Terms." *American Philosophical Quarterly* 2: 193–203.

Armstrong, D. M., 1961. *Perception and the Physical World,* London, Routledge & Kegan Paul.

Ayer, Alfred J., 1940. *Foundations of Empirical Knowledge.* London, Macmillan.

Berkeley, G., 1710. "The Principles of Human Knowledge." In A. A. Luce and T. E. Jessop, eds., *The Works of George Berkeley*, 9 vols. (London: Thomas Nelson and Sons Ltd., 1948–1957), vol. 2, 21–113.

———, 1713. "Three Dialogues between Hylas and Philonous." In A. A. Luce and T. E. Jessop, eds., *The Works of George Berkeley*, 9 vols. (London: Thomas Nelson and Sons Ltd., 1948–1957), vol. 2, 165–263.

Boyd, Richard N., 1981. "Scientific Realism and Naturalistic Epistemology." In: P. D. Asquith and T. Nickles, eds., *PSA 1980,* Vol. 2, (East Lansing: Philosophy of Science Association), 2:613–662.

———, 1983. "On the Current Status of the Issue of Scientific Realism." *Erkenntnis* 19: 45–90; rpt. In: J. Leplin, ed., *Scientific Realism* (Berkeley: University of California Press, 1984), 41–82.

Carnap, Rudolf, 1936–37. "Testability and Meaning." *Philosophy of Science* 3: 420–471, 4: 2–40.

Carrier, Martin, 1991. "What is Wrong with the Miracle Argument?" *Studies in History and Philosophy of Science* 22: 23–36.

———, 1993. "What is Right with the Miracle-Argument: Establishing a Taxonomy of Natural Kinds." *Studies in History and Philosophy of Science* 24: 391–409.

Cartwright, Nancy, 1983. *How the Laws of Physics Lie.* Oxford, The Clarendon Press.

Chisholm, R., 1948. "The Problem of Empiricism." *Journal of Philosophy* 45: 512–17.

———, 1976. *Person and Object.* London, George Allen & Unwin.

Currie, Gregory, and Peter Eggenberger, 1983. "Knowledge of Meaning." *Noûs* 17: 267–279.

David, Marian, 2004. "Theories of Truth." In: I. Niiniluoto, M. Sintonen, and J. Woleński, eds., *The Handbook of Epistemology* (Dordrecht, Kluwer), 331–413.

Devitt, Michael, 1983. "Dummett's Anti-Realism." *Journal of Philosophy* 80: 73–99.

———, 1991. *Realism and Truth*, 2nd rev. ed. Oxford, Blackwell.

———, 2004. "Scientific Realism." In: F. Jackson and M. Smith, eds., *The Oxford Handbook of Contemporary Analytic Philosophy* (Oxford, Oxford University Press), 767–791.

Dummett, Michael, 1978. *Truth and Other Enigmas*. London, Duckworth.

———, 1993. *The Seas of Language*. Oxford, Oxford University Press.

Feyerabend, Paul, 1962. "Explanation, Reduction, and Empiricism." In: H. Feigl and G. Maxwell, eds., *Scientific Explanation, Space, and Time. Minnesota Studies in the Philosophy of Science* (Minneapolis, University of Minnesota Press), 3: 28–97.

———, 1963. "How to Be a Good Empiricist." In: B. Baumrin, ed., *Philosophy of Science: The Delaware Seminar*, (New York, John Wiley & Sons Inc.), 2: 3–40.

———, 1965. "On the 'Meaning' of Scientific Terms." *Journal of Philosophy* 12: 266– 274.

Fine, Arthur, 1986. *The Shaky Game*. Chicago, University of Chicago Press.

Hacking, Ian, 1983. *Representing and Intervening. Introductory Topics in the Philosophy of Natural Science*. Cambridge, Cambridge University Press.

Hanson, Norwood Russell, 1958. *Patterns of Discovery: An Inquiry into the Conceptual Foundations of Science*. Cambridge, Cambridge University Press.

Kitcher, Philip, 2001. "Real Realism." *Philosophical Review* 110: 151–197.

Kuhn, Thomas S., 1962. *The Structure of Scientific Revolutions*. Chicago, Chicago University Press; 2nd rev. ed. 1970.

———, 1977. "Objectivity, Value Judgments, and Theory Choice." In: *idem.*, *The Essential Tension* (Chicago, University of Chicago Press), 320–339.

Ladyman, James, 2002. *Understanding Philosophy of Science*. London, Routledge.

Latour, Bruno, and Steve Woolgar, 1979. *Laboratory Life: The Social Construction of Scientific Facts*. London, Sage.

Laudan, Laurence, 1981. "A Confutation of Convergent Realism." *Philosophy of Science* 48: 19–49.

———, 1984. *Science and Values*. Berkeley, University of California Press.

———, 1990. "Demystifying Underdetermination." *Minnesota Studies in the Philosophy of Science* 14: 267–297.

———, 1998. "Underdetermination." In: E. Craig, ed., *The Routledge Encyclopedia of Philosophy* (London, Routledge), retrieved 17 August 2012 from: www.rep.routledge.com/article/Q112

Leplin, Jarrett, ed., 1984a. *Scientific Realism*. Berkeley, University of California Press.

———, 1984b. "Introduction." In: Leplin (1984a), 1–7.

Levi, Isaac, 1967. *Gambling with Truth*. Cambridge, MA, MIT Press.

Lewis, C. I., 1946. *An Analysis of Knowledge and Valuation*. LaSalle, Open Court.

Maxwell, Grover, 1962. "The Ontological Status of Theoretical Entities." *Minnesota Studies in the Philosophy of Science* (Minneapolis, University of Minnesota Press), 3: 3–27.

McMullin, Ernan, 1983. "Values in science." *PSA 1982* (East Lansing, Philosophy of Science Association), 2: 3–28.

Mill, J. S., 1867. *An Examination of Sir William Hamilton's Philosophy*. London, Longmans Green.

Millar, Alan, 1977. "Truth and Understanding." *Mind* 86: 405–416.

Miller, Alexander, 2003. "The Significance of Semantic Realism." *Synthese* 136: 191–217.

———, 2006. "Realism and Antirealism." In: E. Lepore and B. Smith, eds., *The Oxford Handbook of Philosophy of Language* (Oxford, Oxford University Press), 983–1005.

Popper, Karl R., 1956. "Three Views Concerning Human Knowledge." In: H. D. Lewis, ed., *Contemporary British Philosophy* (London, Allen & Unwin); rpt. in: *idem., Conjectures and Refutations: The Growth of Scientific Knowledge* (London: Routledge & Kegan Paul, 1963), 97–119.

Psillos, Stathis, 1999. *Scientific Realism: How Science Tracks Truth*. New York and London, Routledge.

———, 2006. "Underdetermination." In: D. Borchert, ed., *Encyclopedia of Philosophy*, 2nd rev. ed. (Detroit, Thompson-Gale/MacMillan Reference), 9:575–578.

Putnam, Hilary, 1962. "What Theories Are Not." In: E. Nagel, P. Suppes, and A. Tarski, eds., *Logic, Methodology and Philosophy of Science* (Stanford, Stanford University Press); rpt in: Putnam (1975d), 215–227.

———, 1975a. "Introduction." In: Putnam (1975d), *vii–xiv*.

———, 1975b. "Language and Philosophy." In: Putnam (1975e), 1–32.

———, 1975c. "Meaning of Meaning." In: Putnam (1975e), 215–271.

———, 1975d. *Mathematics, Matter and Method. Philosophical Papers, Vol 1*. Cambridge, Cambridge University Press.

———, 1975e. *Mind, Language, and Reality. Philosophical Papers, Vol 2*. Cambridge, Cambridge University Press.

———, 1978. *Meaning and the Moral Sciences*. London, Routledge & Kegan Paul.

———, 1981. *Reason, Truth and History*. Cambridge, Cambridge University Press.

Quine, W. V. O., 1951. "Two Dogmas of Empiricism." *Philosophical Review* 60.1: 20–43; rpt. in: *idem., From a Logical Point of View* (Cambridge, Mass.: Harvard University Press, 1953), 20–46.

Raatikainen, Panu, 2005. "Conceptions of Truth in Intuitionism." *History and Philosophy of Logic* 25: 131–145.

———, 2010. "The Semantic Realism/Anti-Realism Dispute and Knowledge of Meanings." *The Baltic International Yearbook of Cognition, Logic and Communication, Volume 5: Meaning, Understanding and Knowledge*, 1–13.

———, 2012. "Ramsification and Inductive Inference." *Synthese* 187: 569–577.

Sellars, Wilfrid, 1956. "Empiricism and the Philosophy of Mind." In: H. Feigl and M. Scriven, eds., *Minnesota Studies in the Philosophy of Science* (Minneapolis, University of Minnesota Press) 1: 253–329; rpt. with additional notes in: Sellars (1963), 127–196.

———, 1961. "The Language of Theories." In: H. Feigl and G. Maxwell, eds., *Current Issues in The Philosophy of Science* (New York, Holt, Rinehart, and Winston), 57–77; rpt. in: Sellars (1963), 106–126.

———, 1963. *Science, Perception, and Reality*. London, Routledge & Kegan Paul; rpt. Atascadero, CA, Ridgeview.

Shieh, Sanford, 1998. "On the Conceptual Foundations of Anti-Realism." *Synthese* 115: 33–70.

Smart, J. J. C., 1956. "The Reality of Theoretical Entities." *Australasian Journal of Philosophy* 34: 1–12.

———, 1963. *Philosophy and Scientific Realism*. London, Routledge & Kegan Paul.

Stout, G. F., 1938–39. "Phenomenalism." *Proceedings of the Aristotelian Society* New Series, 39: 1–18.

Strawson, Peter, 1959. *Individuals*. London, Methuen.

Williams, D.C., 1966. *Principles of Empirical Realism: Philosophical Essays*. Springfield, Charles C. Thomas.

Wright, Crispin, 1986. *Realism, Meaning and Truth*. Oxford, Blackwell.

Ylikoski, P., and J. Kuorikoski, 2010. "Dissecting Explanatory Power." *Philosophical Studies* 148: 201–219.

8 Scientific Realism
Independence, Causation, and Abduction

Ilkka Niiniluoto

1 INTRODUCTION

Scientific realism in the narrow sense is the view that scientific inquiry may give us knowledge about unobservable theoretical entities. More generally, realism is a large-scale philosophical program with ontological, semantical, epistemological, theoretical, axiological, methodological, and ethical dimensions. It is important to ask whether realism in these multiple dimensions involves some common criteria or commitments. This paper studies three such interrelated criteria: mind-independence, causal power, and knowledge by abductive reasoning. The functions of these criteria are illustrated by considering the question of what a scientific realist should say about the reality of the past. It turns out that realism about the past effectively rules out many anti-realist philosophical positions, such as subjective idealism, phenomenalism, solipsism, positivism and pragmatism (in some of their varieties), internal realism, and social constructivism.

2 CRITICAL SCIENTIFIC REALISM

Scientific realism as a philosophical position has (*i*) ontological, (*ii*) semantical, (*iii*) methodological, (*iv*) epistemological, and (*v*) theoretical aspects (see Niiniluoto 1999a, Psillos 1999). It holds that (*i*) at least part of reality is ontologically independent of human mind and culture. It takes (*ii*) truth to involve a non-epistemic relation of correspondence between language and reality. An important aim of science is (*iii*) to find true and informative theories. Scientific realism claims that (*iv*) knowledge about mind-independent reality is possible, and that (*v*) the best and deepest part of such knowledge is provided by empirically testable scientific theories which postulate non-observable entities to explain observable phenomena.

Scientific realism is thereby in opposition to doctrines such as subjective idealism, solipsism, phenomenalism, positivism, pragmatism, instrumentalism, skepticism, relativism, and social constructivism.

Critical scientific realism can be distinguished from metaphysical or naive forms of realism by two additional theses. First, according to the principle of *conceptual pluralism,* all inquiry is relative to some conceptual framework. This is what Immanuel Kant argued in his Critical philosophy, but he failed to see that such frameworks can be changed, revised, and enriched—and was thereby driven to a skeptical view concerning knowledge about "things in themselves." According to pluralism, there is no single ideal language, nature's own privileged conceptual scheme as it were, or "Peirceish" in the terminology of Wilfrid Sellars (1968), which would fully describe the world in all of its variety (Niiniluoto 1984, 89; Niiniluoto 1987, 143). Secondly, according to Charles S. Peirce's principle of *fallibilism,* all factual human knowledge is uncertain or corrigible (see Peirce *CP* 5:587). Some of Peirce's pragmatist followers concluded that truth has to be taken as an epistemic notion, but semantic realism can be saved by employing the Tarskian model-theoretical explication of the correspondence theory of truth and by acknowledging that even the best results of science may be false, but still they may be probable, truthlike or approximately true (see Niiniluoto 1987).

Scientific realists are committed to the cultural value of science as a public institution. In the spirit of the Enlightenment, they support the scientific world view and thus oppose various kinds of anti-science movements. On the other hand, scientific realists are divided about the question of moral realism: some of them think that moral values have an independent existence in the realm of ideas or in nature, and that categorical ethical statements have truth values (see *e.g.* Boyd 1988), while others are ethical non-cognitivists or moderate moral relativists (see Niiniluoto 2009).

3 MIND-INDEPENDENCE

For many dimensions of scientific realism, mind-independence is the key condition. More generally, this criterion requires the independence of something from any mind-involving or mind-correlated cultural category, such as perspective, point of view, practice, discourse, linguistic or conceptual framework, scientific paradigm, language-game, form of life, tradition, and style of thinking (*cf.* Pihlström 2009, 12).

Let us start with ontology. While ultimately it is up to scientific inquiry to fallibly tell us what entities there are in the world, ontological realists have outlined many alternative ways of construing the mind-independent reality, or *World 1* in Karl Popper's terminology (see Popper 1972). The physical world has been taken to consist of things, events, or processes. Alternatively, it is constituted by objects, states of affairs, or laws. Among its lawlike features we have causality and spatio-temporal structures. Some scientific realists are realists in the mediaeval sense that they explain properties and predication in terms of universals (see Armstrong 1978), but some of them are nominalists (see Niiniluoto (2002) on Kotarbinski's reism) or trope theorists (see Niiniluoto 2012).

Idealism in its various forms denies ontological realism by claiming that all entities are in some sense in the mind or exist in a mind-dependent way. Hence, for an idealist, statements about mind-independent existence are *false*. In his famous refutation of idealism in 1903, G. E. Moore argued against the principle "*esse est percipi*" that idealists fail to distinguish sensation and its object, or more generally consciousness and its content (see Moore 1922). So a realist denies the identity of reality with our perception or knowledge about reality.

The classical controversy between realists and idealists is continued by debates about the 'constitution', 'construction', or 'structuring' of the world. This idea, which has been developed by the neo-Kantian schools (see Friedman 2000) and pragmatism (see Pihlström 1998), has been a major source of influence on many trends in Twentieth Century philosophy. While traditional *metaphysical realists* assumed that the world has a unique categorial structure, which warrants one true and complete description of the world, Hilary Putnam's (1981) *internal realism* attacked this idea of a "ready-made world" by claiming that objects are relative to descriptions and the structure of the world is created by conceptual frameworks. To avoid subjective idealism, which denies the mind-independent world altogether, internal realism is sometimes formulated with the cookie cutter metaphor: the world is cut into pieces by our concepts. However, this position faces the serious problem that the world without any objects, properties or structure is only a "noumenal jam" of propertyless particulars (see Tuomela 1985), a "world well lost" (see Rorty 1972), a pure abstraction or nothingness (see Niiniluoto 1999a, 217).

One possible conclusion, defended by Sami Pihlström (1998, 57), by appealing to Putnam's later writings, is that statements about mind-independent existence are *meaningless* (*cf.* Niiniluoto 1999a, 7). This view, which in fact rejects both idealism and realism as meaningless, was defended already by Rudolf Carnap's *Aufbau* in 1928 (see Carnap 1967) on the basis of a narrowly empiricist criterion of meaning. Another way of arguing for this kind of anti-realist position is the following: the notion of the mind-independent world is incoherent, since if you try to conceive it, it is no longer independent of mental activity. Or if you try to describe some elements of the non-conceptualized reality, you have thereby interwoven it with your concepts.

To answer this charge of incoherence, the realists have to show that it makes sense to refer to mind-independent things. As a master of semiotics, Peirce in 1904 gave the relevant distinction between the two objects of a sign: its *immediate object* is the object as the sign represents it, and its *dynamic object* is the object in itself (*CP* 8:333, 8:343). The immediate object covers only a particular aspect of the dynamical object which is "really efficient," "unlimited" and what "final study" would show it to be (*CP* 8:183; *cf.* Short 2007, 191–200). Peirce had argued already in 1868 that the Kantian notion of absolutely incognizable things-in-themselves has no meaning, so that whatever is real is cognizable to some degree (*CP* 5:310–311). Hence, things-in-themselves are not unknowable but inexhaustible by our truthlike descriptions. A related important tool for a realist is a modified Fregean theory of reference with a Principle of Charity: a theory refers to those real things which it describes in the most truthlike way (Niiniluoto 1999a, 120–132).

Another approach to the realism debate, based on Kantian *transcendental idealism*, is developed by Pihlström's (2009) "pragmatist metaphysics." Here the term 'ontology' appears to be ambiguous, as it may refer to the *study* of the categorial structure of the world or to the *entities* existing in reality. As a branch of philosophy, ontology is a human activity and therefore not mind-independent. It is then right to point out that the fallible result of ontological inquiry, *our* conception of reality, is dependent on human practices, even moral considerations. Pihlström skillfully applies the Kantian transcendental approach to argue that we humans—and in particular philosophers practicing ontological inquiry—contribute to "making" the "humanly categorized reality." In this transcendental sense, he states that "the world is not absolutely independent of our perspectives and pragmatic purposes," and the world is the "construction" of "the self." This position as such need not be idealist in the anti-realist sense any more than Edmund Husserl's (1970) phenomenological study of the theoretical and practical constitution of the life-world (where the assumption about the external world is put in brackets) or Nicholas Rescher's (1982) "conceptual idealism," which contends that "reality-as-we-picture-it" is "conceptually mind-involving" (*cf.* Niiniluoto 1999a, 209). But speaking about 'the self' in singular involves the danger that this approach collapses into metaphysical objective idealism which assumes a God's-eye-perspective on reality.

In my view, the true ingredient of internal realism and the cookie cutter metaphor is conceptual pluralism: the world can be described or conceptualized with several different linguistic frameworks. However, Putnam was mistaken in his assumption that such conceptual variety would entail the rejection of the non-epistemic notion of truth (Niiniluoto 1984, 178). Attempts to reconcile conceptual pluralism and the correspondence theory of truth, and thereby to reject both metaphysical and internal realism in Putnam's senses, are given by Niiniluoto (1999a) and Kitcher (2001). To maintain the notion of mind-independent existence, one should distinguish *existence* and *identification* (*cf.* Wolterstorff 1987): there exists the mind-independent world (sometimes called THE WORLD) with unidentified objects and facts ('unidentified flying objects' or UFOs in my terminology), and these potential objects and facts can be identified (as 'identified flying objects' or IFOs) by means of several alternative linguistic descriptions. THE WORLD and alternative conceptual frameworks *L* together generate *world-versions* W_L,, where the crucial feature of THE WORLD is its capacity to uniquely determine the truth value of semantically definite statements of *L* in W_L. We choose the language *L*, and THE WORLD 'chooses' the truth values of *L*-statements in the world-version W_L. For the humanly conceptualized world-versions, but not for the mind-independent WORLD, it is correct to say with Putnam (1981, *xi*) that "the mind and the world jointly make the mind and the world." Antonio Dieguez (2011) formulates this kind of "moderate ontological realism" by saying that THE WORLD has a basic intrinsic structure compatible with potentially infinite conceptual classifications but inexhaustible from the point of view of human interests.

The distinction between existence and identification may seem to conflict with Quine's famous dictum "no entity without identity" (see Quine 1969, 23). But Quine's principle concerns our ways of "speaking about objects": when we introduce new entities, identity criteria should be specified.

It should be emphasized that what I have called IFOs in the world-versions W_L are not Kantian "things-for-us," phenomenal or epistemic veils that hide unknowable UFOs from us (Niiniluoto 1999a, 222). As far as THE WORLD itself decides which entities really exist relative to the descriptions of the language *L,* the world-version W_L is typically unknown to us—and thus the proper research domain of a scientist employing the vocabulary of *L*. For example, Paul Dirac described positrons in 1928, and his prediction of their existence was experimentally verified by Carl Anderson in 1932. The hunt of the Higgs particle, introduced by Peter Higgs around 1970, still continues (with some preliminary success) with the Large Hadron Collider experiments at the CERN laboratory.

Some metaphysical systems of objective idealism and dualism postulate the existence of a supernatural and superindividual mind as the ultimate reality. But in order to be testable, statements about mind-independence should refer to the human mind or collectives of such minds. If eliminative materialism is correct, and there are no minds, then trivially everything is mind-independent. All other positions—from reductive and emergent materialism to epiphenomenalism and dualism—have to give a plausible account of the human mind (Popper's *World 2* consisting of subjective experiences, beliefs, and emotions) and the cultural and social products of human action (*World 3* consisting of artifacts, language, abstract entities, and institutions) (see Niiniluoto 2006). A scientific realist can be a realist about Worlds 2 and 3, even though they cannot be mind-independent in the same sense as the physical World 1. Popper himself appealed to the causal powers of such entities (*cf.* the next section). In my view, Peirce's "mediaeval definition" of reality, which he attributed to John Duns Scotus (*CP* 4:28), is relevant here (see Niiniluoto 2006): those things are real "whose characters are independent of what anybody may think them to be" (*CP* 5:405, 5:430). Peirce applied this generalized independence criterion for distinguishing dreaming as a real mental phenomenon (in World 2) from the fictional contents of dreams. By the same criterion, abstract mathematical entities and other public cultural constructions (in World 3) are real in the sense that they are "independent of the vagaries of me and you" (*CP* 5:311).

Mind-independence is also the key idea of the realist theory of truth: the truth or falsity of a statement in language L does not depend on our beliefs or wishes, or on its acceptability under ideal conditions (*cf.* Dewey 1941; Sellars 1968; Putnam 1981; Tuomela 1985), but on its relation to the structure W_L. Such truth conditions are given by Tarski's semantical definition of truth (see Niiniluoto 1999a, 55–64). Peirce's characterization of truth as the limit of inquiry, or the final opinion of the scientific community, can be interpreted in the realist sense (see Niiniluoto 1984, 82, 214). Kitcher (2012) attempts to argue that also the pragmatist account of truth by William James can be understood as a version of the correspondence theory.

Objective truth is the methodological goal of scientific inquiry. Against *instrumentalists,* who treat theories merely as conceptual tools for observational systematization, scientific realists take existential and general theoretical statements as genuine assertions with truth values. Against *social constructivists,* who claim that scientific reality is created or produced by laboratory negotiations and practices (see Latour and Woolgar 1986, 180), scientific realists argue that the objects of the "scientific image" are ontologically but not epistemically primary to the entities of "the manifest image" (Sellars 1968). Thus, as far as inquiry concerns theoretical entities and laws beyond observations, its focus is in the mind-independent reality.

4 CAUSATION AS A CRITERION OF REALITY

If the basic characteristic of reality and truth is their mind-independence, how can we ever gain knowledge about such cognitive targets? Ontological realism may seem to allow anti-realists to devise a weapon against epistemological realism, called the *Inaccessibility of Reality Argument* (IRA) by Kitcher (2001). To answer this challenge, considerations of causation are needed.

A quick answer to Kitcher's problem with IRA is that ontological mind-independence should be distinguished from causal independence (Niiniluoto 1999a, 27, 94, 151). External objects and processes in World 1 are not inaccessible to us: we can be in mutual causal interaction with stones and atoms by means of our perceptions and actions. As Peirce emphasized, "real things" as an "external permanency" "affect our senses according to regular laws" and thus cause beliefs as their effect (see *CP* 5:384, 5:406). This is a feature of their factuality, *haecceity* or "secondness" in Peirce's sense. In experimentation we can causally manipulate nature so that it will give answers to our questions. The scientific method is based upon causal interaction between the object and subject of cognition.

The notion of causality is central to *direct realism, i.e.,* theories of perception which allow us to construe statements of the form "*A* sees *b* as an *F*." Here the epistemic agent *A* sees an external object *b,* which causally influences *A*'s sensation, but the content of perception covers more or less veridically only some aspect *F* of *b* (Niiniluoto 1999a, 91).

For Hume causal connections were merely constant conjunctions of events, and the idea of causal necessity arises from the habit of our mind. For Kant causality is one of the categories that our understanding supplies in its way of thinking about objects. So for both of them causality is in some way mind-involving. In contrast, scientific realists have defended *causality* as a real feature of the mind-independent world. For Peirce, "thirdness" as a lawlike "would-be" is also real. Based upon (deterministic or probabilistic) lawlike dispositions or generative powers of causal mechanisms (*cf.* Bhaskar 1975; Fetzer 1981), causality has been operating in the universe ever since the first seconds after the Big Bang.

Causal relations obtain between events located in space and time. Causes precede their effects, and causal processes determine the direction of time. Hence, a realist view of causation is associated with a realist view of *space* and *time*. While Kant argued that space and time are "forms of intuition," the realist should make a distinction between physical time in World 1 (determined by objective temporal relations and studied by physics), subjective experienced time in World 2 (*e.g.*, Bergson's duration, Husserl's and Heidegger's phenomenology of time consciousness), and theories about time in World 3 (*e.g.*, Newton's classical mechanics, Einstein's theory of relativity) (see Grünbaum 1973, Gale 1978).

Causation and causal powers are elevated to a defining criterion of reality by many philosophers. For Popper (1972), reality consists of ordinary macroscopic objects and everything that is capable of causally influencing them. He applies this criterion to physical processes in World 1, mental states in World 2 with the ability of "downward" mental causation, and public institutional entities in World 3. Ian Hacking (1983) defends the reality of theoretical entities like electrons and positrons by the slogan: "if you can spray them, they are real." According to this principle of causal intervening, causation is a *sufficient* condition of reality: if A causally influences B, then A is real. Another formulation allows the reality of causal powers or potencies: if A has the causal power to influence B, then A is real.

The causality criterion has an important consequence for the debate about the reality of *the past* (*cf.* Niiniluoto 2000). According to *actualism*, everything that is real is actual. An actualist may be an ontological anti-realist about the past or admit that past event *were* real but not any more. Arthur Prior defined *presentism* as the position that (it is always the case that) everything real is actual or present. The causality criterion gives an argument against presentism, since many past events—like the Big Bang, earthquakes, the ice age, organisms, human cultures, and persons—leave causal traces which can be observed today. By these traces—such as cosmic background radiation, geological structures, fossils, documents, pictures, photos, stories, and memories—such past events *are* real, even though they are not present here and now.

Presentism can be criticized, and the reality of the past can be defended, also by Quine's famous criterion of ontological commitment: "to be is to be the value of a bound variable." There are many statements in ordinary language, true by any realist or epistemic theory of truth, which involve quantification over past entities at different points of time, and cannot be translated by suitable temporal operations into paraphrases acceptable for a presentist (*cf.* Sider 1999). While the sentence "A dinosaur lived in the past and weighted 1000 kilos" can be translated into "In the past it was the case that there is a dinosaur which weights 1000 kilos," sentences like "Simo Knuuttila admires Duns Scotus," "An earthquake under the ocean caused the tsunami in Thailand in the next day" and "All human being have been born on earth" (see Needham 1976) are problematic. A scientific realist can also argue that the special and general theories of relativity are incompatible with presentism, since these theories postulate a four dimensional space-time in which objects exist as enduring entities (see Crisp 2003).

A more general argument for the reality of the past can be given by Peirce's independence criterion. Our causal interaction with the past is unidirectional, as there is no retroactive causation from the present to the past. This means that the past is immutable, and thereby real by Peirce's definition. This conclusion concerns even those past events which have lost their causal traces in the present world, and therefore may be forgotten forever. As we shall see in the next section, the reality of such past events is denied by Michael Dummett's (2004) anti-realism about the past.

The future is different from the past in an important sense. There are 'eternal' sentences which are true at any future point of time *t*, among them true temporally definite statements like "The University of Helsinki was founded in Turku in 1640." If such a sentence is true now, it will be true forever. Hence, there are sentences about the future which are true by mathematical or conceptual necessity: "In 2040 the University of Helsinki has its 400th anniversary." However, the situation of contingent events in the future is different. If the world were deterministic, future events would be necessary relative to the present state, so that the future would be real by Peirce's criterion. However, it is more plausible to think that the future is indeterministic in some respects and thus still open between several alternative possibilities (*cf.* Gale 1978). The future course of the world may depend on real potencies or propensities (*cf.* Fetzer 1981) and the choices that human beings make, but the future as such is not yet actual or real, but only possible.

5 ABDUCTION AND REALISM ABOUT THE PAST

The great insight of Peirce's fallibilist analysis of inference was his observation that, besides deduction, science uses two ampliative forms of reasoning: induction and abduction (*cf.* Niiniluoto 1999b). While *induction* makes universal or statistical generalizations from observations, *abduction* is reasoning from effects to causes, or from observational data to hypothetical explanatory theories (*CP* 5:189):

(1) The surprising fact E is observed;
If H were true, E would be a matter of course.
Hence, there is reason to suspect that H is true.

Already perceptual judgments can be understood as "extreme cases" of abduction (*CP* 5:181), since in perception external objects cause our sensations. But against Auguste Comte's positivism, Peirce argued in the realist spirit that science should not be restricted to hypotheses "verifiable by direct observation," but abduction frequently concludes "something which it would be impossible for us to observe directly" (*CP* 2:640). Peirce further pointed out that in science the abductive step is followed by severe observational and empirical tests of the deductive or probable consequences the hypothesis (*CP* 2:634). Our knowledge about theoretical entities ultimately depends upon their observable causal effects. The examples of abduction thus range from compelling everyday observations to the tentative adoption of theoretical hypotheses in science by virtue of their explanatory and predictive power. The success of such theoretical abductive arguments is denied by empiricist anti-realists (see Laudan 1984, van Fraassen 1989, Kukla and Walmsley 2004) and defended by scientific realists (see Psillos 1999, Leplin 2004, Niiniluoto 2004).

Abduction of the form (1) can be strengthened, if the theory H is able to give a unifying explanation of mutually independent facts (see Niiniluoto 2008). For critical epistemological realists, it is interesting to study the idea that abductive inference (1) can be reformulated by taking its conclusion to concern the *truthlikeness* of a hypothetical theory on the basis of its success in explanation and prediction (see Niiniluoto 2005). This modification of abduction is also relevant to the so called "ultimate argument for scientific realism" (*cf.* Musgrave 1988). Following the 1950s, when scientific realism became a tenable position after the dominance of empiricism and instrumentalism, several philosophers of science (among them J.J.C. Smart, Hilary Putnam, Grover Maxwell, and Richard Boyd) have defended realism as the best hypothesis which explains the practical (empirical and pragmatic) success of science. The ability of scientific theories to explain surprising phenomena and to yield correct empirical predictions and effective rules of action would be a "cosmic coincidence" or a "miracle" unless they refer to real things and are true or at least approximately true or truthlike (see Psillos 1999, Niiniluoto 1999a).

Peirce's examples of abduction included historical inferences. In the same way as electrons manifest themselves in Wilson's cloud chamber, past events manifest themselves in their traces and documents today. For example, the fact that Napoleon Bonaparte once lived is no longer "susceptible of direct observation," but we believe it because "its effects (such as the histories, the monuments *etc.*) are observed" (*CP* 2:714). Such inferences are known as *retroductions.* By the same principles, we can infer that stones and trees existed in the past, and were composed of atoms and cells, respectively. As there is no retroactive causation backwards in time, the past is cognitively accessible to us only by means of retroduction.

The enormity and profundity of the problem posed to anti-realists or radical social constructivists by the reality of the past is put into proper perspective by Martin Rudwick (2005), who examines in detail the social and historical construction of the discipline of geohistory. This study, which shows that human history is just a slight, recent blip in the historical record of the earth, contributed to shifting the sense of 'history' away from its original, purely descriptive sense to its fundamentally chronological sense of historical change subject to causal inquiry and explanation.

Realism about the past allows us to formulate an argument against all views which claim that all reality is dependent upon or relative to human minds, language, culture, or practices. This "argument from the past" (see Niiniluoto 1999a, 40) points out that we have very strong evidence from many fields of contemporary science that there was a point of time t_0 in the history of our planet when no human being had yet appeared through evolution. We can estimate that t_0 is 5 million years BCE. This gives us a powerful test of mind-independence: whatever existed before t_0 must be ontologically independent of human mentality and human cultural constructs such as concepts. This refutes all philosophical doctrines—such as subjective idealism, phenomenalism, solipsism, positivism and pragmatism (in some of their varieties), internal realism, and social constructivism—which literally claim that nothing exists independently of the human mind or that all objects are our constructions.

It is important to give a correct formulation of known truths about the past. For this purpose, it is useful to distinguish causal and conceptual construction. Shaping material and technological artifacts is an example of *causal construction,* while identification of kinds of concrete and abstract entities by linguistic descriptions and definitions is a typical method of *conceptual construction.* The concept of 'dinosaur' was created by scientists in the nineteenth century after the discovery of fossils, but this linguistic act did not retroactively causally produce dinosaurs in the past. The sentence "Dinosaurs exist" did not exist 100 million years ago, so it was not true at that time. The following counterfactual statement can be claimed to be true: if the world had been described by current language 100 million years ago, then the sentence 'Dinosaurs exist' would have been true. But the justification for this sentence is based on the truth of the sentence 'Dinosaurs existed 100 million years ago'. By Tarski's T-equivalence, this is equivalent to the fact that dinosaurs existed 100 million years ago. Thus, this historical statement refers to dinosaurs as IFOs or identified objects in the world version W_L conceptually created by the language L of the paleontologists, but dinosaurs as UFOs or really existing past entities in THE WORLD are sufficient as the truth-makers to determine its truth. Scientists who started to talk about dinosaurs influenced our world view, but they did not change the past by causally constructing new objects in the past. The realist is not only claiming that dinosaurs today belong to our world version or our current scientific theory, which makes claims about the past, but more strongly that dinosaurs in flesh and blood lived, mated, interacted, and died as real entities in the past (see Niiniluoto 1999a, 220).

The argument from the past does not refute objective idealism which takes past objects to be ideas in God's mind. For example, Nicholas Malebranche's occasionalism claims that God is the only true cause and we see external objects by means of ideas in God (see Schmeltz, 2009). A similar view was held by George Berkeley. Such metaphysical positions are immune to empirical criticism, but also devoid of empirical content and explanatory power. Perhaps the argument does not affect those who think that we 'construct' objects only in some metaphorical sense (whatever that may be). Pihlström (1998, 73) suggests that historical truths about dinosaurs living on earth before any language-using being appeared are not relevant to the dispute between realism and pragmatism (or internal realism). But clearly it is a problem for those like Nelson Goodman who insist that we literally are "worldmakers" or "starmakers" (*cf.* McCormick 1996). This was Putnam's (1990) initial position in his internal realism, but, as Pihlström acutely observes, he soon changed his view to an "increasingly realist" position. It is also a problem for those who think that talk about such past objects is incoherent or meaningless.

Shaun O'Dwyer (2010) argues that, in spite of his naturalism, John Dewey's version of pragmatism lacks the conceptual resources to move beyond the contemporary realism/anti-realism debate about the past. Referring to Dewey's opposition to Arthur Lovejoy's conception of entities which are "existentially transcendent of the knowing experience" (*cf.* Morgenbesser 1977), and Dewey's denial of epistemically inaccessible true or false propositions, O'Dwyer finds Dewey's position to have "too many affinities" with Dummett's (2004) anti-realism about the past.

For a scientific realist, there are both known and unknown truths about the past. The former are warranted by abduction, the latter are true by their correspondence to the way the world was. Bertrand Russell's (1940) example, directed against Dewey's (1941) epistemic notion of truth as warranted assertability, was the proposition "It snowed on Manhattan Island on the first of January in the year 1 AD." By the law of excluded middle, or the principle of bivalence, this proposition is true or false, even though we have no evidence, and are never likely to have any, to decide the alternative. Another illustration concerns the number of times Winston Churchill sneezed in 1939. Assuming that sneezing is defined sufficiently clearly, the statement 'Churchill sneezed 100 times in 1939' has a truth value depending on facts about him during that year. A common thread in Dewey's notion of warranted assertability, Dummett's (1982) semantical anti-realism, and

Putnam's (1990) notion of truth as ideal acceptability, is the denial of such "recognition-transcendent" truths. Dummett's conclusion, inspired by verificationism and intuitionist philosophy of mathematics, is that such undecidable propositions are meaningless. Nicholas Jardine (1986) suggests that counterfactual evidence obtained by time-travel to the past should be allowed. However, as we have no problems in understanding statements about snowing in 1 AD. or sneezing in 1939, these attempts to avoid skepticism seem to be exaggerated caution (Niiniluoto1999a, 107). While many realists support "human memory organizations" (museums, archives, and libraries) in their efforts to preserve cultural products from times past, especially their information content in World 3, they may very well admit that there are large parts of past and contemporary reality which are and will remain unknown to us.

The debate about realism is also significant for the methodology of historiography. Franklin Ankersmit (1989) defends the instrumentalist view that the past as unobservable cannot be "the proper object" of historical investigation, but the historian's task is to construct narratives which are coherent with the traces the past has left us. The realist, instead, sees the historians—just as their counterparts in natural science—as truth-seekers whose retroductive descriptions and interpretations should correspond to the actual course of past events.

REFERENCES

Ankersmit, Frank, 1989. "The Use of Language in the Writing of History." In: H. Coleman, ed., *Working with Language* (Berlin, Mouton de Gruyter), 57–81.
Armstrong, David, 1978. *Universals and Scientific Realism,* 2 vols. Cambridge, Cambridge University Press.
Bhaskar, Roy, 1975. *A Realist Theory of Science.* Leeds, Leeds Books.
Boyd, Richard, 1988. "How to be a Moral Realist?" In: G. Sayre-McCord, ed., *Essays on Moral Realism* (Ithaca, NY, Cornell University Press), 181–228.
Carnap, Rudolf, 1967. *The Logical Structure of the World and Pseudoproblems in Philosophy.* Berkeley, University of California Press.
Crisp, Thomas M., 2003. "Presentism." In: M. J. Loux and D. W. Zimmerman, eds., *The Oxford Handbook of Metaphysics* (Oxford, Oxford University Press), 211–245.
Dewey, John, 1941. "Propositions, Warranted Assertability, and Truth." *The Journal of Philosophy* 38.7: 169–186.
Dieguez, Antonio, 2011. "Kitcher's Modest Realism: The Reconceptualization of Scientific Objectivity." In: Gonzalez (2012), 141–169.
Dummett, Michael, 1982. "Realism." *Synthese* 52: 55–112.
———, 2004. *Truth and the Past.* New York, Columbia University Press.
Fetzer, James, 1981. *Scientific Knowledge: Causation, Explanation, and Corroboration.* Dordrecht, D. Reidel.
Freidman, Michael, 2000. *A Parting of the Ways: Carnap, Cassirer, and Heidegger.* La Salle, Ill., Open Court.
Gale, Richard, ed., 1978. *The Philosophy of Time: A Collection of Essays.* New Jersey, Humanities Press.

Gonzalez, Wenceslao, ed., 2011. *Scientific Realism and Democratic Society: The Philosophy of Philip Kitcher.* Amsterdam, Rodopi.
Grünbaum, Adolf, 1973. *Philosophical Problems of Space and Time*, 2nd enlarged edition. Dordrecht, D. Reidel.
Hacking, Ian, 1983. *Representing and Intervening.* Cambridge, Cambridge University Press.
Husserl, Edmund, 1970. *The Crisis of European Sciences and Transcendental Phenenomenology.* Evanston, Ill., Northwestern University Press.
Jardine, Nicholas, 1986. *The Fortunes of Inquiry.* Oxford, Oxford University Press.
Kitcher, Philip, 2001. "Real Realism: The Galilean Strategy." *The Philosophical Review* 110.2: 151–197.
———, 2011. "Scientific Realism: The Truth in Pragmatism." In: Gonzalez (2011), 171–189.
Kukla, André and Walmsley, Joel, 2004. "A Theory's Predictive Success does not Warrant Belief in the Unobservable Entities it Postulates." In: C. Hitchcock, ed., *Contemporary Debates in Philosophy of Science* (Oxford, Blackwell), 133–148.
Latour, Bruno and Woolgar, Steve, 1986. *Laboratory Life: The Construction of Scientific Facts.* Princeton, Princeton University Press.
Laudan, Larry, 1984. *Science and Values.* Berkeley and Los Angeles, University of California Press.
Leplin, Jarrett, 2004. "A Theory's Predictive Success can Warrant Belief in the Unobservable Entities it Postulates." In: C. Hitchcock, ed., *Contemporary Debates in Philosophy of Science* (Oxford, Blackwell), 117–132.
McCormick, Peter, ed., 1996. *Starmaking.* Cambridge, Mass., MIT Press.
Moore, G. E., 1922. *Philosophical Studies.* London, Paul, Trench, Truber.
Morgenbesser, Sidney, ed., 1977. *Dewey and His Critics.* New York, The Journal of Philosophy.
Musgrave, Alan, 1988. "The Ultimate Argument for Scientific Realism." In: R. Nola, ed., *Relativism and Realism in Science* (Dordrecht: Kluwer), 229–252.
Needham, Paul, 1976. "The Speaker's Point of View." *Synthese* 32: 309–327.
Niiniluoto, Ilkka, 1984. *Is Science Progressive?* Dordrecht, D. Reidel.
———, 1987. *Truthlikeness.* Dordrecht, D. Reidel,
———, 1999a. *Critical Scientific Realism.* Oxford, Oxford University Press.
———, 1999b. "Defending Abduction." *Philosophy of Science (Proceedings),* 66: S436–S451.
———, 2000. "Onko menneisyys todellista?" [Is the Past Real?]. In: S. Pihlström, A. Siitonen, and R. Vilkko, eds., *Aika* (Helsinki, Gaudeamus), 246–261.
———, 2002. "Kotarbinski as a Scientific Realist." *Erkenntnis* 56: 63–82.
———, 2004. "Truth-seeking by Abduction." In: F. Stadler, ed., *Induction and Deduction in the Sciences* (Dordrecht, Kluwer), 57–82.
———, 2005. "Abduction and Truthlikeness." In: R. Festa, A. Aliseda, and J. Peijnenburg, eds., *Confirmation, Empirical Progress, and Truth Approximation* (Amsterdam, Rodopi), 255–275.
———, 2006. "World 3: A Critical Defense." In: I. Jarvie, K. Milford, and D. Miller, eds., *Karl Popper: A Centenary Assessment, Volume II: Metaphysics and Epistemology* (Aldershot, Ashgate), 59–69.
———, 2007. "Evaluation of Theories." In: T. Kuipers, ed., *Handbook of Philosophy of Science: General Philosophy of Science—Focal Issues* (Amsterdam, Elsevier), 175–217.
———, 2008. "Abduction, Unification, and Bayesian Confirmation: Comment on Schurz." In: C. Dermont, L. Keiff, and H. Rückert, eds., *Dialogues, Logic and Other Strange Things* (London, College Publications), 365–370.
———, 2009. "Facts and Values—A Useful Distinction." In: S. Pihlström and H. Rydenfelt, eds., *Pragmatist Perspectives* (Acta Philosophica Fennica 86. Helsinki, Societas Philosophica Fennica), 109–133.

———, 2012. "On Tropic Realism." In: L. Haaparanta and H. J. Koskinen, eds., *Categories of Being* (Oxford, Oxford University Press), 439–452.
O'Dwyer, Shaun, 2010. "Pragmatism and Anti-Realism about the Past." *Transactions of the Charles S. Peirce Society* 46.3: 401–422.
Peirce, Charles S., 1931–1956. C. Hartshorne, P. Weiss, and A. Burks, eds., *The Collected Papers of Charles Sanders Peirce,* 8 vols. Cambridge, MA, Harvard University Press.
Pihlström, Sami, 1998. *Pragmatism and Philosophical Anthropology: Understanding Our Human Life in a Human World.* New York, Peter Lang.
———, 2009. *Pragmatist Metaphysics: An Essay on the Ethical Grounds of Ontology.* London, Continuum.
Popper, Karl R., 1972. *Objective Knowledge: An Evolutionary Approach.* Oxford, Oxford University Press.
Psillos, Stathis, 1999. *Scientific Realism: How Science Tracks Truth.* London, Routledge.
Putnam, Hilary, 1981. *Reason, Truth and History.* Cambridge, Cambridge University Press.
———, 1990. *Realism with a Human Face.* Cambridge, Mass., Harvard University Press.
Quine, W.V.O., 1969. *Ontological Relativity and Other Essays.* New York, Columbia University Press.
Rorty, Richard, 1972. "The World Well Lost." *The Journal of Philosophy* 69: 649–665.
Rudwick, Martin, 2005. *Bursting the Limits of Time: The Reconstruction of Geohistory in the Age of Revolution.* Chicago, University of Chicago Press.
Russell, Bertrand, 1940. *An Inquiry Concerning Meaning and Truth.* London, Allen & Unwin.
Schmeltz, Tod, 2009. "Nicholas Malebranche." In: E. Zalta, ed., *The Stanford Encyclopedia of Philosophy* (Winter 2009 Edition). http://plato.stanford.edu/archives/winter09/entries/malebranche/
Sellars, Wilfrid, 1968. *Science and Metaphysics.* London, Routledge & Kegan Paul.
Short, T. L., 2007. *Peirce's Theory of Signs.* Cambridge, Cambridge University Press.
Sider, Theodore, 1999. "Presentism and Ontological Commitment." *The Journal of Philosophy* 96: 325–347.
Tuomela, Raimo, 1985. *Science, Action and Reality.* Dordrecht, Reidel.
van Fraassen, Bas, 1989. *Laws and Symmetry.* Oxford, Oxford University Press.
Wolterstorff, Nicholas, 1987. "Are Concept-Users World-Makers?" In: J. E. Tomberlin, ed., *Philosophical Perspectives I: Metaphysics* (Atascadero, Ridgeview), 233–277.

9 Cognitive Semantics and Newton's Rule 4 of Experimental Philosophy

Scientific Realism without Empiricism

Kenneth R. Westphal

1 INTRODUCTION

Gareth Evans' classic article, "Identity and Predication" (1975), provides the basis for a specifically *cognitive* semantics of singular reference which has important implications for epistemology and for philosophy of science. Here I argue that this semantics of singular cognitive reference directly and strongly supports Newton's Rule 4 of (natural, experimental) Philosophy in ways which support Newton's realism about gravitational force. I begin with Newton's Rule 4 of Philosophy and its role in Newton's justification of realism about gravitational force (§2). I then briefly summarize Evans' account of predication and explicate its implications for the semantics of singular cognitive reference (§3). I then argue that this semantics of singular cognitive reference is embedded in and strongly supports Newton's Rule 4, and that it rules out not only Cartesian physics (*per* Harper) but also Cartesian, infallibilist presumptions about empirical justification generally (§4). I then argue that this semantics of singular cognitive reference reveals a key defect in Bas van Fraassen's anti-realist "Constructive Empiricism," even in its recent (2008) version, and in many common objections to realism (§5).

2 NEWTON'S RULE 4 AND HIS CAUSAL REALISM

Newton's Rule 4 of Philosophy states:

> In experimental philosophy, propositions gathered from phenomena by induction should be considered either exactly or very nearly true notwithstanding any contrary hypotheses, until yet other phenomena make such propositions either more exact or liable to exceptions. (Newton 1999, 796; 1871, 389)

Newton directly adds:

> This rule should be followed so that arguments based on induction may not be nullified by hypotheses. (*Ibid.*)

Newton's Rule 4 requires any competing scientific hypothesis to have, not merely empirical evidence in its favor, but sufficient evidence with sufficient precision either to make an accepted scientific theory or law "more exact" or to restrict it by demonstrating specified "exceptions" to it. Rule 4 is central to Newton's methodology, in ways I now indicate.[1]

Recent scholarship, especially by William Harper (2011), shows that Newton was significantly more sophisticated about scientific method and explanatory success than contemporary philosophers of science,[2] that his methodology and ideal of explanatory success justified his realism about gravitational force and that (when provided the relevant data and theoretical analysis) they also justify the shift from Newtonian mechanics to General Relativity, as Einstein himself realized. Moreover, Newton's methodology vastly reduces the problem of the "underdetermination" of theory by observation (Harper 2011, 194–219, 238–56, 372–8) and is used today in cosmology (Harper 2011, 394–6). Harper's *tour de force* is a major contribution to contemporary philosophy of science, as well as to history and philosophy of science.

Here I focus on one key issue and one central instance of it, which show that Newton's Rule 4 centrally embeds two important epistemological implications of the Thesis of Singular Cognitive Reference (§§3, 4), and how these implications support his causal realism about gravitational force (§§4–5).

When rejecting mere hypotheses, Newton famously states that:

> [W]hatever is not deduced from the phenomena must be called a hypothesis; and hypotheses, whether metaphysical or physical, or based on occult qualities, or mechanical, have no place in experimental philosophy. In this experimental philosophy, propositions are deduced from the phenomena and are made general by induction. The impenetrability, mobility, and impetus of bodies, and the laws of motion and of the law of gravity have been found by this method. And it is enough that gravity really exists and acts according to the laws that we have set forth and is sufficient to explain all the motions of the heavenly bodies and of our sea. (Newton 1999, 943; 1871, 530; *cf. Opticks,* 401–2)

This passage has been persistently misread by generations of philosophers—starting at least with Berkeley and Hume—for two main reasons. First, it has been widely assumed that by "deduction" from the phenomena Newton must mean logical deduction, though this makes his view hopeless because statements can only be deduced logically from other statements, not from experiences nor from natural phenomena (nor from anything non-propositional). Second, it has been widely assumed that by "induction" Newton must mean simple numerical induction, so that his theory is challenged by Hume's problem of induction. Both assumptions are incorrect.[3]

Newton uses the term 'deduction' in a broader sense roughly equivalent to 'justify' by evaluating empirical evidence; recall the forensic use of the term 'deduction' from evidence. The question then is, what sort of "justification" Newton proposes to derive from natural phenomena. An especially important example of Newton's "deduction from the phenomena" is provided by Harper's reply to the concern that Newton appears to assume as a hypothesis, rather than to prove on the basis of phenomena, that the inverse-square law of mutual gravitational attraction holds generally, and not merely for those few spaces in the cosmos occupied by bodies we have observed in our solar system (Harper 2011, 28–31, 137–42).

Three aspects of Harper's response to this concern suffice for present purposes:

1) Newton's method seeks converging measurements by various independent means of causal parameters, where:
    i) Systematic dependencies identified by a theory make the phenomenon to be explained measure the value of the theoretical parameter which explains it.
    ii) Alternatives to the phenomenon would carry information about alternative values of that same parameter.
2) This feature of Newton's method highlights the importance of the links between Newton's three distinct ways of measuring centripetal force and acceleration fields.
3) Newton's Rules of Philosophizing—and centrally Rule 4—support generalizing the causal parameters thus measured. (Harper 2011, 257–64, 361–64)

Newton's analyses and proofs are, as Harper shows, very rich, subtle, and thorough; here only some of their rudiments may be illustrated briefly. One important example of the kind of "systematic dependencies" mentioned in (1) is Newton's recognition of the further significance of Kepler's "Areal" Law (law of areas). Kepler determined that the (roughly triangular) area swept by a planet orbiting the Sun is constant, although the planet follows an elliptical orbit in which it accelerates when approaching the Sun and decelerates when receding from the Sun. This is Kepler's Areal Law:

> The line joining the planet to the Sun sweeps out equal areas in equal times as the planet travels around the ellipse.

Newton realized that this constancy indicates precisely an orbit about the centre of motion of the Sun and the planet, because an increasing areal rate would place the focal point of the planet's orbit outside and 'ahead' of the Sun, while a decreasing areal rate would place the focal point outside and 'behind' the Sun. The former would result in an expanding, the latter in a contracting orbit; either case represents orbital degeneration rather than stability. Newton's observational data (which included Kepler's and Brahé's) clearly indicated orbital stability. The stability with which planetary orbits satisfy Kepler's areal law indicates that their orbits *measure* precisely an inverse-square acceleration field directed towards the Sun (Harper 2011, 109–20).

This same result—precisely an inverse-square acceleration field—is measured independently by determining whether there is orbital precession, that is, whether planets follow the same orbit repeatedly, or whether the location of an orbit's aphelion and perihelion (its most distant and closest points to the Sun; called 'apsides') shift by rotating about the Sun, either 'forwards' or 'backwards' with respect to the direction of orbital rotation, with subsequent orbits.[4] Absence of such rotation or precession measures precisely an inverse square force of acceleration; a different rate of diminution of field strength would produce either positive or negative orbital precession (Newton 1999, 802; 1871, 395; Harper 2011, 120–6).

These two crucial steps, undertaken by Newton for the independent cases of six planets and two distinct aspects of their motions, *are* Newton's (initial) deduction from planetary orbital phenomena of the existence of an inverse-square acceleration field of force radiating from the Sun, in contrast to any other rate of diminution. Extrapolating from these sets of orbital phenomena and their univocal measurement of an inverse-square attractive force to a field of such force radiating from the Sun *is* Newton's initial "generalization by induction" of the consequences he has deduced from the orbital phenomena (Harper 2011, 44–5, 128–9, 135–46, 257–84).

As Harper notes, Newton has additional data on the motions of bodies which provide further precise measurements of the inverse-square attraction of gravitational force: Comets, the four moons of Jupiter, the Earth's moon, the rotation of Jupiter and the Sun about their common centre of motion, and a vast range of terrestrial phenomena, including pendula, free-fall, and (quaintly enough) floating magnets. Indeed, the entirety of *Principia,* Book 3, Newton's "System of the World," *is* his proof of universal gravitation, all based on multiple, precise, agreeing measurements of the inverse-square gravitational field provided by many diverse phenomena of motion. All of these were further bolstered in 1759 by Clairaut's successful, precise prediction of the return of Halley's comet.

Harper (2011, 355–68) explains very nicely how Newton's appeal to his First Law can be used to extend his Third Law in order to show that Jupiter's tendency to move toward the Sun—that is, its tendency to orbit the Sun rather than to move away from the Sun on a tangent—counts as an attraction between Jupiter and the Sun. Newton's First and Third Laws are these:

Law 1. Every body continues in its state of resting or of moving uniformly in a straight line, except insofar as it is driven by impressed forces to alter its state. (Newton 1999, 416; 1871, 13)

Law 3. To an action there is always a contrary and equal reaction; or, the mutual actions of two bodies upon each other are always equal and directed to contrary parts. (Newton 1999, 417; 1871, 13)[5]

Measuring an attractive force between, *e.g.*, Jupiter and the Sun, requires using Newton's definitions of the quantities of absolute, accelerative and motive centripetal force (Harper 2011, 86–94); these are his Definitions 6–8:

> Def. 6. The *absolute* quantity of centripetal force is the measure of the same, greater or less in proportion to the efficacy of the cause propagating it from the centre through the encircling regions. (Newton 1999, 406; 1871, 4)
>
> Def. 7. The *accelerative* quantity of centripetal force is the measure of the same, proportional to the velocity which it generates in a given time. (Newton 1999, 407; 1871, 4)
>
> Def. 8. The *motive* quantity of centripetal force is the measure of the same proportional to the motion which it generates in a given time. (Newton 1999, 407; 1871, 5)

Note that Newton expressly defines *measures* of *quantities* of force. His definitions have been widely misread as defining *forces,* a misreading central to effacing Newton's realism about gravitational force and to reducing his concerns merely to those of empirical adequacy.

Harper (2011, 375–8) shows how Newton identifies systematic dependencies which enable orbital phenomena to provide measurements of the Sun's gravitational field. Each of these measures is supported by Newton's method of successive approximations (*cf.* Smith 2002a, 2002b). Each of Newton's measures begins with an approximation of the physical situation which is used to calculate an approximate measure of the target value. At each stage of calculation and recalculation, divergences between the initial approximation and the actually observed phenomenon count as *theory-mediated* secondary phenomena, which are to be explained by reiterated use of the same explanatory resources. With an approximate result in hand, Newton progressively eliminates approximations by reiterated use of the very same explanatory resources to achieve ever more accurate, ever less idealized measures of the target value. Reiterated deployment of the same theoretical apparatus produces ever more precise *and* converging measures of the target value, thus supporting very robustly Newton's claim thereby to measure a real value. The progressive elimination of approximations can and often did lead, not only to much greater precision, but in several central cases, Newton's results stand in formal contradiction to his initial approximations. This important feature of Newton's method cannot be explicated by hypothetico-deductive (H-D) methods (*cf.* Harper 2011, 126–42).

More important is the point that the success of these successive approximations *in each case* of Newton's vast array of *independent* measures of the inverse-square rate of gravitational attraction greatly bolsters the strength of his conclusions based on the agreement *among* all of these measures of the inverse-square field of gravitational attraction. The wide variety of *agreeing* measures of the inverse-square attraction of gravitational force provide a very robust measurement of that force. Harper stresses that this is one of Newton's key ideals of theoretical success, to provide "convergent accurate measurement of causal parameters by the phenomena they are taken to explain" (Harper 2011, 104–7, 194–200). This is a vastly stronger ideal of theoretical success than empiricist descriptive, predictive, and retrodictive accuracy (across the data set), for three main reasons.

First, Law 3, the equality of action and reaction, or the mutual equality of attractions (or repulsions) between two bodies, is required to disentangle the masses from the weights of any two orbiting bodies. Disentangling these two characteristics is required in order to use their motions to measure the force of their attraction, whatever it may be. Newton's Law 3 has vastly more empirical support than any alternative assumption that the strength of attractive forces varies pair-wise among bodies, primarily because it alone provides for convergent agreeing measures of the relative masses within our solar system.

Second, Newton's gravitational theory famously integrated a vast range of celestial and terrestrial phenomena within a common, comprehensive explanatory theory. This explanatory integration provides more than just comprehensiveness: By using the same theory to explain this vast range of phenomena, Newton's *Principia* is able to use this vast range of phenomena to provide accurate, convergent agreeing measures of the strength of gravitational attraction and its inverse-square rate of diminution across our solar system, including some comets. For example, both the orbit of the earth's moon and the length of a terrestrial seconds pendulum near sea level provide accurate agreeing measures of the force of the earth's gravity (Harper 2011, 180–6, 195–203, 215–7).

Relying upon the empiricist criterion of descriptive, predictive, and retrodictive accuracy across the data set (*i.e.*, 'empirical adequacy') cannot rule out hypotheses that different material bodies have different powers of attractive force. Nor can it rule out the suggestion that the inverse-square law holds only for those distances and regions of space for which we have observational data. Nor can it disentangle the masses from the weights of orbiting bodies in ways achieved by Newton's use of Law 3, which is crucial to Newton's entire set of astronomical measures of the inverse-square ratio of gravitational force to distance. Using Newton's Law 3 provides for converging, agreeing, precise measures of one and the same attractive force among celestial and terrestrial bodies, *and* provides grounds for seeking to explain deviations from their predicted motions by using the very same theoretical and observational resources to search for other bodies affecting their motions. This strategy is central to Newton's extremely successful method of progressive elimination of initial idealizations, which results in convergent agreeing measures of the inverse-square power of gravitational attraction.

Third, using Law 3 also enables Newton to measure the relative masses of bodies with satellites, including the Sun, the Earth, Jupiter, and Saturn. Success in solving this very difficult analytical problem provides further confirmation of Law 3 by showing that it is implied by the observed phenomena Law 3 is used to measure. Therefore, Newton's Third Law *is* "deduced from the phenomena," though it is deduced from them indirectly rather than directly; it is not simply postulated, *à la* H-D methods (Harper 2011, 274–9, 355–64).

Understanding Newton's realism about gravitational force requires distinguishing it from causal agnosticism. According to causal agnosticism, causal structures generate observed events and regularities, though we cannot know what those causal structures are. This issue was hotly debated by Newton and his contemporaries, especially Leibniz (Janiak 2007). Newton is not a causal agnostic about gravitational force. Newton concludes that

> . . . it is enough that gravity really exists and acts according to the laws that we have set forth and is sufficient to explain all the motions of the heavenly bodies and of our sea. (Newton 1999, 943; 1871, 530; *cf. Opticks*, 401–2)

Newton was a realist about gravitational force; he was agnostic only about how gravitational force operates as a physical cause.

The progressive increase in accuracy required by Newton's method and ideal of explanatory success significantly exceeds the requirements of other accounts of theoretical adequacy current in philosophy of science. Newton's procedure may recall Glymour's "boot-strap" analysis. However, Harper has shown that Newton's methodology and explanatory ideal are stronger and more adequate than Glymour's boot-strap account, and that they overcome problems confronting that account.[6] Indeed, Harper (2011, 378–85, 392) shows that Newton's methodology and ideal of explanatory success apply to—and strongly support—the shift from Newtonian mechanics to General Relativity: on the basis of the relevant evidence and theoretical analysis, *Newton's* methodology and ideal of explanatory success strongly support General Relativity (*contra* Kuhn 1970, 94, 102, 107–8). In brief, Newton understood both the demands upon and the achievements of physical science better than have most philosophers and historians of science up to the present day. In these regards, Newton's causal realism is empirically supported far more than is empiricist anti-realism, if one insists on treating these two views as 'hypotheses'.

Bas van Fraassen (2002, 129; 2004a, 130–1), too, hails Newton's Rule 4—at least sometimes; elsewhere he (2007, 365) dismisses Rule 4 in the same breath as the traditional empiricist principle of sensory evidence (*sola experientia*). It may be expected that the interpretation of Rule 4 shall prove controversial within history and philosophy of science, as empiricists respond to Harper's (2011) findings. I shall argue (§4) that the semantic core of Newton's Rule 4, as Harper understands it (rightly, I submit), is supported directly and decisively by the semantics of singular cognitive reference (§3), so that objections to Rule 4 based solely upon considerations drawn from history and philosophy of science cannot undermine either Newton's or Harper's interpretation and use of Rule 4.[7]

3 PREDICATION AND THE SEMANTICS OF SINGULAR COGNITIVE REFERENCE

In "Identity and Predication" Evans (1975) argues for the following constraint on any theory of meaning regarding the predicates of a language, which I shall call "The Evans Constraint":

> THE EVANS CONSTRAINT: It is unacceptable to discern predication of *A*'s in sentences of a given kind unless those sentences are such that their systematic mastery requires mastery of the identity conditions of *A*'s. (Evans 1975, 355–6)

The rationale for The Evans Constraint is that only objects which satisfy certain specifiable conditions can be properly characterized using the predicate in question. Specifying the conditions objects must meet to satisfy any particular predicate requires also specifying objects which fail to meet those same conditions (349, 352);[8] in this regard, predication is differential or discriminatory. More specifically, predicates pertain to aspects, to parts, or to whole objects (349, 351, 351 n. 9); predication within a language is structured accordingly.[9] Relevant aspects, parts, or whole objects pertain to the identity conditions of the objects discussed within a language and their features. Distinguishing relevant aspects, parts, or whole objects belongs to the mastery of those identity conditions (354). These remarks are all variations of the principle of the nonidentity of discernables (353). This principle pertains to whatever speakers of the target language can discern, that is, can discriminate or distinguish, by noting their spatio-temporal regions and boundaries, including whole individuals and their aspects, parts, or features. Mastery of those identity conditions is required for mastery of the relevant predicates; this is the Evans Constraint.

Coordinating qualities or characteristics with spatio-temporal regions or boundaries is central to Evans' thesis about predication, which he states in these terms:

> . . . the line tracing the area of [ascriptive] relevance delimits that area in relation to which one or the other, but not both, of a pair of contradictory predicates may be chosen. And that is what it is for a line to be a boundary, marking something off from other things. (Evans 1975, 352, *cf.* 350–3)

It is central to Evans' analysis that specifying the relevant boundary for the use of either member of a pair (or set) of contrary (mutually exclusive, though not necessarily "contradictory") predicates is only possible by specifying the region relevant to the characteristic in question, and *vice versa,* where this region will be either co-extensive with or included within the spatio-temporal region occupied by some spatio-temporal particular. More generally, Evans demonstrates—even if he only implicitly argues—that predication requires conjointly specifying the relevant spatio-temporal region and some manifest characteristic(s) of any particular we experience or identify. These conjoint specifications may be rough and approximate; the key point is that spatio-temporal designation and ascription of manifest characteristics are *conjoint, mutually interdependent,* specifically *cognitive* achievements. This result I shall call "The Evans Thesis."

That these discriminatory abilities are specifically *cognitive* abilities is underscored by how the Evans Thesis highlights the cognitive *in*sufficiency of the descriptions theory of reference. According to Russell's theory of descriptions, a description designates or refers to whatever individual(s) it describes, if there be any. I do not dispute descriptions as linguistic referring devices, nor with explicating conceptual content via descriptions. These points belong to philosophy of language (or by extension also to philosophy of mind). Whatever may be the status of Russell's theory of descriptions within philosophy of language, it is insufficient for epistemology. This is my present point. According to the descriptions theory of reference, our statements refer to whatever is described when we analyze the meanings of our terms or statement into explicit descriptions. The problem with this approach within epistemology is that, no matter how specific or extensive a description may be, no description by itself determines whether it is empty, determinate, or ambiguous because it describes no, only one, or instead several individuals. Which may be the case is not a function of the description alone: it is equally a function of what there is (or is not). The inclusion of a definite article (such as 'the' or 'the one and only') within an attributive phrase does not, because it cannot, settle this issue because no definite article can ensure that the phrase in which it occurs may be either empty or ambiguous; this was, after all, Russell's problem (*ca.* 1905) about "the present King of France."

To *know* any one spatio-temporal particular requires both correctly ascribing characteristics to it *and* locating it within space and time. Integrating both of these is required for predication, and also for knowledge of—or even for error about—that individual. Only through singular sensory presentation *and* competent use of conceptions of 'time', 'times', 'space', 'spaces' (spatial regions), 'individual', and 'individuation' can we locate any object or event—together with (at least some of) its aspects, parts, and characteristics—within space and time. Only through ostensive designation can we *ascribe* the predicates used in our (perhaps implicit) description *to* any *one,* putatively known particular. Therefore, predication is required for singular, specifically *cognitive* reference to any sensed, spatio-temporal particular. Only through this kind of predication as a cognitive achievement can anyone specify (even approximately) the relevant spatio-temporal region (putatively) containing the particular one purports to designate ostensively—by specifying its occupant, the (putatively) known particular. Only in this way can we note, specify, or determine precisely *which* spatio-temporal region to designate, in order to grasp *this* (intended, ostended, presented) particular, and to ascribe to it any manifest characteristics, all of which is required to achieve any knowledge (whether presumptive or actual) of that particular.

Thus determinate, significant cognitive claims or judgments are possible for us only through conjoint spatio-temporal designation of, and predicative ascription of characteristics to, any particular. Recognizing any particular object or event requires conceptually identifying both the region it occupies and at least some of its manifest characteristics. This, in brief, is the cognitive significance of the Evans Thesis. How precisely we may identify either the region or the features of any particular is a further issue; approximations suffice, provided they are sufficiently accurate to discriminate between particulars. Much of this point was made already by Donnellan (1966), when he noted that a partially incorrect definite description can nevertheless be used successfully to refer unambiguously to a particular.[10] Without at all denying the importance of descriptions within philosophy of language (or *mutatis mutandis* philosophy of mind), analyses of the meanings of our terms, or of the contents of our concepts or descriptive phrases do not, because they cannot, suffice for epistemology. Only by analyzing the cognitive dimensions of predication can we understand how the terms or concepts we use in our judgments, claims, statements, or assertions can have specifically *cognitive* significance, in addition to their linguistic meaning or conceptual content.

This point about cognitive reference is necessary, though not sufficient, for knowledge, in part because it says nothing about justification, but more importantly: this point about cognitive reference is necessary for either truth or falsehood, or for approximation or accuracy. Successful cognitive reference, as specified here, is necessary also to evaluate the truth, falsehood, or accuracy of a judgment, statement, or claim, and it is necessary to assess its justification or justificatory status. This point about cognitive reference is neutral about whether an epistemology is formulated in terms of judgments, statements, or claims. It is also neutral about how linguistic meaning or conceptual content are understood or analyzed, *e.g.*, as descriptions, classifications, or intensions (in contrast to extensions). 'Cognitive reference', so understood, is necessary for any substantive claim, judgment, or statement to have cognitive status, as a putative truth (in contrast to an arbitrary utterance or sentence). To summarize this point I state the following

> Thesis of Singular Cognitive Reference: Terms or phrases have 'meaning', and concepts have content (intension), as predicates of possible judgments, although (in substantive domains) no proposition has specifically *cognitive* significance unless and until it is incorporated into

a candidate cognitive judgment which that person refers to some actual particular(s) *S*/he localizes (at least putatively) within space and time.

This Thesis has three important implications for epistemology, including history and philosophy of science.

One very important consequence of this Thesis of Singular Cognitive Reference is that it shows that justificatory infallibilism is in principle *irrelevant* to the non-formal domain of empirical knowledge. Strictly speaking, formal domains are those which involve no existence postulates. Strictly speaking, the one purely formal domain is a careful reconstruction of Aristotle's Square of Opposition (Wolff 1995, 2000, 2009, 2012). All further logical or mathematical domains involve various sorts of existence postulates, including semantic postulates. We may define 'formal domains' more broadly to include all formally defined logistic systems (Lewis 1930/1970, 10). Whether we construe formal domains narrowly or broadly in either way, deduction suffices for justification within any formal domain because deduction *constitutes* justification within any formal domain. Indeed, a domain *is* a formal domain only insofar as deduction constitutes justification within it. However, only within formal domains is justification constituted by provability.

The relevance of any such logistic system to any non-formal, substantive domain rests not upon formal considerations alone, but also upon substantive considerations of how useful a specific logistic system may be within a non-formal, substantive domain (Lewis 1929, 298; *cf.* Carnap 1950). The use of any specified logistic system within any non-formal domain does not *suffice* for justification within that domain. The use of that system within any non-formal domain requires further justificatory resources, not limited to formal deduction. This holds too for the use of that system in justifying any particular claim within its domain of application. Consequently, within any substantive domain, fallibilism is no skeptical capitulation, not because infallibilist standards of justification are too stringent, but because *in principle* they are inappropriate to any and to all substantive domains. Conversely, within any substantive domain, a mere logical possibility as such has *no* cognitive status and so cannot serve to "defeat" or to undermine (refute) an otherwise well-grounded line of justificatory reasoning within that domain.

The domain of (putative) empirical knowledge includes spatio-temporal objects and events; accordingly, empirical knowledge is a non-formal domain. Consequently, the Thesis of Singular Cognitive Reference rules out the ideal of infallible justification (*scientia*) within the entire non-formal domain of empirical knowledge. Recognizing that only fallibilist accounts of justification are tenable within the non-formal domain of empirical knowledge is no concession, and certainly no capitulation, to skepticism.

A second important implication of the Thesis of Singular Cognitive Reference is that it secures the key aim of verification empiricism *without* invoking verification empiricism! The Thesis of Singular Cognitive Reference holds independently of whether the concepts we use in cognitive judgments (in non-formal, substantive domains) are *a priori, a posteriori,* or mixed. The cognitive-semantic point of this Thesis is that, whatever may be the conceptual content or linguistic meaning of our claims, judgments, or propositions, they have no cognitive status unless and until they are referred to particulars we have (presumptively) localized within space and time. This requirement is a necessary condition for the truth-evaluability of our claims (*etc.*), and it is a necessary condition for us to know enough about our claims and whatever about which we make those claims to discover and thereby to determine their truth value, or their accuracy, or their use as approximations. It is also necessary (though not sufficient) for our assessing the justification of our cognitive claims about those particulars.

Third, the Thesis of Singular Cognitive Reference shows that philosophy of language and philosophy of mind may augment, but cannot supplant, epistemology, for two main reasons. First, specifically cognitive reference requires fulfilling more stringent requirements than linguistic reference or whatever "reference" may be involved in conceptual content (predicates, intensions). "Mental" content may encompass cognitive content, but cognition requires justification, and cognitive justification is an achievement, one which cannot be reduced merely to what one may say or think about one's own (or about another's) cognitive-justificatory success, either in general, or on any particular occasion.

This way of putting the point may recall Charles Travis' (2005) excellent analysis, "A Sense of Occasion." His (2005, 302–3, 313) appeals to what we register in or of our situation on the basis of clear-headed alert sensory experience, and to contextually specific relevant alternatives to any presently perceived state of affairs, either coincides with, or is underwritten by, an unacknowledged account of cognitive information channels such as Dretske's (1981); Dretske's account undergirds the generality (the "reasonably systematic" holding; 309) of Travis's Austinian point about the occasion-sensitivity of what factively *means* (indicates) what is the case. (This is not to usurp, but rather to undergird, Travis' analysis; conversely, Dretske's account of information channels must (and does) link nicely to Travis' astute sense of the sense of cognitive occasion.) Consider, too, that the Thesis of Singular Cognitive Reference undergirds O. K. Bouwsma's (1949) brilliant deflation of Cartesian skepticism. To put the main point briefly, cognitive justification integrates internalist and externalist factors, such as the generally reliable functioning of our cognitive capacities, including our sensory-motor and attentional systems. (Obviously, these latter pertain to our capacities to localize particulars within space and time.)

To avoid possible misunderstanding, note that the Thesis of Singular Cognitive Reference does not rule out second-hand "knowledge by description" based upon reliable testimony or written reports; it only establishes some basic cognitive conditions upon the acquisition of empirical knowledge, by identifying basic conditions under which alone synthetic statements have specifically cognitive status within any non-formal domain. Much more can be said to support the Thesis of Singular Cognitive Reference, and I have done so elsewhere.[11] I have said enough here, however, to distinguish specifically *cognitive* reference from issues within philosophy of language or philosophy of mind about mental or propositional content, semantics, or theory of linguistic reference. I now argue that the Thesis of Singular Cognitive Reference is embedded centrally within Newton's Rule 4 of (experimental) Philosophy, that it strongly supports Newton's Rule 4 (§4), and specifically that it supports the role of Rule 4 in justifying Newton's causal realism about gravity against van Fraassen's Constructive Empiricism (§5).

4 COGNITIVE SEMANTICS, NEWTON'S RULE 4, AND ANTI-CARTESIANISM

Understanding the significance of singular cognitive reference for scientific knowledge requires removing an expository simplification made in the preceding section. I have formulated singular cognitive reference in terms of localized spatio-temporal perceptible 'particulars'. The term 'particulars' commonly connotes medium-sized dry goods, though its use can be much broader. Singular cognitive reference pertains to spatio-temporal particulars construed very broadly, to include any kind of particular we may locate in space and time, whether these be individual physical objects such as planets, an orbital system, fields of force (such as gravity or magnetism), or any distinct, identifiable natural phenomenon or process, *e.g.*, an aurora borealis, or electrons jumping a gap in a tunneling electron microscope. This matters in connection with Newton's gravitational theory, because he sought to explain, not individual facts about various motions of any one celestial body, but the general phenomena of the regularity of orbital motions, of the rate of free fall near the earth's surface, the periodicity of pendula, and the uniformity of gravitational attraction throughout our solar system—and presumptively throughout the universe until proven otherwise, *per* Rules 3 and 4. All of these general natural phenomena can be and have been located within space and time. Hence they satisfy a key requirement of the semantics of singular cognitive reference, which allows sensory presentation via observational instruments. What bearing, then, does the Thesis of Singular Cognitive Reference have on Newton's Rule 4?

Newton's Rule 4 embeds the core point of the Thesis of Singular Cognitive Reference. Newton's main point in Rule 4 is to distinguish between hypotheses which do and those which do not compete with, or provide an *alternative* to, an established theory or law. In making this distinction, however, Rule 4 also distinguishes between hypotheses with cognitive status and those lacking such status, which count instead as suggestions, proposals, or as yet untested suppositions. I don't wish to be stipulative, but philosophers, especially those favoring H-D methodology, are prone to use the term 'hypothesis' promiscuously, so that almost any idea about how an event might occur counts as a 'hypothesis'. Such promiscuity was also common among Seventeenth and Eighteenth Century scientists, many of whom unhesitatingly described both Newtonian gravitational theory and Cartesian vortex theory as scientific "hypotheses." The specific contrast drawn by Newton's Rule 4 between competing and non-competing scientific hypotheses is rooted in a more general contrast also implied by Rule 4. Because Rule 4 requires of any competing hypothesis that it have evidence in its favor, it requires a competing hypothesis be referred to localized, identified physical particulars; without such reference, there can be no evidence supporting that hypothesis. And *per* above (§3), without such reference to localized particulars, no hypothesis is so much as a candidate for truth-evaluation, nor for evaluation of its accuracy, its merits as an approximation, or its justification. Rule 4 thus requires competitor hypotheses, not only to be logically possible, it further requires them to have cognitive status, insofar as they are supported by at least some favorable empirical evidence, and so must be referred in some definite manner to localized spatio-temporal particulars. Consequently, Newton's Rule 4 rules out Cartesian epistemology, which restricts rational justification to logical deduction from premises which survive scrutiny by the *malin génie, i.e.* infallibilism (*scientia*). Precisely for this reason, Rule 4 and Harper's interpretation of it will be contested by philosophers of science who presume infallibilism about empirical justification; in the next section (§5) we shall see that this includes contemporary empiricists. Hence it is important to see that Newton's rejection of infallibilism about empirical justification, implicit in Rule 4, is sound.

Obviously, nothing about the extent of evidence sufficient to justify a scientific hypothesis, nor to justify its status as a competitor to an established theory or law, is implied by the Thesis of Singular Cognitive Reference. However, because Rule 4 requires that there *be* positive empirical evidence for a competing hypothesis, it embeds the core point of singular cognitive reference: to *be* a *cognitive* claim—also within the sciences—requires referring that claim to localized spatio-temporal particulars, which alone can provide relevant evidence (*pro* or *contra*). This reference to spatio-temporally localized particulars which alone can provide empirical evidence is required by the central point of Rule 4, that this evidence must differentially favor the proposed competitor. (Otherwise the evidence cited in its support would equally well support the established hypothesis, and so would provide no evidence specifically favoring a proposed theoretical alternative; so far as that evidence would show, that proposal is not an *alternative*.) Hence the Thesis of Singular Cognitive Reference directly supports the requirement embedded in Rule 4 that to *be* a competing scientific hypothesis requires having at least some positive evidence in its favor. Without such evidence, the proposed empirical alternative merely states a proposal with no cognitive status because it is not referred to identified, localized particulars; it would be merely a proposal, a suggestion, and not a scientific hypothesis with cognitive status.

To this sound point of cognitive semantics Newton's Rule 4 adds the altogether credible methodological requirement that a *competing* scientific hypothesis have sufficient differential evidence favoring it either to render an established hypothesis "more exact" or to restrict its scope by demonstrating specified "exceptions" to it. Newton's justification for his Rule 4 is methodological: the *Principia* shows that, by adopting this methodological rule, unprecedented advances in natural science can be achieved.[12] This is not trivial: Newton's *Principia* is *inter alia* a sustained treatise on measurement theory, and on actual measurements of gravitational forces of attraction.

The Thesis of Singular Cognitive Reference directly and strongly supports the cognitive-semantic requirement embedded in Newton's Rule 4. By anchoring one core point of Newton's Rule 4 in a sound semantics of singular cognitive reference, the Thesis of Singular Cognitive Reference shows that Newton's Rule 4 cannot be countered simply on grounds specific to history and philosophy of science. Instead, criticizing or rejecting Newton's Rule 4 and his use of it would require the much more ambitious task of criticizing and rejecting a central cognitive and semantic precondition of commonsense knowledge and hence a central epistemological precondition of any sound philosophy of science.

Harper (2011, 212–4, 341–6, 361–4) rightly notes that Newton's Rule 4 directly opposes Cartesian physics, which restricted itself to logically possible explanations, *e.g.*, cosmic vortices to account for planetary orbits. In its condemnation of Copernicus, the Roman Church decreed that natural scientists could only propose *possible* explanations of natural phenomena, not actual explanations. Descartes complied and officially regarded his explanatory models as merely possible explanations of natural phenomena. Newton's Rule 4 rejects merely possible explanations as scientifically irrelevant; this is one key point of his infamous *hypotheses non fingo*. Newton's examples of the mere "hypotheses" he condemns and rejects make plain that here he rejects mere proposals lacking specific empirical evidence. Merely possible alternative accounts defeat cognitive justification only if strict logical deduction is both necessary and sufficient for justification. Newton's Rule 4 rejects the ideal of justificatory infallibilism, and thus also the sufficiency of a mere logical possibility to defeat the cognitive justification of a scientific theory (or of one of its components).

It is important to note, however, that neither the Thesis of Singular Cognitive Reference nor Newton's Rule 4 reject the H-D model of inquiry or explanation. They do, however, set an important condition for the cognitive status of any specific use of H-D methods: until positive evidence is provided to justify a hypothesis, at least partially, that hypothesis has no cognitive status, and cannot defeat the justification of any evidentially supported theory or law within its (purported) domain. Newton's Rule 4 further requires of any presumptive alternative hypothesis—whether derived in accord with his own ideal of multiple independent agreeing measures, or by using H-D methods—that to be an *alternative* hypothesis, a hypothesis must either improve upon the precision of the relevant established theory, or it must delimit the scope of that theory by demonstrating specified exceptions to it.

Because Newton's Rule 4 embeds the Thesis of Singular Cognitive Reference, Newton's Rule 4 likewise rules out Cartesian epistemology, which restricts rational justification to logical deduction from premises which survive scrutiny by the *malin génie, i.e.* infallibilism. Precisely for this reason, Rule 4 and Harper's interpretation of it will be contested by philosophers of science who presume infallibilism about empirical justification. We shall now see that this includes contemporary empiricists (§5).

5 COGNITIVE SEMANTICS *VERSUS* CONSTRUCTIVE EMPIRICISM

5.1 Van Fraassen's Constructive Empiricism

In *The Empirical Stance* (2002) and more recently in *Representing Science: Paradoxes of Perspective* (2008), Bas van Fraassen renewed his efforts, inaugurated in *The Scientific Image* (1980; '*SI*'), to expound and recommend his philosophy of science, designated 'Constructive Empiricism', an anti-realist position defined by two central theses:

1) Science aims to give us theories which are empirically adequate;
2) acceptance of a theory involves as belief only that it is empirically adequate. (*SI* 12)

Constructive Empiricism emphasizes the pragmatics of language and a key distinction between believing a scientific theory to be true, and merely accepting a theory in view solely of its empirical adequacy, insofar as it implies with sufficient accuracy all the observable events within its domain.[13] He himself calls his position 'anti-realist' (*SI* 12); its fundamental principles remain constant amidst subsequent refinements. I reconsider van Fraassen's original (1980) exposition of Constructive Empiricism because it contains both its primary justification, and four fundamental, neglected flaws, both substantive and methodological, which undermine Constructive Empiricism in all its versions (including his 2008).[14]

The Thesis of Singular Cognitive Reference refutes infallibilism (*scientia*) about cognitive justification in non-formal domains (§§3, 4). This result may appear to have no bearing on Constructive Empiricism, because, *e.g.*, van Fraassen (2002, 1–30) so often stresses scientific caution about the truth of theories. However, van Fraassen's core distinction between merely accepting a scientific theory and believing it to be true is an instance of a common epistemological strategy of regarding a weaker belief as better justified than a stronger one, if they are based on the same evidence (*etc.*):

> . . . the assertion of empirical adequacy is a great deal weaker than the assertion of truth. . . . (*SI* 69)

Van Fraassen repeatedly appeals to this premise to justify his rejection of scientific realism, both about any one scientific theory and within philosophy of natural science generally.

5.2 Van Fraassen's Core Argument

Van Fraassen's core argument may be put thus:

1) 'Empirical adequacy' is adequacy to describe, predict, retrodict (and systematize) the observable events within the domain of a scientific theory, throughout the history of the universe. (*SI* 12, 45; 2008, 317)

2) Any belief that an empirically adequate scientific theory is also true can only be justified by whatever observations justify that theory's empirical adequacy.

3) The Logical Law of Weakening: If $(A \supset B)$, then $((A \,\&\, C) \supset B)$.

∴ 4) If two beliefs are based upon, are equally adequate to, and are justified by the same evidence, the stronger of those two beliefs is *less* well justified by that evidence than is the weaker (less committal) belief.

5) To be true, a scientific theory's fundamental concepts, principles, laws, or explanatory model(s) refer to and accurately describe actual structures or aspects of nature, including any (putative) unobservables within its domain.

6) Realism regarding any scientific theory is the claim that the theory is (very nearly) true about the objects or events within its domain, whether observable or unobservable.

7) Empirical adequacy is much weaker than and does not involve the (putative) truth of any scientific theory, law, explanation, or hypothesis regarding (putative) unobservables within its domain.

∴ 8) Unless the domain of a scientific theory contains only observables, realism about any empirically adequate scientific theory is always and in principle a much stronger claim than is the claim that the theory is empirically adequate.[15]

∴ 9) Constructive Empiricism is better justified than Scientific Realism, as an interpretation of any particular scientific theory which includes nonobservables within its domain, and as an interpretation of natural science generally.

This line of reasoning has become familiar to the point of appearing to many to be self-evident. However, Premises (3) and (4) cannot support van Fraassen's conclusions.

5.3 Van Fraassen's 'Law' of Weakening

Van Fraassen's justification of Constructive Empiricism appeals centrally to what he calls 'the Law of Weakening' (here Premise 3). Indeed, he argued that this contrast in strength or weakness of beliefs is simply a matter of logic.[16] In connection with "the Law of Weakening" van Fraassen noted that

> . . . the 'if . . . then' [in English] is not correctly identified with any of the sorts of implication traditionally discussed in logical theory, for those obey the Law of Weakening:
>
> 1. If A then B; hence: if A and C then B.
>
> But our conditionals, in natural language, typically do not obey that law:
>
> 2. If the match is struck it will light; hence (?): if the match is dunked in coffee and struck, it will light;
>
> the reader will think of many other examples. The explanation of why that 'law' does not hold is that our conditionals carry a tacit *ceteris paribus* clause:
>
> 3. If the plant had not been sprayed (*and all else had remained the same*) then it would not have died.
>
> The logical effect of this tacit clause is to make the 'law' of Weakening inapplicable. (*SI* 114–5; brackets and italics original, underscoring added)

NB: *The* ceteris paribus *clause, tacit in any causal-explanatory conditional statement, entails that van Fraassen's logical 'Law' of Weakening is irrelevant to* all *explanatory domains.* As van Fraassen here notes, because the logical 'Law of Weakening' holds *only* of systems of strict conditionals, it is therefore irrelevant to any domains which employ *ceteris paribus* clauses (whether explicitly or implicitly). His illustration is truth-functional for ease of presentation, but he correctly notes (as quoted) that *none* of "the sorts of implication traditionally discussed in logical theory" correctly capture our ordinary language conditionals. Hence the logical law of weakening is *irrelevant* to issues about scientific explanation, because causal explanations employ, ineliminably if implicitly, *ceteris paribus* clauses (Goodman 1946, Hempel 1988).[17] Nor does restricting one's commitment to a theory's empirical adequacy dispense with *ceteris paribus* clauses (see below). Thus van Fraassen's appeal to the logical Law of Weakening (Premise 3), involved in his key distinction between accepting a scientific theory and believing it to be true, is based upon an infallibilist presumption about empirical justification, namely, that whatever is required for justification within a logistic system holds as such also in non-formal domains. Such infallibilist presumptions are exposed as *irrelevant in principle* to the non-formal domain of empirical knowledge by the Thesis of Singular Cognitive Reference, by Newton's Rule 4 *and* by the *ceteris paribus* clauses implicit (if not explicit) in any causal-explanatory conditional statement. Consequently, van Fraassen cannot use the 'logical' Law of Weakening to justify his Constructive Empiricist account of any particular scientific theory, nor to criticize any realist interpretation of a scientific theory. Nor can van Fraassen use his logical law of weakening to justify his Constructive Empiricism in general, nor to justify his rejection of scientific realism in general.

Note further that van Fraassen's "Law of Weakening" is not a principle of formal *logic* because logical relations either hold or fail to hold, they do not come in degrees or 'strengths', whether distributively or collectively. 'Weakening' pertains to degree, strength, or kind of justificatory support.

Appreciating the character and scope of this result requires distinguishing van Fraassen's claim about conditional statements in explanatory contexts from what might be thought to be a similar result reached by Brandom (1981), who demonstrated this semantic paradox of material implication: Determining the truth-values of all conditional sentences within a truth-functional language *also* determines all the truth values of all of the simple (categorical) sentences of that language. This result is absurd, because merely conditional truths should not determine categorical truths. Consequently, material implication cannot render 'if . . . then' in ordinary usage. Brandom (1981, 130) notes that modal forms of conditional sentences, such as C. I. Lewis' strict implication, do not generate this paradox. His result about material implication and his observation about modal conditionals are correct. However, van Fraassen makes a different, very important point about modal forms of conditional sentences (all of which are strict logical implications): that *no* logic of conditionals can capture 'if . . . then' in explanatory contexts (whether commonsense, scientific, or forensic), because explanatory usage of 'if . . . then' presumes at least implicitly a *ceteris paribus* clause (*cf.* Goodman 1946, Hempel 1988), so that even modal forms of conditional statements (such as strict implication) cannot correctly render uses of 'if . . . then' within explanatory contexts.

Van Fraassen's "Law of Weakening," as a (purported) *logical* principle (*SI* 118), pertains as such only to strictly formal domains. However, the 'beliefs' relevant to Premise 4 are scientific beliefs, either about the empirical adequacy or the truth of scientific theories. Van Fraassen uses Premise 3 (the Law of Weakening) within the non-formal domain of philosophy of science to justify his Premise 4. Accordingly, Premise 4 is not, and cannot be, justified simply by logical principles alone; certainly not by simple instantiation of Premise 3. This is a key example of an infallibilist presupposition, of presuming that what is either required, or sufficient, for justification within strictly formal domains holds as such also in non-formal domains.

Infallibilist assumptions appear in all criticisms of, and all alternatives to, realism—both commonsense and scientific—which appeal to 'logical gaps' between evidence and any relevant realist claim or view, as if logical gaps automatically are cognitive gaps because they are justificatory gaps. To the contrary, logical gaps as such count as justificatory gaps only within strictly formal domains, domains which can be defined by the sufficiency of strict deduction for justification within those domains (above, §3). However, empirical knowledge, both commonsense and scientific, is a non-formal domain. Consequently, logical gaps *per se* are not justificatory gaps within the domain of empirical knowledge.[18]

5.4 Explanatory Parsimony

There is, of course, an important rule of evidence, most familiar as Ockham's Razor, according to which, of two explanations equally adequate to the same phenomenon, the less ontologically committal is better justified than the more ontologically committal explanation. This is the principle of simplicity, or of explanatory parsimony. It is not, however, a principle of logic, nor can it be justified simply by principles of logic. How and when the principle of simplicity can be used to assess competing causal explanations is a complex and delicate matter (Sober 1975), but one point is uncontroversial: The principle of explanatory parsimony becomes relevant only *subsequent* to determining that two alternative causal explanations are equally adequate to the relevant domain and evidence. Explanatory parsimony pertains to competing *explanations;* as such it does not pertain to the distinction between any one scientific explanation and *its* empirical adequacy. Van Fraassen cannot support his Constructive Empiricism by independent appeal to the principle of explanatory simplicity—in effect, treating Premise 4 as a statement of Ockham's Razor—because explanatory parsimony pertains to two otherwise equally adequate explanations of some domain. Harper has shown, however, that "empirical adequacy" is insufficient for disentangling the masses from the weights of planets in the ways central to the achievements of Newton's mechanics (above, §2.3).[19] Consequently, scientific realism and Constructive Empiricism do not appeal simply to the empirical adequacy of scientific theories (Premises 2, 5; above, §5.2).

5.5 Empirical Adequacy

Van Fraassen repeatedly claims—even in 2008—that empirical adequacy—even for Newton—only concerns what in fact occurs in nature:

> When Newton claims empirical adequacy for his theory, he is claiming that his theory has some model such that *all actual appearances are identifiable with (isomorphic to) motions* in that model. (This refers of course to all actual appearances throughout the history of

the universe, and whether in fact observed or not.) (*SI* 45, *cf.* 46; 1991, 372; 2008, 317).

> Remember that empirical adequacy concerns actual phenomena: what does happen, and not, what would happen under different circumstances. (*SI* 60; *cf.* 61)

> . . . the precise definition of empirical adequacy . . . relates the theory to the *actual* phenomena (and not to anything which *would* happen if the world *were* different, assertions about which have, to my mind, no basis in fact but reflect only the background theories with which we operate). . . . (*SI* 64[20])

However, regarding, *e.g.*, the stability of the apsides of each planetary orbit, where any rotation of the apsides (precession) would indicate some rate of diminution of attractive force other than precisely an inverse square ratio to distance, Newton stressed quite the opposite. That the gravitational force between any two primary planets varies by the inverse square of the distance between them is stated in Proposition 2, Theorem 2, of *Principia,* Book III, as follows:

> *The forces by which the primary planets are continually drawn away from rectilinear motions and are maintained in their respective orbits . . . are inversely as the squares of their distances from its center.* (Newton 1999, 802; 1871, 395)

Concerning his justification of this second part of the theorem, the inverse square law, Newton states:

> . . . this second part of the proposition is proved with the greatest exactness from the fact that the aphelia are at rest. For the slightest departure from the ratio of the square would (by book 2, prop. 45, corol. 1) necessarily result in a noticeable motion of the apsides in a single revolution and an immense such motion in many revolutions. (Newton 1999, 802; 1871, 395; *cf.* Harper 2011, 116)

As Harper repeatedly and rightly stresses, Newton's causal-explanatory gravitational theory gives pride of place to certain subjunctive conditional statements. Such subjunctive conditionals are central to Newton's methodology of devising analyses of motions which enable those motions to measure the strength of a force, which requires distinguishing the actual value measured from other values. Having noted van Fraassen's infallibilist fallacy, it is important to stress that the systematic dependencies Newton formulates as subjunctive conditional statements, are formulated as mathematically and physically precise continuous functions. That Newton's functions are continuous is not the key point here, but rather that his subjunctive conditionals are mathematically and physically defined: they are not creatures of modal logic, and are not subject to the vagaries of ill-defined "accessibility relations" between possible worlds, nor of philosophers' "modal intuitions," including not infrequent "modal skepticism."

Van Fraassen's empiricist focus upon solely what does happen in nature, and his rejection of counterfactuals about what would happen *in nature* under identifiably different conditions, is flatly inconsistent with Newton's mechanics. (Restricting modality to a theoretical *model* mislocates it and mistakes much physical theory.) In this crucial, elementary regard, van Fraassen's Constructive Empiricism is plainly *in*adequate to its purported domain, which includes, centrally, Newton's classical mechanics. NB: In *Representing Science* (2008, 317–9) van Fraassen reaffirms exactly the same Constructive Empiricism, and does so directly in connection with Newton's classical mechanics; in *Representing Science* he restricts and revises his Constructive Empiricism only to accommodate statistical theories.[21] This fundamental flaw in Constructive Empiricism should not have been neglected for thirty three years.

5.6 Pragmatics *versus* Semantics?

These points raise a host of further issues which cannot be examined here. Let me conclude with these observations. I have claimed that van Fraassen's appeal to "weakening" is basic to his central arguments for Constructive Empiricism. One might suppose that his main argument concerns his highly contextual, entirely "pragmatic" account of explanation. Van Fraassen's claim that explanation is pragmatic rather than cognitive rests upon the argument examined and criticized above.[22] In addition, van Fraassen seeks to "transpose" Morris's semiotic distinctions among syntax, semantics, and pragmatics into philosophy of science (*SI* 89). In just this vein he much more recently claims:

> As to knowledge, all the philosophical puzzles that pertain to it specifically seem best transposed to philosophy of language, to investigation of the grammar, the logic, and most of all the pragmatics rather than [*sic*] semantics, of the term 'know'. (van Fraassen 2007, 346; *cf.* 2000, 275)

These are precisely the kind of claims mentioned and criticized above, of seeking to supplant epistemology with philosophy of language or philosophy of mind (or both). This strategy cannot and does not work, because philosophy of language or philosophy of mind do not suffice, whether singly or conjointly, to account for cognitive justification. Indeed, they do not suffice even for the semantics of singular cognitive reference (above, §§3, 4). Most fundamentally, van Fraassen's contention that truth is solely a matter of semantics whereas explanation is wholly a matter of pragmatics neglects the single most important lesson of Gettier's (1963) paper, namely, that to understand cognitive justification requires attending to the candidate cognizant subject—the infamous *S*'s—actual circumstances and actual cognitive processes. All of Gettier's counter-examples turn on such factors. All such factors fall into Morris' category of 'pragmatics', but this does not show—nor did Morris as a semiotician seek to show—that because relations between any statement, the speaker, and his or her circumstances are 'pragmatic', that they are not cognitive.

6 CONCLUSION

Van Fraassen noted (1980, 19) that "the major questions of epistemology" cannot be settled "*en passant* in philosophy of science." Indeed. Yet developing a philosophy of science on faulty epistemological preconceptions is ill-fated from the outset. For reasons examined here I conclude that this is the misfortune of van Fraassen's Constructive Empiricism, whether in 1980 or in 2008. That the fundamental flaws in Constructive Empiricism identified here have gone unnoted for thirty three years indicates clearly that much contemporary philosophy of science requires fundamental re-examination of its central epistemological presuppositions. We shall not understand empirical sciences until we rescind the notion that empiricism has a monopoly on empirical knowledge.

NOTES

1. Here I follow Harper (2011), who kindly allowed me to study his (2011) prior to its publication.
2. *E.g.*, Newton's criteria of theoretical adequacy are not subject to Christensen's objections to Glymore's boot-strap account (Harper 2011, 133–6).
3. Both mistaken assumptions result from the deductivist view of scientific explanation, which is presupposed by both Hume's 'Problem' and Goodman's 'Riddle' of Induction, and was central to Logical Positivism and Logical Empiricism up to *circa* 1980; see Suppe (1977), Grünbaum and Salmon (1988), Kyburg (1988), Salmon (1989). That observation statements do not follow logically from observations (experiences) was stressed by Hempel (1935) and Schlick (1935); it recurs in Davidson's (2001, 141, 143) view that only a belief can justify another belief.
4. By Kepler's Areal Law and Newton's Bk 1, Propositions 1 and 2 and their Corollaries (Harper 2011, 76–83, 156–9), this force is directed to the focus of an elliptical orbit occupied by the sun.
5. To recall, Newton's Second Law concerns the composition of distinct forces, often referred to as the "Parallelogram of Forces."
6. These problems are due to Christensen (1983, 1990); Harper (2011, 133–6) responds on Newton's behalf.
7. Though Newton stated Rule 4 only in the third edition of the *Principia*, its use is evident in the earlier editions, and is implicit in Kepler's laborious though ultimately successful and significantly more precise determination of planetary orbits and the three laws he discovered about them. It is also implicit in Galileo's law of free fall, in particular, his discovery that acceleration varies directly

with time (squared) rather than with distance. For further critical discussion of Harper (2011), see Huggett *et al* (2013).

8. In this section, otherwise unattributed parenthetical page numbers refer to Evans (1975).
9. Evans contends that predication within a language is structured in ways which govern the construction and assent conditions of compound sentences containing negation (whether internal or external) and conjunction in ways Quine's indeterminacy theses cannot accommodate. I believe Evans' case against Quine is successful, but these issues are not presently germane.
10. Recall his example of using the description 'The man in the corner drinking a martini', used in a context in which it univocally referred to a particular man who happened to be drinking water from a martini glass.
11. *E.g.*, Westphal (2008b), (2011b). Some readers may wonder whether the present analysis commits me to some of Putnam's reasons for espousing "internal realism." To the contrary, I have argued elsewhere that Putnam failed to make his case for internal realism (Westphal 1997, *xxiii–xxvii*), and that Carnap's attempt to scuttle framework-independent issues about realism (to which Putnam's case for internal realism centrally appealed) fails for reasons strictly internal to Carnap's account (Westphal 1989, 47–67). (Putnam's original argument for internal realism disregards the distinction between formal and non-formal domains, and disregards what may be called Kaplan's *Caveat,* that in any use of formal modeling, it is imperative to distinguish carefully between genuine features of the domain so modeled and mere artifacts of the model).
12. Harper (2011) explicates brilliantly Newton's use of Rule 4 in the *Principia.* I submit that Newton's use of Rule 4, as explicated by Harper, provides ample scientific, methodological justification for Rule 4. However, philosophers in the empiricist tradition will ask, not what follows from, nor what can be based upon, Rule 4, but rather, what if anything justifies Rule 4 antecedently? To this question I have not found an answer in Harper's research, which is why I propose the present answer.
13. I discuss van Fraassen (2002) in Westphal (2006), §4. (The quotation on 138–9 of my (2006) is cited incorrectly; I have not yet been able to relocate its source. Nevertheless, the passage formulates the attitudes, values, and beliefs van Fraassen (2002, 37, 47; *cf.* 62, 152) ascribes to the empirical stance.)
14. I examine Constructive Empiricism in much greater detail in Westphal (2014).
15. The parenthetical qualification is required because if a scientific "hypothesis is solely about what is observable . . . empirical adequacy coincides with truth" (*SI* 72).
16. He identifies "Weakening" as a "logical law" like Contraposition (*SI* 118).
17. The familiar notion that in 'mature' (physical) sciences causal locutions are replaced by transformations (whether strict or statistical) neglects the distinction between the formulation of laws of nature and their use in explaining any phenomenon, and that such use reintroduces *ceteris paribus* clauses and, in many typical cases, specific causal relations, in addition to initial conditions and parameters (or 'boundary conditions') of the relevant physical system. By themselves, laws of nature explain nothing.
18. I discuss this point in greater detail in Westphal (2010–11); I develop the positive alternative in Westphal (2011a) and (2013). I critically examine global perceptual skepticism in Westphal (2004, §§62–3; 2011b).
19. For more general critique of "empirical adequacy" as the goal of physics, see Hüttemann (1997). Harper (2011, 389–94) shows that Laudan's confutation of convergent realism does not hold against Newton's methodology.
20. Likewise van Fraassen states: "My view is that physical theories do indeed describe much more than what is observable, but that what matters is empirical

adequacy, and not the truth or falsity of how they go beyond observable phenomena" (*SI* 64); "When the hypothesis is solely about what is observable . . . empirical adequacy coincides with truth" (1980, 72); ". . . we must define empirical adequacy directly, without an empirical detour: all the actual observable phenomena fit the empirical substructures in a certain one of these models" (*SI* 84).

21. For further critical reflections on van Fraassen (2008), see Okruhlick (2009). Fortunately, much of van Fraassen's *Representing Science* is, by design (2008, 3), independent of his Constructive Empiricism. An excellent attempt is made by Muller (2005) to show that Constructive Empiricism can support the modalities involved in 'observability' and in scientific explanations. In my (2014) I argue that his attempt is unsuccessful.
22. See *SI* 4, 89–90, 115–6, 129–30; 2000, 273–4; 2007, 337, 340, 349; 2008, 3, 17, 25, 82, 136.

REFERENCES

Bouwsma, O. K., 1949. "Descartes' Evil Genius." *The Philosophical Review* 58.2: 141–151.

Brandom, Robert, 1981. "Semantic Paradox of Material Implication." *Notre Dame Journal of Symbolic Logic* 22.2:129–132.

Christensen, David, 1983. "Glymore on Evidential Relevance." *Philosophy of Science* 50: 471–481; rpt. in: L. Sklar, ed., *The Philosophy of Science: A Collection of Essays* (London: Taylor & Francis, 2000), 119–129.

———, 1990. "The Irrelevance of Bootstrapping." *Philosophy of Science* 57: 644–662; rpt. in: L. Sklar, ed., *The Philosophy of Science: A Collection of Essays* (London: Taylor & Francis, 2000), 130–148.

Donnellan, Keith, 1966. "Reference and Definite Descriptions." *The Philosophical Review* 77.3: 281–304.

Dretske, Frederick, 1981. *Knowledge and the Flow of Information*. Cambridge, Mass., MIT Press.

Evans, Gareth, 1975. "Identity and Predication." *Journal of Philosophy* 72.13: 343–63; rpt in: *idem., Collected Papers* (Oxford: Oxford University Press, 1985), 25–48.

Fraassen, Bastian C. van, 1980. *The Scientific Image*. Oxford, The Clarendon Press.

———, 2001. "Constructive Empiricism Now." *Philosophical Studies* 106:151–170.

———, 2002. *The Empirical Stance*. New Haven, Yale University Press.

———, 2004a. "Preçis of *The Empirical Stance*." *Philosophical Studies* 121:127–132.

———, 2004b. "Replies to Discussion on *The Empirical Stance*." *Philosophical Studies* 121.2:171–192.

———, 2006. "Representation: The Problem for Structuralism." *Philosophy of Science* 73.5:536–547.

———, 2007. "Reply: From a view of science to a new empiricism." In: B. Monton, ed., *Images of Empiricism: Essays on Science and Stances, with a Reply from Bas van Fraassen* (Oxford, Oxford University Press), 337–383.

———, 2008. *Scientific Representation: Paradoxes of Perspective*. Oxford, Oxford University Press.

Goodman, Nelson, 1946. "The Problem of Counterfactual Conditionals." Rpt. in: *idem., Fact, Fiction and Forecast*, 4th ed. (Cambridge, Mass.: Harvard University Press, 1983), 1–27.

Grünbaum, Adolf, and Wesley Salmon, eds., 1988. *The Limitations of Deductivism*. Pittsburgh, University of Pittsburgh Press.

Harper, William L., 2011. *Isaac Newton's Scientific Method: Turning Data into Evidence about Gravity and Cosmology*. New York, Oxford University Press.

Hempel, Carl G., 1935. "On the Logical Positivists' Theory of Truth." *Analysis* 2,4:50–59.

———, 1965. *Aspects of Scientific Explanation and Other Essays in the Philosophy of Science*. New York, Free Press.

———, 1988. "Provisos: A Problem Concerning the Inferential Function of Scientific Theories." In: A. Grünbaum and W. Salmon, eds., *The Limitations of Deductivism* (Pittsburgh, University of Pittsburgh Press), 19–36.

Huggett, Nick, George E. Smith, David Marshall Miller and William Harper, 2013. "On Newton's method." *Metascience*. DOI: 10.1007/s11016-013-9745-y

Hüttemann, Andreas, 1997. *Idealisierung und das Ziel der Physik. Eine Untersuchung zum Realismus, Empirismus und Konstrukivismus in der Wissenschaftstheorie*. Berlin, de Gruyter.

Janiak, Andrew, 2007. "Newton and the Reality of Force." *Journal of the History of Philosophy* 45.1: 127–147.

Kaplan, David, 1975. "How to Russell a Frege-Church." *The Journal of Philosophy* 72.19:716–729.

———, 1989. "On Demonstratives." In: J. Almog, *et al*, eds., *Themes from Kaplan* (New York, Oxford University Press), 481–563.

Kuhn, Thomas, 1970. *The Structure of Scientific Revolutions*, 2nd ed. Chicago, University of Chicago Press.

Kyburg, Henry Jr., 1988. "The Justification of Deduction in Science." In: A. Grünbaum and W. Salmon, eds., *The Limitations of Deductivism* (Pittsburgh, University of Pittsburgh Press), 61–94.

Lewis, Clarence Irving, 1929. *Mind and the World Order: Outline of a Theory of Knowledge*. New York, Charles Scribner; rpt. New York: Dover, 1956.

———, 1930. "Logic and Pragmatism." In: G. P. Adams and W. P. Montague, eds., *Contemporary American Philosophy* (New York, Macmillan), 31–50; rpt. in: *idem*. (1970), 3–19.

———, 1970. J. D. Goheen and J. L. Mothershead, Jr., eds., *Collected Papers of Clarence Irving Lewis*. Stanford, Stanford University Press.

Newton, Isaac, 1878. *Philosophiae naturalis principia mathematica*, 3d rev. ed. London (1726); rpt. Glasgow, Maclehose; with corrections.

———, 1952. *Opticks*. New York, Dover; rev. ed. 1979.

———, 1999. *The Principia: Mathematical Principles of Natural Philosophy*. I. B. Cohen & A. Whitman, trs., with assistance from J. Budenz. Berkeley, University of California Press.

Okruhlik, Kathleen, 2009. "Critical Notice, *Scientific Representation: Paradoxes of Perspective*." *Canadian Journal of Philosophy* 39.4: 671–694.

Russell, Bertrand, 1905. "On Denoting." *Mind* 14.4:479–493.

Salmon, Wesley, 1989. *Four Decades of Scientific Explanation*. Pittsburgh, University of Pittsburgh Press.

Schlesinger, George, 1975. "Confirmation and Parsimony." In: G. Maxwell and R. M. Anderson, Jr., eds., *Induction, Probability, and Confirmation. Minnesota Studies in the Philosophy of Science* 6 (Minneapolis, University of Minnesota Press), 324–342.

Schlick, Moritz, 1935. "Facts and Propositions." *Analysis* 2.5:65–70.

Smith, George E., 2002a. "The Methodology of the *Principia*." In: I. B. Cohen and G. Smith, eds., *The Cambridge Companion to Newton* (Cambridge, Cambridge University Press), 138–173.

———, 2002b. "From the Phenomenon of the Ellipse to an Inverse-Square Force: Why Not?" In: D. Malament, ed., *Reading Natural Philosophy: Essays in the*

History and Philosophy of Science and Mathematics (LaSalle, Ill., Open Court), 31–70.

———, 2007. "Closing the Loop: Testing Newtonian Gravity, Then and Now." Unpublished lecture, available at: www.stanford.edu/dept/cisst/visitors. html

Sober, Elliot, 1975. *Simplicity.* Oxford, Oxford University Press.

Suppe, Frederick, ed., 1977. *The Structure of Scientific Theories,* 2nd rev. ed. Champaign-Urbana, Ill., University of Illinois Press.

Travis, Charles, 2005. "A Sense of Occasion." *Philosophical Quarterly* 55.219:286–314.

Westphal, Kenneth R., 1989. *Hegel's Epistemological Realism.* Philosophical Studies Series, 43; Dordrecht and Boston, Kluwer.

———, 1997. "Frederick L. Will's Pragmatic Realism: An Introduction." In: Frederick L. Will, *Pragmatism and Realism,* ed. by K. R. Westphal (Lanham, Md., Rowman & Littlefield), xiii–lxi.

———, 2004. *Kant's Transcendental Proof of Realism.* Cambridge, Cambridge University Press.

———, 2006. "Science and the Philosophers." In: H. Koskinen, S. Pihlström, and R. Vilkko, eds., *Science: A Challenge to Philosophy?* (Frankfurt/Main, Lang), 125–52.

———, 2008. "Contemporary Epistemology: Kant, Hegel, McDowell." In: J. Lindgaard, ed., *John McDowell: Experience, Norm and Nature* (Oxford, Blackwell), 124–151.

———, 2010–11. "Analytic Philosophy and the Long Tail of *Scientia:* Hegel and the Historicity of Philosophy." *The Owl of Minerva* 42.1–2: 1–18.

———, 2011a. "Urteilskraft, gegenseitige Anerkennung und rationale Rechtfertigung." In: H.-D. Klein, ed., *Ethik als prima philosophia?* (Würzburg, Königshausen & Neumann), 171–193.

———, 2011b. 'Self-Consciousness, Anti-Cartesianism and Cognitive Semantics in Hegel's 1807 *Phenomenology*'. In: S. Houlgate and M. Baur, eds., *The Blackwell Companion to Hegel* (Oxford, Wiley-Blackwell), 68–90.

———, 2013. "Rational Justification & Mutual Recognition in Substantive Domains." *Dialogue: Canadian Journal of Philosophy/Revue Canadienne de philosophie* 52:1–40; DOI: 10.1017/S0012217313000796

———, 2014. "Hegel's Semantics of Singular Cognitive Reference, Newton's Methodological Rule Four and Scientific Realism Today." In: *Philosophical Inquiries* 2.1:9–65.

Wolff, Michael, 1995. "Was ist formale Logik?" In: C. Fricke, P. König, and T. Petersen, eds., *Das Recht der Vernunft. Kant und Hegel über Denken, Erkennen und Handeln* (Stuttgart, Frommann-Holzboog), 19–31.

———, 2000. "Kantische Urteilstafel und vollständige Induktion: Nachtrag zu meiner Kontroverse mit Ulrich Nortmann." *Zeitschrift für Philosophische Forschung* 54.1:86–94.

———, 2009. *Abhandlungen über die Prinzipien der Logik,* 2nd rev. ed. Frankfurt am Main, Klostermann.

———, 2012. "Viele Logiken—Eine Vernunft. Warum der logische Pluralismus ein Irtum ist." *Methodos* 7:73–128.

10 Naturalism without Metaphysics

Jonathan Knowles

1 INTRODUCTION

Naturalism is a very broad category in philosophy today, and difficult to define. If one looks for a lowest common denominator, one might perhaps say that all naturalists are committed to rejecting supernatural entities and modes of explanation. As a definition, this is not very enlightening, since the idea of the supernatural is obviously correlative to that of the natural. Nevertheless, when 'supernatural' is glossed in terms of things like ghosts, elves, 'biofields', astrology, and the like, I suppose most contemporary philosophers would count as naturalists in this sense. Such a 'minimal' naturalism would not seem to be very interesting within current philosophical debates. Many have pointed out that it clearly falls short of anything like a *metaphysical* naturalism, such as would see the totality of facts (properties, entities) as reducible or at least supervenient upon those of, say, a completed fundamental physics. Elsewhere I have argued (Knowles 2008, 2010) that even minimal naturalism entails commitments to a certain priority for (natural) scientific endeavors over others, which in turn puts constraints on one's non-(natural) scientific theorizing, so that various kinds of what I have called 'non-scientific' and others have called 'liberal' naturalism—which promote the autonomy of non-scientific modes of understanding without admitting anything supernatural—are unstable and plausibly collapse into what I call *scientific* naturalism, the view that science—natural science—is our *unique* source of what can be considered our most *fundamental* knowledge and systematic explanation.

Many liberal naturalists are also 'anti-metaphysicians' or, as some put it, 'neo-pragmatists', in the sense that they renounce as futile or meaningless metaphysical questions about what *really* can be said to exist and what cannot: questions about the nature of a completely objective reality, as it is independent of any mind or system of representation. Perhaps the clearest example of such an anti-metaphysical, liberal naturalist is Richard Rorty; other philosophers often regarded as subscribing to similar views are Wittgenstein, Hilary Putnam, Robert Brandom, John McDowell, and (perhaps) Huw Price,[1] whose views are discussed more fully below. I find myself largely in sympathy with these thinkers' anti-metaphysicalism, but the liberal naturalism they espouse is, I believe, unstable. In this paper I investigate the issue of whether a *scientific naturalist* can also be an *anti-metaphysician*—a combination not widely defended, at least explicitly.[2] Philosophical belief being what it is, I approach this task with an open mind, though also, in accord with philosophical method, in the spirit of trying to see what *can* be said on behalf of such a position. In particular, I will examine whether this combination is at least consistent and not wholly implausible. My central concerns are two questions: Does science *itself* suggest a metaphysical conception of its results and practice? Can scientific naturalism (henceforth 'SN') maintain its credentials while refraining from metaphysical theses about the 'true nature of reality'? I shall argue, at least provisionally, for negative and positive answers, respectively, to these questions.

2 CRITIQUE OF METAPHYSICAL NATURALISM

First I should indicate why I reject a metaphysical kind of naturalism. Insofar as I hold scientific naturalism and reject liberal naturalism, many will wonder why I demur at metaphysical naturalism, insofar as this would seem to give SN precisely the backing it needs to defend itself definitively against its liberal detractors. If science delineates the mind-independent facts in terms of which we must understand any other facts (mind-independent or otherwise), then surely it must have a fundamental status in relation to other modes of understanding.

In order to see why I reject such a view of science, I need first to say what I mean by 'metaphysical naturalism' (henceforth 'MN'). As I understand it, this is the view that there is an objective reality with a determinate nature that we can meaningfully seek to capture in our thought and talk about the world, and that natural science is best-placed to delineate this nature, at least at a fundamental or base level. MN is thus clearly a *kind* of scientific naturalism, though it is not my scientific naturalism, *i.e.* SN.[3] The central tenets of MN are:

a) *Realism:* There is a way (or there are ways) the world is completely independently of how we think about it; there are *mind-independent facts* (properties, entities, *etc.*).[4]

b) *Representationalism:* We can form a conception of these mind-independent facts in thought and/or language: we can represent or *mirror* reality.

c) *Naturalism:* Science (*i.e.* natural science—a qualification I omit hereafter) has turned out to be the most adept way of giving us this conception, at least at a fundamental level—that is, in such a way that all other facts can be made sense of in scientific terms.[5]

Sometimes this whole package is referred to as *metaphysical realism* ('MR'), but I will reserve that epithet here for the combination of (a) and (b) alone, that is, without the assumption of naturalism. (For the record, nothing of what I have to say will impact on MR to the extent this might be deployed non-naturalistically.)

Sometimes MN is seen as equivalent to *physicalism:* the view that the most basic facts will be furnished by a completed physics. Many thinkers develop MN under something like the assumption of physicalism (*e.g.* Jackson 1998). I believe there is a tight connection between the very idea of a fundamental set of objective facts that ground all others, and the central ideas behind contemporary physicalism. In principle, one can imagine other sciences than physics contributing to the ultimate facts, but this would be more a question of how we define 'physics' than a divergence from the basic idea. Now, as we shall see in the following section, I think there are good reasons for not being a physicalist insofar as *actual* science does not mandate anything like physicalism as an overarching metaphysics. Science should surely be a main source of inspiration and constraint for any naturalistic position. However, I take it also to be an important and interesting task to show that MN is not viable as a position in its own right, and thus not something that might be recommended as a framework within which someone could say we *should* conduct science.

In addition to the question about physicalism, some readers may wonder why clause (b) should belong to MN (or to MR). Isn't the key idea that there is an objective reality with some definite nature, regardless of any ideas about how or even *whether* we represent such a reality (*cf.* Devitt 1991, 2010, 2013; Searle 1995, 199–225)? I don't believe so; the very idea of mind-independent facts (properties, entities) makes sense only assuming that we stand to these in something like a representational relation. This doesn't mean that we must be actually able to represent all the facts, but rather that they are in principle 'representable'. I defend this commitment briefly at the end of this section, and otherwise proceed on the assumption that this is the correct understanding of MN (and MR).

MN has been attacked by many philosophers over many years. My rejection of it does not depend on explicitly idealist alternatives to realism, nor on skeptical claims to the effect that science cannot in fact tell us about ultimate reality. Instead, I sympathize with views that question MN's notion of mind-independent facts due to problems with representationalism. This large topic cannot be handled in one paper, let alone in part of one. Accordingly I shall present and defend one important recent challenge to representationalism, due to Huw Price.[6] I will also, as noted, defend the relevance of representationalism to the very idea of mind-independent facts.

Price's critique of representationalism begins with the notion of a substantial *reference* relation holding between words and things in the world (*cf.* Price 2004, Price & Menzies 2009). Most parties to the debate would find this notion unproblematic, since reference relations are typically regarded as constituting at least an important part of anything worth calling a representational relation. Understanding representation in terms of reference, MN[7] is committed, claims Price (as I understand him), to the following:

a) There are substantial reference relations between our words or concepts and things in the world.
b) There are more or less determinate facts of the matter as to exactly what our thoughts or words refer to in the world, and how they so refer.
c) These are genuine, currently open, empirical questions.
d) A theory of that to which our thoughts or words refer, and how they do so, will also apply to the concept of reference itself.

These commitments follow from MN's commitment to naturalism, *i.e.* to science providing us with *all* the basic facts, and from the reasonable assumption that for MN semantic facts are genuine facts, albeit ones about which we are at present relatively ignorant.[8] Price presents three arguments (or groups of arguments) against MN: two directly challenge (a)–(c), the third challenges the internal consistency of (a)–(d).

The first argument is simply that, so far as scientific theory of reference is concerned, deflationary theories need to be considered a live option (Price 2004, 79). Deflationism about truth has many sympathizers and though deflationary theories of reference may be less widely discussed, they cannot be dismissed out of hand. Such theories precisely deny that facts about what words refer to involve genuine word-world relations, *a fortiori* that there is any empirical uncovering of this relation to carry out. This is not so much an argument against MN as a reminder that its view of reference cannot be taken for granted, yet it is an important point to register in the overall dialectic.

Price's second argument derives from Stephen Stich and his skeptical ruminations on the status of eliminativist arguments, of the kind to which Stich himself was attracted in relation to notions like belief and desire in his early work (*cf.* Stich 1996; Price 2004, 80–1; Price & Menzies 2009, 195–7). Eliminativism about, *e.g.*, belief often takes the form of arguing that there is nothing in the world that satisfies a certain 'Ramsey sentence' for belief.[9] Stich points out that this only shows there are no such things as beliefs, if we take the Ramsey sentence to determinately specify the reference relation for our concept 'belief'. However, what grounds are there for thinking this, given the existence of other theories of reference (including deflationism) and, avers Stich, the bleak prospects for empirical resolution amongst them? The latter claim is further buttressed by the thought that we could only ever establish such a reference theory subsequent to deciding *what* beliefs are to begin with, and whether indeed there are any. For Stich the upshot is that reference is no help in deciding issues about eliminativism. For Price the upshot is more radical: a substantive notion of reference should, arguably, not be countenanced at all, since it is of no help in adjudicating questions of science. Hence MN falters.

The third argument is more an argument group. The considerations in this group are also more principled in that they aim to show that there is a kind of self-referential inconsistency involved in the idea that MN can provide the same kind of treatment of reference relations as it can of other natural phenomena (assuming that reference is one such). I have identified three such considerations in Price's writings. The first is Paul Boghossian's (1989, 1990; *cf.* Price 2004, 79–81; Price & Menzies 2009, 197–8) argument against (metaphysically) naturalistic theories of reference, which goes as follows: On the presuppositions of such theories, it must be not just an open question exactly *what* reference is, but indeed *whether* there is any such phenomenon at all. In other words, the very idea of such an empirical theory entails the genuine possibility of irrealism about reference. That, however, is incoherent, for we would then have no way of expressing the conclusion that the term or concept 'reference' fails to refer.

The second argument is presented most explicitly in Price & Menzies (2009, 194–5). They claim here that metaphysical naturalists—at least those subscribing to the so-called 'Canberra plan' for naturalism outlined by Jackson (1998)—need to identify *prima facie* non-natural entities and properties as the *relata* of substantial semantic relations (such as reference) in order to understand how they might, with the help of Ramsey sentences codifying our folk theory of them, be identified with underlying natural properties. They then argue that this procedure cannot be applied to the semantic terms themselves, in particular not to reference, since the Ramsey procedure depends upon a distinction between theoretical terms and observation terms, and thus precludes anything playing both roles. Hence metaphysical naturalism is necessarily incomplete and to the extent that it aims to be complete, it is thus self-undermining.

Price's (2004, 82–3) final argument in this last group is similar in spirit to Boghossian's. Under the presuppositions of MN, different empirical theories of reference must be capable of conflicting with one another. However, under those same presuppositions they turn out not to be able to conflict. To illustrate, Price asks us to imagine an empirical theory T that says:

(1) *Reference* is relation R

and another theory S that says:

(2) *Reference* is relation R*

They thus appear to conflict. However, according to T and S what is the case is in fact, respectively:

(3) 'Reference' stands in relation R to R

(4) 'Reference' stands in relation R* to R*.

These claims do not conflict. So reference isn't, after all, a substantial, empirically scrutable relationship, contrary to the presuppositions of MN, at least so far as reference is itself subject to the same kind of theorizing, as MN supposes. Hence MN is incoherent.

How might a supporter of MN respond to these objections? I think the most obvious response would be to question whether seeking a naturalistic theory of reference must be understood in terms of what our concept 'reference' refers to; if not, then much of the force of these arguments would appear to evaporate. Indeed, Stich himself wants to say that we may still be able to investigate whether there are beliefs, if *belief* is something like a natural kind; his point is only that we should stop trying to launch this inquiry from a linguistic starting point. Price (2004, 84–7) claims that a metaphysical naturalism that just looks to the world, rather than starting with language, will not have the resources to argue for anything like physicalism or indeed for any kind of reductive naturalism, at least at a general level.[10] I will argue below, on independent grounds, that a position which seeks to divorce the notion of mind-independent fact from the idea that there are substantial relations of reference is not viable. However, as far as the arguments just reviewed are concerned, a supporter of MN might say she can proceed with her naturalistic project with a substantial notion of reference that has been investigated as a 'natural kind', in the way just suggested (or perhaps that the two projects might be developed in tandem).

To halt this line of response, we need to ask whether reference *is* the kind of phenomenon that might be investigated as a natural kind. The answer would seem to be that it is not.[11] A common view is that a natural kind is one which supports 'twin-earth intuitions': if astronauts from earth were to find something that looked, smelled, tasted, and generally behaved like, say, water on a distant planet, it could still be reasonable or at least intelligible to report back that they wondered whether it is water, and not some superficially indistinguishable substance. Hence water is plausibly a natural kind. But how would we react to such a report for something like reference? The astronauts have found in the natives' language a term that plays more or less exactly the role in their language that 'reference' does for us in ours—could they then reasonably add "But we are unsure whether the term expresses *reference*?" Intuitively this seems highly odd. There is, it would seem, nothing *more* to reference than (dispositions to) patterns of use amongst speakers of the language. So intuitively it is not a natural kind with a discoverable essence, in the way water is. But if not then it seems we are left, in seeking to understand reference, with having to ask what our concept 'reference' refers to, if anything, and Price's arguments against MN will apply.[12]

No doubt this and related discussions will continue to rage for some time. The immediately preceding argument, for instance, is not conclusive, insofar as it builds on intuitions. Perhaps to see the truth of MN we need to shake ourselves free of such poor intuitions. However, it seems clear there are substantial worries here for MN that need to be countered by its supporters.

I now consider an objection to the whole drift of the previous arguments, namely that the kind of realism involved in MN is simply not beholden to any semantic theses, such as representationalism. Both Devitt (1991, 2010, 2013) and Searle (1995, 199–225) have forcefully argued that realism is not beholden to theses about how language or thought relates to the world, such as the correspondence theory of truth. Their writings make clear the latter is just one particular, egregious example and that neither would they countenance that MN has anything to do with theory of reference.[13] They also argue, convincingly, that idealistic alternatives to realism are implausible. Within the context of naturalism, this amounts to saying that a view reasonably called 'metaphysical naturalism' can be maintained independently of commitment to representationalism.

I think this conclusion mistaken. To my mind, what Devitt and Searle (and others with similar views) fail to appreciate is a distinction between metaphysical realism (MR, understood as above) and what I will call 'common sense realism' ('CR'). I concur with Devitt and Searle in the idea that there are dinosaurs, chairs and electrons existing independently of us and that there are truths about these things, many of which we don't know or may never know; moreover, this is not something which requires a correspondence theory of truth or any thesis about reference to defend; it is consistent, *e.g.*, with a deflationary theory of truth and reference. However, the view just sketched only amounts to what I am calling CR. Devitt and Searle want to argue for more than this: they want to say that there are *mind-independent* facts: that there is an *objective* reality completely independent of thought—and that is something quite different. Now by CR it is true that *dinosaurs exist* (*i.e.* existed) *independently of* (*i.e.* long before) *us*. By CR it is also either true or false that *dinosaurs became extinct because a large meteor hit the earth*, even though we may never have good grounds for knowing which. However this does not amount to countenancing the idea of 'mind-independent' or 'concept-independent' facts. Indeed the first statement above follows fairly straightforwardly from an everyday understanding of what dinosaurs are—that is, it gives (partial) expression to our *concept* of what dinosaurs are (were). The second connects dinosaurs with a speculation about their fate through an understanding of a range of further concepts. Dinosaurs certainly didn't need humans to exist, but this does not license the idea of facts about them—including that they *were* dinosaurs—as something distinct from claims made in terms of our concepts. As we ordinarily think about it, it is precisely the latter, our concepts and the understanding they embody, that furnish us with our facts, at least when they combine to express true thoughts (to use a helpful Fregean term). I stress that this is not meant to rule out wanting to see facts as more than just true thoughts; rather my aim is merely to show that CR does not license the idea of such things. Something more needs to be said.

I should register that Devitt and Searle admit that the ways we 'carve up' the world are to an extent (or in a certain sense) up to us and our concepts; however, they do not see this as compromising their metaphysical realism. For Searle this takes the form of saying that although which general concepts or categories we use is up to us, whether things fall under them is not; this is rather a matter of objective fact (1995, 160*ff.*). However, insofar as each thing requires individuation (as the particular *kind* of thing it is), it is hard to see how this gets us to a concept-free reality. Let us say we choose to divide up the world (*inter alia*) in terms of the green things and the red things; that a post box falls into the latter and not the former category, that it *is* red, is then not up to us. But that the post box can be said to fall into the category of red things at all is dependent on our agreeing that *it is a red thing;* this is also a fact, and one we frame in terms of our concepts. Of course, we don't decide what is true, but the world of facts being essentially conceptualized doesn't entail this.

Devitt goes a step further than Searle and sees categories as part of objective reality: along with (most of) the categories we generally take to be instantiated, there are categories consisting of, as it might seem to us, such arbitrary bedfellows as my left big toe, Tony Blair's smart phone and the Arc de Triomphe (*cf.* Devitt 2013, 108–9). We might have used different concepts from those we in fact do, and thereby picked out different categories, dividing up the things in the world in quite different ways; which categories we land upon depends on our explanatory goals and purposes. What we are not free to do, says Devitt, is decide whether things fall into the different categories we pick out. However, this line does not advance the cause of mind-independent facts. The non-standard categories Devitt talks about are just as dependent on our concepts as the ones we are more accustomed to, even if we have no explicit name for them, for they must be defined in relation to terms we already understand. The line also falls afoul of the previous problem that 'things' are not self-individuating, which means we need our concepts to make sense of them—as falling into whatever 'objective' category we choose to focus on.

What this little discussion shows is that to make sense of the idea of mind- or concept-*independent* facts we do—surely—need a substantial representational theory, involving something like the classical notion of reference and correspondence truth: we need to think of something independent of thought that *makes* our thoughts true, by virtue of *corresponding* to them (somehow or other) in structure and content. I conclude that metaphysical realism, hence metaphysical naturalism, is dependent upon representationalism, and that MN therefore is vulnerable to the kinds of objection raised against its notion of reference above.

3 SCIENCE AND METAPHYSICS

We have now seen why, in spite of wanting to be a scientific naturalist, I do not accept MN and the support it might thereby give to my preference for science. This discussion has proceeded independently of anything specific about the nature of science. But of course if science—*real* science—were in fact to be such as to *mandate* something like MN, then something in my constellation of commitments would have to give way; perhaps one might see it as showing that something of the argumentation in §2 had gone wrong. In this section, I argue that there is no reason to regard science as committed to or even tending towards anything like metaphysical naturalism (MN). This should give some initial succor to the idea of naturalism without metaphysics, though as we shall see the issue needs to be prosecuted a good deal further before we can return a full bill of health to my preferred conjunction.

Why might science be seen as somehow wedded to the picture of MN? A commitment on its part to physicalism would be one possible motivation. If physics is a kind of foundation for science and everything else, in the sense that all (genuine) facts, properties or entities are, at bottom, physical facts, properties, or entities, then arguably we need a framework for making sense of this idea; and MN seems well placed to provide such a framework. However, as alluded to in §2, physicalism in the strict sense is not the only view of science that would motivate MN, insofar as properties investigated by other sciences might also belong among the basic ones (though this seems more a terminological issue than a substantive one). What mandates MN in particular then is a certain (what I will call) *reductive/physicalistic* view of science: to make sense of how different disciplines asymmetrically depend on others, we need to invoke the idea that certain kinds of facts can *make true* certain statements upon which those facts would obviously bear. The facts which are the relevant truth-makers will themselves be fairly straightforwardly reflected in—represented by—the statements of fundamental science, while the other statements will somehow be analyzed in terms of these. Therefore, *if* science is wedded to reductive physicalism (henceforth 'RP'), then it is also likely to involve MN.

However, that antecendent is extremely doubtful. My reason for saying this is *not* that RP is by definition an extra-scientific thesis, one which fails to figure in physics or any other science. In one sense that is true, but it seems clearly possible that some such commitment might be *implicit* in physics or science generally in virtue of its (or their) dominant explanatory practices. It is important for my view that science be able to support claims which are not explicitly part of science, in particular, the claim of my own SN, that scientific knowledge is uniquely fundamental. Though controversial, I believe that this view is implicit in scientific practice itself. This is not to say that it is the view of most or even many scientists, though that may well be true too; rather it reflects the fact that scientific explanatory practice does not presuppose extra-scientific input nor acknowledge as equally valid distinct explanatory programs. (This view allows for philosophers to contribute to developing and resolving more abstract and *récherché* aspects of scientific theory). These ideas are not central to the arguments presented here, but it may allay the worry that the very idea of SN is somehow mixed up with metaphysics from the outset.

Returning to RP, my objection to it is substantive, but simpler. It is just that science does not in fact cleave to explanatory practices that are identifiably physicalistic or reductionistic. As thinkers such as John Dupré (1994), Noam Chomsky (*e.g.* 1995), Richard Rorty (*e.g.* 1991), and others have done much to demonstrate, science is *by default* pluralistic, seeking not to carve nature at some assumed physical or any other joints but simply to provide the best theories we can about phenomena of interest, be this the movement of stars and planets, the behavior of sub-atomic particles, the development of biological organisms, the nature of linguistic competence, or whatever else we turn our scientific gaze towards. Furthermore, the idea that we need to *vindicate* discourses at various levels by reference to lower ones simply has no echo in real science. This is not to say that science has no truck with reduction, but as Bjørn Ramberg (2004, 43) has put it in his interpretation of Rorty, this should be viewed as one among several *meta-tools* of science, not a defining characteristic of the very enterprise.

A further aspect of this picture is that science is not concerned with the fate of commonsense categories in the way some advocates of RP have supposed. Not content to subject the discourses of the various special sciences to demands of vindication by physics, thinkers like Jackson see the fate of everyday categories like free will, solidity, color, and so on as susceptible to resolution by the findings of science. This too, I would contend, is fundamentally at odds with how science itself would see the significance of its findings. Science starts out with various commonsense assumptions to identify phenomena of interest and moves on from there to ever deeper, more systematic levels of understanding. What becomes of the initial characterizations is of no particular interest to science. Are objects still rightly called solid now that we know they consist mainly of empty space? Are any things really *yellow,* since yellow is in some way an observer-relative category? From a scientific perspective these are idle questions (and thus probably idle questions *tout court,* since no one would deny that lemons are yellow). I should add immediately that this does not mean that scientists are not interested in such categories at all—color, say, from the perspective of perceptual psychology, or free will from the perspective of neuroscience. Rather, it is the metaphysical naturalist's concern for *what these categories are,* and *whether they exist,* that I think is belied by scientific practice.

These points give support to the idea that science has no particular interest in metaphysics—at least if we think of metaphysics in terms of distinctive kinds of question about reality and existence posed by modern philosophy. However, though I take this to be significant, the problem I have set myself to resolve in this piece is not quite so easily put aside. Metaphysics, like the notorious lump in the carpet, has an uncomfortable tendency to reappear after you have trodden it flat.

4 PRICE'S SUBJECT NATURALISM AND THE RE-EMERGENCE OF METAPHYSICS

I think this worry can be appreciated by considering further the position of Huw Price, who I take it would find the above anti-reductionistic conception of science congenial. Price, however, recommends, not merely my scientific naturalism (SN), but what he calls *subject* naturalism (Price 2004). Subject naturalism starts, logically at any rate, with a scientific account of language and thought. According to Price, such an enquiry suggests—for reasons canvassed in §2—that, contrary to MN, our basic semantic categories, such as truth and reference, are not best understood as substantive relations to an extra-linguistic reality of truth-making facts, but rather as categories that allow us to use our language in certain ways. Following from this, concepts such as 'cause', 'value', and 'probability' are to be understood, not, as for MN, by understanding which physical facts make true the statements involving them, but in terms of their being expressions of psychological states underlying practical goals. Subject naturalism (henceforth 'SuN') thus enunciates a kind of *pragmatist* position, combining semantic minimalism, functional pluralism about language, expressivism and a particular take on truth—though in complex ways I cannot detail here (Price 2011 collects relevant pieces). In this section, I first highlight certain central aspects of Price's SuN and explain how these can appear to be commitments of any scientific naturalist who rejects MN. I then explain how SuN lets metaphysics in through the back door, and how this again leads to incoherence.[14] In the following (and final) section of the paper, I sketch an escape route for the scientific naturalist from those apparent commitments.

An aspect of SuN very important to Price is that it is ultimately a piece of natural science—a kind of linguistics, psychology or anthropology—albeit pitched at a high level of abstraction. Of all the things we might want to understand, language and concept use, in their multifarious varieties, are certainly a central range of phenomena. According to MN, language functions in the manner outlined by the representationalist paradigm. According to Price, there are good reasons for doubting this hypothesis. But the phenomena themselves—language and linguistic practices—remain, and we need to say something about them. Price's idea is that we should theorize about language in terms of it giving expression to certain basic psychological reactions we have to the world. The same applies, importantly, to scientific language, including causal locutions. In this way SuN induces a kind of generalized or *global* expressivism. For Price, such an expressivism also involves a way of demurring at metaphysics—demurring at saying what *really* exists, what is *really* true and so on (which is not to say that nothing is ever true or false simpliciter, or that truth has no value beyond justification).

There are two things I want to say about Price's global expressivist, subject naturalist package. Firstly, I think it is a highly attractive project for one who rejects MN, one which seems to promise much explanatory understanding of how language relates to the world around us. In that sense, it is good science, or would seem to be. However, secondly, in spite of what Price claims, it seems that his position must, in the final analysis, involve metaphysical commitments if it is to avoid straightforward incoherence—though I shall further suggest that these commitments themselves are dubiously coherent. His position is meant to be *global* expressivism. Hence ultimately we must understand not just everyday talk of values and tastes, and not just scientific talk of causes, space and time, but also *the very expressivist view itself*—an integral part of the science of language—in expressivist terms. But how exactly should we understand expressivist talk *expressivistically?* What could that even mean? We seem to lose our grip on what we are saying if we go along this line. If we think we are doing serious science and not just some kind of intellectual game-playing, it thus seems we must understand the expressivist part of SuN in a non-expressivist way. In doing this, however, we let metaphysics back in. For example, if Price holds that somewhere along the line we will find a particular scientific discourse which serves as the non-expressivist framework for giving expressivist theories for all other discourses, then he owes us an account of how we are to understand this discourse. The natural candidate here is representationalism, which throws us back into the problems of MN. On the other hand, if he claims that there is no such discourse, but rather that all discourses must be understood expressivistically as responses to experience of a world whose intrinsic nature we cannot grasp, then although this avoids MN, we are still deeply mired in metaphysics, now something like the Kantian *Ding an sich*—something we are compelled to posit but of which we can form no positive conception. This is, arguably, at least as incoherent as anything MN is committed to.

Perhaps one might argue, in response to this, that there need be no fundamental level at all, or alternatively that we can simply demur at theorizing about talk about this fundamental level. The former line leads away from naturalism, and in any case makes it difficult to see how expressivism can be anything more than one among several models—without truth commitment of any kind, even a deflationary one—that we use to reflect on the relations between different discourses (*cf.* Knowles 2011, 79). The latter line might seem tempting, but it leads to asking why we ever should have felt the need to understand the function of language at any level. (I return to this point below.)

What this brief tour of SuN shows is that avoiding metaphysics by stressing anti-reductionism is not so easy as it may appear. Furthermore, the metaphysical consequences of SuN are not at all attractive. Thus it may seem that my scientific naturalism is caught between a rock and a hard place, and that it is the real culprit behind this mess.[15]

5 SCIENCE FOR NATURALISM WITHOUT METAPHYSICS

What, then, to do? What I aim to do here is modest. In particular, I do not offer a complete or (I imagine) completely convincing account, but one I hope is at least consistent and not wholly implausible.

First we need to find a way of reasonably demurring at the project of theorizing about thought and language, in the way Price's subject naturalism takes to be incumbent upon us. I think this, in and of itself, is not so difficult, again from the perspective of science. Although Price sees his SuN as a part of natural science, it is far from clear that it actually makes contact with anything in the scientific study of language as it exists today, let alone with any more general study of mind. It would not be an unreasonable stand in the science of language to follow Chomsky's repeated animadversions against semantics, in the broad sense of the study of how language relates to the world, on the grounds that no one has really got beyond a position which could be described as pre-scientific. (This would apply to some of Chomsky's more recent, pragmatist-sounding forays into the field—see note 15–as well as more traditional truth-theoretic or 'realist' semantics.)

SN still faces two serious pressures, however. On the one hand, the overall position can begin to sound a lot like Rorty's (*e.g.* 1979, 1998). For Rorty too we should stop scrutinizing the mind/language–world interface in the hope of providing an answer to our philosophical problems. The problem with co-opting Rorty for naturalism is of course that we then seem hard pressed to say why *science* has any privileged status with respect to other discourses we seem to find enriching, appealing, edifying, or whatever—*useful* to use the ultimate pragmatist category.

The other source of pressure comes from considering whether as good *scientists* we really can just demur at metaphysics. It is one thing to say that there is no prospect of a serious science of semantics, and that is important. However, it is another thing to say that there is no prospect for a serious science of the *mind*—something which is arguably belied by much ongoing work in cognitive science, as well as being in principle difficult to reconcile with the broad understanding of ourselves provided by evolutionary biology. Moreover, if this science of mind, nascent or real, acknowledges some kind of *interface* between organism, or brain, on the one hand, and world, on the other, then one might think some kind metaphysics will inevitably emerge.

In response to the first pressure, I think a case can be made for saying that scientific discourses are in a way more fundamental than non-scientific ones, even without introducing metaphysical considerations.[16] To begin to see this, we should first register that there should be no inclination to follow Rorty (1986; 1998, 19–42) in jettisoning truth as a distinctive norm in our thinking. Embracing semantic minimalism does not, as several authors have pointed out, promise a reduction of truth to justification. Rorty (1986, 334–5) is partly aware of this problem, but insists that to say some claim is justified but not necessarily true only alludes to possible future contexts of justification for that claim—it *cautions* us. However he neglects to tell us what to make of cases where a claim is neither justified *nor* unjustified, for then we can still say it is either true or false. This means we can say that we are often concerned with *finding out* what is true, rather than with any other more specific practical purpose, and that this knowledge is something we should use as the basis (or part thereof) for practical decision-making. This is an important and demarcatory idea behind science.

Can we further give some account of why discourses of politics, value, art appreciation, and the like are not *as* truth-apt, or at least as cognitively fundamental, as those of science? Well, perhaps that is not so hard, at least in broad outline. Is it likely that we will ever come to useful agreement about questions of politics, value, and morality, or even what the fundamental questions are? Do we have any real signs of theoretical progress, of the kind we have in natural science? Arguably we do not. History, to take another, slightly different example, though dealing with many previously unknown facts is hard-pushed to find theoretical foundations that can be seriously tested. Perhaps, then, we can say that though in daily language and in human or social science, discourses such as these function well enough to allow us to use the truth predicate in them, nevertheless they have no way of being extended to theoretical, deeply explanatory contexts. Hence *systematic truth,* or the search for this—theorizing that enters into a genuine dialectic with what is as yet unknown—seems to be the prerogative of science. I understand these are controversial issues, but the idea that there may be some way of giving priority to natural science over other discourses in questions of fundamental knowledge strikes me as not entirely hopeless even when one has renounced metaphysics.

The other source of pressure on my view—from the idea that cognitive science will lead to metaphysical thinking through its investigation of how mind relates to world—raises even more intricate issues and deserves a much more extensive treatment than I can offer here. But a few words will, I hope, give a shape of the kind of answer I favor. My basic line of attack derives again from stressing what science itself has to say about these matters, rather than some philosophically idealized conception of science. The first step is indeed already suggested by the pluralistic picture of science defended above (§3). Thus although science may look at many different phenomena, including those relating to what we traditionally think of us as 'the mind', it is arguable that it does not engage and has no vocabulary for engaging with the idea of how *mind relates to world,* couched at this level of generality, any more than it does for thinking of reality as a whole, in the way physicalists suggest it does.

One might think that modern biology inevitably forces such a general question upon us: We are evolved creatures that arose out of the primordial ooze, shaped by its ever-changing influences upon us over the eons. Our cognitive faculties must then at some level be seen as adaptations to this world, ones that in some cases more, in others less, reflect the way things are in physical reality. But does modern biology really mandate this picture—mandate seeing ourselves and our thoughts, as well as those of animals, as set over against and representing a wholly mind-independent world, more or less adequately? It strikes me that this picture is in fact scientifically suspect. What an organism experiences, however basically, is *always already* a world *for it,* filled in and specified in relation to categories that only make sense in relation to *its* particular subjectivity and needs. There is never a cognitive meeting between 'things-in-themselves'—that is, the things described by physics—and a similarly 'transcendental' organism.

These thoughts relate to recent alternatives to the traditional 'receptivity' view of perception and perceptual consciousness. Inspired by the phenomenological work of Merleau-Ponty (1962), these challenge the idea of perception as the deriving of an internal representation of outer physical things through the processing of impoverished retinal images, seeing it instead as an embodied and world-involving activity (*cf.* Noë 2004, O'Regan 2011). To the extent perception has objects, these need to be viewed as what Gibson (1979) called *affordances:* entities whose very existence only makes sense in relation to the needs and actions of the perceiving being and thus are directly perceptible. On this kind of view, known as *enactivism,* information in the environment has to be understood in relation to the standing abilities of organisms, and its pick-up in relation to their activities across time. Perception and action are thus at best nominally distinct, while affordances are theoretical posits that serve to illuminate the various different organism-environment interactions—ultimately, in a way that one can hope will mesh with phenomenological inquiry into the basic structures of consciousness (see Varela *et al* 1991; Thompson 2007).

Much more needs to be said than I can say here about this kind of view and its relation to empirical work, and how it relates to the idea of representation, both in perception and in thought. But lest one is tempted to see a lapse here into a kind of idealism or anti-realism about the external world (which no naturalist can accept), remember that the overarching point is that there is no perspective, at least in science, which allows us to see ourselves as cognitively pitched over and against a purely physical world—in terms of which affordances might then be somehow 'fitted in' to a world of more basic posits. There is therefore no case for saying that the theoretical entities of biology, here affordances, are less 'real' than those of physics, and one will only think otherwise if still in the grip of metaphysical naturalism.[17] In sum, I think one can seriously question whether biological science, and hence cognitive science, courts the (in)famous 'sideways-on' view of reality that tends to feed metaphysics.

A possibly more troubling aspect of my package is the role it assigns to language—on which subject I now conclude with some (even more!) speculative and compressed remarks. Though perceptual consciousness might be amenable to a systems-theoretical biological account, it is harder to see how language—with which we theorize everything, include such systems, and language itself—can be. For an anti-representationalist like myself, language does not of course 'describe' or 'capture' experience or reality; but it is nevertheless a capacity we humans have, that on which all genuine, *i.e.* truth-evaluable, thought depends. Is this to remain a wholly unexplained explainer—or else something, metaphysical naturalism having been cast aside, that will necessitate some kind of neo-Hegelian, non-naturalistic account of assertion and content, perhaps along the lines sketched in

Brandom (1994)? I think the answer here is unclear, but at least it is not as clear as many might have thought that naturalism is ruled out by such considerations. The game of 'giving and asking for reasons', for example, might simply be viewed as one of the natural things our particular species goes in for, as Price (2010) avers. Alternatively one might try to use the Chomskyan idea of thought as tied to (internal) linguistic *structure* and this in turn as somehow the upshot of a particular structure or functioning of the mind/brain (a view most fully developed by McGilvray 1999). Insofar as this account has scientific credentials and prescinds from talk of *semantic* content, is not clear to me that any element of metaphysics, anti-naturalism, or paradox is involved it.[18]

6 CONCLUSION

In this paper I have tried to illuminate some of the intricacies involved in defending a non-metaphysical but non-anodyne form of naturalism. The discussion has ranged from a critique of metaphysical naturalism and Price's alternative subject naturalism to discussions about the proper scientific understanding of perceptual consciousness. Though one might reasonably wonder about the cogency of the idea that science without metaphysical backing can still be said to have some kind of privileged status over other discourses, I hope to have shown that the position is at least consistent, given a proper understanding of the goals of science, and of biological and cognitive science. Much more must be said about everything I have touched upon, but sometimes it is worth taking a step back to see how broader areas fit together—for if they do not, we can hope to save ourselves a lot of unnecessary work.[19]

NOTES

1. Price is certainly a neo-pragmatist; the question concerns his classification as a liberal naturalist, which he seems to leave (perhaps deliberately) open. Section 4 has some material relevant to this issue.
2. This includes, as I indicated in the previous note, Price, as well as, I think Quine. Quine, the arch naturalist, does clearly in a sense also want to bypass metaphysics insofar as he avows a thoroughgoing pragmatism, but at the same time he seems pulled towards it as witnessed by his stipulation of more or less *a priori* physicalistic constraints on adequate scientific explanations, as well as various forays into issues of ontology. Nevertheless, I do see Quine as one of the chief precursors to the idea of anti-metaphysical naturalism. The idea has also been investigated in recent literature on the Carnap-Quine debate, *inter alia* by Price (*cf. e.g.* his 2007).
3. One might actually question whether MN is properly naturalistic, insofar as it might seem to be wedded aprioristically to a certain conception of the nature of reality (*viz. physicalism*). For further discussion of this point, see below and §3.

4. I will talk in terms of facts for simplicity's sake. There are no doubt many questions in detail about how one should construe MN's 'reality', but I believe they do not impact on my discussion, in which the important idea is that this reality is *mind-independent.*
5. I should stress that MN is consistent as well with all sorts of local realisms and anti-realisms about discourses other than the (basic) scientific one(s) (anti-realism about the ethical, say).
6. Of course the most famous attack of the kind I am concerned with is Putnam's, who, taking his lead from Quine, argues that there can be no unique reference relation to an independently construed reality, and that nothing within the naturalist's picture, such as causal relations, can assuage this since talk of these will be subject to the same crippling indeterminacy (*cf.* Putnam 1981). Though yet another discussion of Putnam's much-discussed argument would no doubt be of great interest, on this occasion I note merely that its cogency and significance are still hotly contested. Some of Price's arguments below are reminiscent of Putnam's.
7. Price talks of 'object naturalism' rather than 'metaphysical naturalism', but any differences are insignificant as far as I can see.
8. Assuming MN depends for its idea of mind-independence on representationalism —which, I remind, will be defended below—it would not seem plausible for it to adopt anything short of a full-blooded (naturalistic) realism about representational facts.
9. Ramsey (1929) argued that the content of a physical theory is captured in a sentence, later dubbed its 'Ramsey sentence', obtained by taking an axiomatization of a theory and replacing all the theoretical terms with observation variables and existentially quantifying over those variables. A Ramsey sentence of a theory is thus a statement in higher-order logic stating that the theory has a model consistent with a fixed interpretation of its observation terms. Ramsey sentences treat theoretical terms as disguised definite descriptions. The original theory and its Ramsey sentence both imply the same observational sentences; hence a theory's Ramsey sentence captures its factual content (*sub specie* a positivist account of physical theories).—*Ed.*
10. Such a general argument, says Price, has as a premise something to the effect that 'all referents are physical' (*cf.* Price & Menzies 2009, 11). I will not elaborate upon or evaluate this aspect of Price's view here.
11. I am indebted to Jussi Haukioja for the following point. For further discussion, in a distinct though related context, see Haukioja & Cohnitz (2013).
12. Some might feel inclined to deny there can be any kind of empirical question about reference as a substantial, empirically scrutable relation if it is not a natural kind. I take this as standing in need of substantiation, and hence that the arguments of Price above are still germane. Even if we accepted it, it would seem to weaken rather than embolden MN.
13. For Searle, realism is the view that there *is* a reality independent of all human representations; he is otherwise happy with talks of facts, and with truth as correspondence to these (albeit this is further commitment, beyond realism). Devitt (2013, 103) characterizes his realism in these terms: "Tokens of most commonsense, and scientific, physical types exist independently of the mental." Devitt (2013, 105–7) also holds a correspondence theory of truth, but not in the traditional sense as one which sees truth as a relation between claims and facts. It may then be that Devitt does not want to countenance an ontology of *facts* in a certain sense of the word. However, I do not see this as significant to our debate. Thus he (2013, 106) writes that "[t]he realism dispute arises from the age-old metaphysical question 'What *ultimately* is there and *what is it like?*'" He also thinks there are mind-independent categories, as we

shall see. Thus he clearly holds that there are objects and categories of certain objective, thought-independent kinds, which implies an individuative power in the world, as it were. This is enough for MN's (and MR's) idea of objective facts.

14. The outline presented below summarizes ideas in Price (2004) and (2011, 3–33). A fuller exposition of the critique is given in Knowles (2011).
15. I should add as a conclusion to this section that I don't think it is necessary to pursue precisely *Price's* (allegedly) anti-metaphysical project to end up in kind of the bind he does. Chomsky (1995) has, for example, recommended a view of word meaning which seems to amount to pretty much the same kind of pragmatist position as Price's, broadly speaking, and thus also to face similarly awkward metaphysical quandaries.
16. The following two paragraphs summarize a discussion in Knowles (forthcoming). Note they take up a further stage of the dialectic between scientific and non-scientific (or liberal) naturalism than that alluded to in the introduction (Knowles 2008, 2010), for the latter did not presuppose any particular take on the possibility of a metaphysical grounding of science (something now firmly rejected).
17. One might wonder how we can think about the entities of physics on this view if we are not cognitively related to them. Here we need to remember the underlying anti-representationalism of the view I am developing, which does not see theoretical claims as made true in virtue of truth-makers. Physical theories are thus in a way mere posits, though no less real than anything else for that. For further discussion see Knowles (2013, §4).
18. I am assuming that this core Chomskyan 'internalist' conception of language is not beholden to the further pragmatist claims made by Chomsky himself; *cf*. footnote 15.
19. I thank Jussi Haukioja for discussion of the issues on reference in §2. Thanks also, for their comments and questions, to the audience at the conference on realism in Helsinki, June 2011, at which an earlier version of this paper was presented. Special thanks go the editor of the volume, Ken Westphal, for suggestions of stylistic and other kinds which have greatly improved the final product.

REFERENCES

Boghossian, P., "The Rule-following Considerations." *Mind* 98.392: 507–549.

Brandom, R., 1994. *Making it Explicit*. Cambridge, Mass., Harvard University Press.

Chomsky, N., 1995. "Language and Nature." *Mind* 104.413: 1–61.

Devitt, M.,1991. *Realism and Truth*, 2nd ed. Princeton, Princeton University Press.

———, 2010. *Putting Metaphysics First*. Oxford, Oxford University Press.

———, 2013. "Hilary and Me: Tracking down Putnam on the Realism Issue." In: M. Baghramian, ed., *Reading Putnam* (Oxon, Routledge), 101–120.

Dupré, J., 1994. *The Disorder of Things*. Cambridge, Mass., Harvard University Press.

Haukioja, J. and D. Cohnitz, 2013. "Meta-externalism *versus* Meta-internalism in the Theory of Reference." *Australasian Journal of Philosophy* 91.3: 475–500.

Jackson, F., 1998. *From Metaphysics to Ethics*. Oxford, Oxford University Press.

Knowles, J., 2008. "What is Naturalism? Towards a Univocal Theory." *SATS: Nordic Journal of Philosophy*, 9.2: 28–57.

———, 2010. "Two Kinds of Non-scientific Naturalism." In: M. Milkowski and K. Talmont-Kaminski, eds., *Beyond Description: Naturalism and Normativity* (London, College Publications), 157–175.

———, 2013. "Challenges to Cartesian Materialism: Understanding Naturalism and the Mind-World Relation." In: M. Milkowski and K. Talmont-Kaminski, eds., *Regarding Mind, Naturally* (Cambridge, Cambridge Scholars Publishing), 182–202.
———, forthcoming. "Rortian Realism." *Nordic Studies in Pragmatism.*
McGilvray, J., 1999. *Chomsky: Language, Mind, and Politics.* Oxford, Wiley-Blackwell.
Noë, A., *Action in Perception.* Cambridge, Mass., MIT Press.
O'Regan, K., 2011. *Why Red Doesn't Sound like a Bell: Understanding the Feel of Consciousness.* Oxford, Oxford University Press.
Price, H., 2004. "Naturalism without Representationalism." In: M. De Caro and D. Macarthur, eds., *Naturalism in Question* (Cambridge, Mass.: Harvard University Press), 71–88.
———, 2007. "Quining Naturalism." *Journal of Philosophy,* 104.8: 375–405.
———, 2010. "One Cheer for Representationalism." In: R. Auxier and L. Hahn, eds., *The Philosophy of Richard Rorty* (Library of Living Philosophers 32; Chicago, Open Court), 269–289.
———, 2011. *Naturalism without Mirrors.* Oxford, Oxford University Press.
———, and Menzies, P., 2009. "Is Semantics in the Plan?" In: D. Braddon-Mitchell and R. Nola, eds., *Naturalism and analysis* (Cambridge, Mass., MIT Press), 183–200.
Putnam, H., 1981. *Reason, Truth and History.* Cambridge, Cambridge University Press.
Ramberg, B., 2004. "Naturalizing Idealizations." *Contemporary Pragmatism* 1.2: 1–63.
Ramsey, Frank P., 1929. "Theories." Rpt. in: *idem.*, R. B. Braithwaite, ed., *Foundations of Mathematics and Other Logical Essays* (Paterson, NJ: Littlefield and Adams, 1960), 212–236.
Rorty, R., 1979. *Philosophy and the Mirror of Nature.* Cambridge: Cambridge University press.
———, 1986. "Pragmatism, Davidson and Truth." In: E. Lepore, ed., *Truth and Interpretation* (Oxford, Blackwell), 333–355.
———, 1991. *Objectivity, Relativism and Truth.* Cambridge, Cambridge University press.
———, 1998. *Truth and Progress.* Cambridge, Cambridge University press.
Searle, J., 1995. *The Construction of Social Reality.* London, Penguin.
Stich, S., 1996. "Deconstructing the Mind." In: *idem.*, *Deconstructing the Mind* (New York, Oxford University Press), 3–89.
Thompson, E., 2007. *Mind in Life.* Cambridge, Mass., Harvard University Press.
Varela, F., E. Thompson and E. Rosch, 1991. *The Embodied Mind: Cognitive Science and Human Experience.* Cambridge, Mass., MIT Press.

Part III

Pragmatism and Realism

11 Majesty of Truth and the Moral Sentiment

Emerson's and Peirce's Ethico-Ontological Realism

Heikki A. Kovalainen and Douglas R. Anderson

1 INTRODUCTION

Since the chief aim of the present essay is to explain how Emerson and Peirce pertain to ethico-ontological realism, we begin with a few words on what kind of ethics is hereby meant. Peirce's philosophy, including both his classification of the normative sciences as well as his writings on religion, suggests not only that logic be based on ethics and aesthetics, but also that religious—in particular, Christian[1]—ethics will fill the spot reserved for ethics. In this chapter, however, we will not examine Peirce's discussion of Christianity as such; rather, Christian ethics will be understood as an attitudinal sensibility and orientation *exemplifying* Peirce's ideal of ethically guided inquiry. It ought to be added that since many of Peirce's best-known formulations pertaining to his realism, and in particular, his definition of the term 'real', stem from his writings on Christianity; a careful student of Peircean realism ought not to ignore these texts.

Even when attention has been paid, in the secondary literature, to Peirce's remarks on religion, they have customarily been taken as subordinate to his astute discussions of pragmaticism and the methodology of the sciences.[2] To the contrary, we agree with Michael Raposa (1989, 4) that "Peirce's philosophy as a whole seems to have been shaped and informed by certain religious beliefs and ideas. . . ." In particular, as we have argued in detail elsewhere,[3] Peirce's very conception of the parallelism between Christian ethics of charity and the scientific spirit first makes possible genuine non-egoistic inquiry into the world. As commentators such as Pfeifer (1981), Orange (1984), Anderson (1995), and Short (2000) have proposed, the spirit of self-sacrifice attributed by Peirce to Christianity goes to the very heart of how he understands scientific inquiry. In several of his published essays, especially in "What is Christian Faith?" (1893) and "A Neglected Argument for the Reality of God" (1908), Peirce more than once underscored how a religious attitude is not only something *supportive* of scientific inquiry, but even its precondition. In the latter essay, for example, he maintained that good reasoners should inquire "in scientific singleness of heart," trading on a mediaeval Christian phrase to bring science and a religious attitude into proximity (*EP* 2:436).

Insofar as the affinity between Emerson and Peirce will be a theme in the present chapter, let us next cite the key pieces of evidence lending initial credibility to the comparison. To begin with the negative or somewhat neutral evidence, Peirce knew but did not always unreservedly admire Emerson's work. Unlike James and Dewey, who were both intimately and documentedly influenced by Emerson's essays, Peirce's historical encounter with Emerson was more indirect. Although he and his family were, as Peirce puts it, "within hearing to the Transcendentalists,"[4] Peirce's published works show only little explicit philosophical reference to Emerson's ideas. His pre-college manuscripts, however, demonstrate him to have read Emerson by that time (Smyth 1997, 52), and in a much-quoted autobiographical note in the "Law of Mind," he grudgingly admitted to having caught something of the Transcendentalist virus (*CP* 6:102). One important way in which Peirce makes use of Emerson in his corpus is his repeated references to Emerson's poem, "The Sphinx": "I am thy spirit, yoke-fellow, / Of thine eye I am eyebeam . . . Thou art the unanswered question; / Couldst see thy proper eye, / Always it asketh, asketh; / And each answer is a lie" (*CW* 9:24).[5]

Even more positive evidence may be found; permit us to cite a few facts that have gone unnoticed in the secondary literature. Emerson and Peirce, to begin with, were both lecturing at Harvard in 1869 and 1870: Peirce delivered his lectures "On British Logicians" in 1869–1870, while Emerson—anticipating Peirce's well-known title *Law of Mind*—gave a lecture titled "Metre of Mind—Law of Mind" in 1870.[6] In the lectures given at Harvard under the title "The Natural History of the Intellect" (1870–71), Emerson (*CW* 12:17) worded what Peirce (*CP* 6:158) virtually reiterated some thirty years later, that "at last Matter is dead Mind."[7] Following Kruse (2010, 387), it must be added that the Emersonian echoes figuring in Peirce's work are ultimately no coincidence; there are indeed texts in which Peirce *identified himself precisely with New England Transcendentalism.* In a rejoinder to Paul Carus's criticism of the pragmatist's work, in an article published in *The Monist,* Peirce maintained: "I carefully recorded my opposition to all philosophies which deny the reality of the Absolute," citing his original article by saying that "the one intelligible theory of the universe is that of objective idealism, that matter is effete mind" (*CP* 6:605). In an unpublished draft of the same paper, finally, Peirce went as far as to state that "I frankly pigeon-hole myself as a modified Schellingian, or New England transcendentalist" (MS 958; 1892–93).

These confessions from Peirce, of course, raise a number of questions that ought to be addressed briefly. He appears to run together, firstly, the Emersonian and the Schellingian influences in his work. This is natural inasmuch as Emerson, too, drew important elements from Schelling, whose work the essayist knew and studied at an early stage of his career.[8] Despite these self-conceived comparisons, it goes without saying that neither Emerson's nor Schelling's standards of scientific accuracy were up to the pitch of Peirce's own. To cite one important difference between Emerson and Peirce, there is an obvious discrepancy in their respective styles of philosophizing, which also explains some of their substantial disagreements. Peirce had, as is well known, extensive experience both in mathematics and the empirical sciences, and he occasionally insisted that philosophy ought to be dull and dry. Emerson, in turn, conducted his studies in theology and worked three years as a Unitarian minister before resigning the post and setting out on the unlikely path of a public American intellectual. However, such stereotypes have only limited applicability. To content ourselves here with just two pieces of contrary evidence, Emerson was also immersed in the contemporaneous sciences of his day, and Peirce showed remarkable religious sensibilities, not to mention his high philosophical sympathies for the arts and artists.

Thirdly, it is somewhat striking that Peirce claimed, alongside his references to Emerson and Schelling, to be defending a version of *objective idealism.* Here one must, however, exercise high caution with the terms, since Peircean 'objective idealism' means nothing like what, say, Hilary Putnam would tag 'metaphysical idealism'. In contrast, Peirce's objective idealism virtually aligns with his *scholastic realism;* they are both doctrines according to which ideas or *universals* do not exist in an other-worldly realm but as objectively real 'generals'. In the secondary literature, to be sure, Peirce's realism has customarily been discussed in conjunction with his theory of scientific inquiry, and consequently, *scientific realism* has often been associated with Peircean thought. Here we prefer to talk about Peirce's *realism* instead of idealism, if only because the earlier term causes less confusion in the context of contemporary philosophy. One need not be a Peirce scholar, indeed, to be acquainted with the basic tenets of his best-known article, "The Fixation of Belief," where Peirce's survey of different methods of arriving at beliefs lands ultimately in the "new conception . . . that of reality" (*EP* 1:120). To overcome skeptical doubts, he noted, "it is necessary that a method should be found by which our beliefs may be determined by nothing human, but by some external permanency—by something upon which our thinking has no effect" (*ibid.*). But Peirce's "external permanency" is not something fixed once and for all; to the contrary, truth, reality, and *even the scientific method itself* will not be fixed until inquiry itself is completed. Since this, however, is mostly an ideal goal, Peirce's realism implies a strong version of revisionism. And since the Peircean concept of *reality,* as the ideal limit of inquiry, is essentially non-human, this raises the question whether Peirce's realism is allied with his understanding of religion, in particular, his conception of God.

Contrary to the narrow-minded readings of Peirce's realism, then, his position can be shown to carry important religious overtones as well. The second important idea emerging from Peirce's religious texts, indeed, is what might be termed *Christian realism,* according to which God is *real,* and as such a *real general* may be taken to align with the ideal aim of scientific inquiry. Here a terminological clarification will be useful, since Peirce says precisely that God *is real* rather than God *exists,* which latter would, on his view, connote reacting with the other like things in the environment (*CP* 6:495). He records the vernacularity and vagueness of the term 'God', but finds that it is precisely in such a vagueness that the term is well understood (*CP* 6:494). 'The real', in turn, is defined by Peirce as "that which holds its characters on such a tenure that it makes not the slightest difference what any man or men may have *thought* them to be . . .; but the real thing's characters will remain absolutely untouched" (*CP* 6:495).[9] While reality is thus defined in virtue of its being independent of any particular person's thought, this is not to say that Peirce's sense of reality would imply ultimate mind-independence: "whatever is meant by any term as 'the real' is cognizable in some degree," he states in "Some Consequences of Four Incapacities," and "so is of the nature of a cognition, in the *objective* sense of that term" (*CP* 5:310–311; emphasis added). Peirce's *reality,* then, is never absolutely unknowable, yet is thought-independent, *viz.* persisting regardless of whether it is being thought or not.

It is precisely in this two-fold nature of Peircean realism that he might be read, as we will argue, as sharing key features with the moral realism of R. W. Emerson. We will be suggesting, in particular, that both Emerson and Peirce exemplify what might be termed *ethico-ontological realism,* comprising two central claims: (1) the natural universe (or, if one prefers, the *empirical world*) is what is ultimately real, yet its precise nature cannot be definitely determined by science as it presently stands, but only *ideally* by future science. (2) Scientific research consists in ethically guided self-controlled inquiry, which means that science, both in theory and in practice, has not only logical but also ethical (and even aesthetic) underpinnings. We begin the discussion with an elaboration of Peirce's notion of inquiry, helping to set his realism in context. Thereafter we introduce the reader to Emersonian realism vis-à-vis some of his later work and the essay "Nominalist and Realist." The Peircean stripe of realism will then be juxtaposed with the Emersonian notion of the *moral sentiment.* Best sense of this complex notion, so it will be argued, may be made with the help of comparative reflections on Peirce's conception of *reality.*

2 PRACTICE OF INQUIRY AND THE NORMATIVE SCIENCES

By 1903, as is well known, Peirce had developed his account of the normative sciences in which he held logic to rest, in its theoretical orientation, on ethics and aesthetics. As a mode of self-controlled conduct, logic aims at "logical goodness, or *truth*" (*EP* 2:204). A few years later he went a step further when he hypothesized "Reason" to be an ultimate aesthetic ideal and then pointed out what this would practically entail for us:

> The one thing whose admirableness is not due to an ulterior Reason is Reason itself comprehended in all its fullness, so far as we can

> comprehend it. Under this conception, the ideal of conduct will be to execute our little function in the operation of the creation by giving a hand toward rendering the world more reasonable whenever, as the slang is, it is "up to us" to do so. In logic, it will be observed that knowledge is reasonableness; and the ideal of reasoning will be to follow such methods as must develop knowledge the most speedily. (*EP* 2:255)

When Peirce turned to the *practice* of inquiry—and particularly to the practices of abduction which grow out of perception—he argued that certain ways of being were beneficial to the inquirer. To begin with, he maintained that overall an inquirer needed to establish an other-directed, non-egoistic attitude in order to conduct inquiry effectively. In the "Further Consequences of Four Incapacities," under the heading "The Social Theory of Logic," he went as far as to state that "logic . . . rigidly requires, before all else, that no determinate fact, nothing which can happen to a man's self, should be of more consequence to him than everything else," continuing in a quasi-Christian vein, that he "who would not sacrifice his own soul to save the whole world, is illogical in his inferences, collectively" (*CP* 5:354).[10] These easily misunderstood remarks imply, on a modest reading, no more than that the ideal goal of inquiry is impersonal and that the inquirer must thus for the sake of truth forego personal goals. "No man," Peirce claimed, "can be logical who reckons his personal well-being as a matter of overwhelming moment" (MS L, WJ, 3–13–87).

As a corollary, Peirce argued that in observation or perception as well as in the development of hypotheses, an inquirer needs to develop an effectively receptive attitude. Attempts to dominate our perceptions and ideas would, for Peirce, lead us astray: "It is not by dealing out cold justice to the circle of my ideas that I can make them grow, but by cherishing and tending them as I would the flowers of my garden" (*CP* 6:289). Such a receptive attitude is crucial to science precisely because, for Peirce, "all truth is ascertained by observation" (*W* 4:246) and all inquiry both begins and ends in perception. Moreover, Peirce believed, as did Emerson, that this receptive attitude is not easily achieved. "Attentive experience," he argued, is achieved "usually with some, often with great, effort" (*CP* 2:605). One reason why it is difficult to attain is because elements of experience provided to us through receptivity are constantly present to us, and thus we are not customarily awakened to their reality (*CP* 5:42).

Thus Peirce suggested that scientists look to artists and poets to see how perception and observation are done best. On this view, the artists, compared to scientific men, are "much finer and more accurate observers than they are, except of the special minutiae that the scientific man is looking for" (*CP* 1:315). Peirce developed this idea at length in his 1903 Harvard lectures on pragmatism, and especially in his lecture on phenomenology. He underscored, above all, "that rare faculty, *the faculty of seeing what stares one in the face,* just as it presents itself, unreplaced by any interpretation . . ."

(*EP* 2:147; emphasis added). This Emersonian dimension of his work has generally been downplayed, but he took it as central to his entire discussion of the process of abduction and inquiry. To achieve a poetic, receptive mood which "approaches the state in which the present appears as it is present," scientists will need to learn the kind of humility manifest in the observation of artists. "The act of observation," Peirce claimed, "is the deliberate yielding of ourselves to that *force majeure*—an early surrender at discretion, due to our foreseeing that we must, whatever we do, be borne down by that power at last" (*CP* 2:581).[11] In short, for Peirce, good science requires that a "craving to know" must overbalance "all prejudice, all vanity, and all passion" (*CP* 1:34).

As we see, then, Peirce believed that an inquirer who sought short-term goals rather than the ideal goal of truth would sooner or later run into difficulties. Specifically, inquiry driven by self-interest and self-importance—an inquiry of egotism—would override any attempts to remain self-aversive. It would instigate a forgetfulness of our fallibility. One tends to forget, indeed, how Peirce's epoch-making observation that *the method of science is itself a scientific result* finds wording in his texts on religion:

> That which constitutes science, then, is not so much correct conclusions, as it is a correct method. But the method of science is itself a scientific result. It did not spring out of the brain of a beginner: it was a historic attainment and a scientific achievement. So that not even this method ought to be regarded as essential to the beginnings of science. That which is essential, however, is the scientific spirit, which is determined not to rest satisfied with existing opinions, but to press on to the real truth of nature. (*CP* 6:428)

Here we must consider one important facet of what Peirce has to say about the *scientific spirit:* it presses on the real truth of nature, in other words, nature "talks back" and thought is representational and not merely constructive in the way Richard Rorty has repeatedly suggested.[12] Peirce, working in his Neo-Platonic way, dismissed modernist suggestions that we can completely control our ideas and learn through such intense control. Both modernists and post-modernists, from a Peircean perspective, overestimate the importance of human minds—modernists seem to believe we can dominate nature through knowing with certainty and post-moderns, including neo-pragmatists, believe we make the 'world' through our vocabularies. For both Emerson and Peirce, the story is more modest. They agree that the human organism, in being a product of nature, has a natural ability to know nature. Without what Peirce routinely called an attunement between our ideas and God's, science could not get started. "I infer," he argued, "in the first place that man divines something of the secret principles of the universe because his mind has developed as a part of the universe and under the influence of these same secret principles . . ." (*CP* 7:46). Again, in his clarificatory remarks following "The Neglected Argument for the Reality of God," he essentially suggested that the conduct of scientific research, while not thinking any thoughts of God, can still catch "a fragment of His thoughts, as it were" (*CP* 6:502). The trick is to allow this attunement—Galileo's light of nature—room to operate, and this we cannot do by dictating either to nature or to our ideas what their relationship should be. The modesty appears in the fragmentariness of our knowing: we know things, but we do so without certainty, finality, and fixity.

Kruse has already adeptly developed the relationship between Emerson's notion of imagination and Peirce's conception of musement in the process of abduction. She notes that musement in the development of hypotheses involves the free play of ideas in the imagination. In a letter to William James on Christmas 1909, Peirce noted how he finds his best work in philosophy to be in Logic, summarizing this as his "proof that the warrant of Hypothesis is mainly that unless the human intellect is of the kind of the Creative Spirit, it is useless to attempt any discovery of positive truth" (MS L, WJ, 12-25-09). What we would add to Kruse's assessment is that at this level too there is an element of Emersonian receptivity. Abduction—hypothesis development—is not merely creative, though it can be that; it is also the allowing of ideas to find their natural place in accounting for things. For instance, Peirce pointed to Poe's "Murders of the Rue Morgue" in which the detective allows ideas to reveal themselves as "smoothly fitting keys" (*EP* 2:437). This permissiveness with ideas, as Carl Hausman (1974) has noted, involves a cherishing love for ideas. Achieving such an attitude towards one's ideas is, again, no easy task. Indeed, the attitude is better achieved when one acknowledges, as Peirce does, that we do not own ideas—they do not occur as features of singular brains. Ideas have lives and histories of their own and are, for Peirce, public realities.

3 EMERSONIAN REALISM: ROOTS AND MISUNDERSTANDINGS

To prepare for a proper comparison between Emerson and Peirce, permit us next to explain how Emerson might be understood as an ontological realist. In the secondary literature, to be sure, he has more often than not been read as a metaphysical idealist. Thus his various accents on the primacy of the empirical world have been customarily downplayed, while those quasi-Platonic overtones appearing to deny the reality of the external world have been accentuated.[13] We believe, however, such an interpretation to be severely one-sided. Consider, for instance, a key essay from his later chef d'oeuvre *The Conduct of Life* (1860), "Fate," one of his most pervasively Proto-Peircean pieces. "Certain ideas are in the air. We are all impressionable, for we are made of them; all impressionable, but some more than others, and these first express them" (*CW* 6:24). Ideas do not exist in an other-worldly realm, but *incarnate* themselves "in the healthiest and strongest," and thus the great man is "most imbued with the spirit of the time," in other words, "the impressionable man" (*CW* 6:7, 24). Anticipating the Peircean dictum of *us being in thoughts rather than they in us,* Emerson noted repeatedly how the unity of laws operating in nature is "not in us so much as we are in it" (*CW* 6:14).[14] More than once, he appealed to the contemporaneity of scientific discoveries as evidence for the general ideas prevailing in the surrounding reality. "The truth is in the air, and the most impressionable brain will announce it first, but all will announce it a few minutes later" (*CW* 6:24).

In general terms, 'fate' is a master term for the later Emerson by which he voiced precisely the extramental character of worldly events. For Peirce, scientific inquiry was sooner or later *fated* to broach the ideal limit of external reality, while Emerson's later philosophy, in particular "Fate," spoke about the ways in which fate holds sway over the conduct of our lives. Emerson like Peirce, found that the laws of nature operate in the empirical world and cannot be denied. To be precise, while the determination of events could be—and indeed, ought to be—thought, thought has no power to alter the course of natural events, *unless* the events being thought of originate in the human will. Further examples of Emerson's realism may be drawn from his later lectures, where 'fate' again stands for an Emersonian emblem of an extramental reality:

> A perception came into my mind, by no chance, by no illusion: it is there by a fate, the same or much older than that in geology or gravitation. . . . The fundamental fact is the correspondence of the human being to the world,—his correspondence and impressionability,—so that every change in that writes a record on the mind. The mind yields sympathetically to the tendencies or laws which stream through things, and make the order of nature . . . What we require in a good mind is sensibility to truth. (*LL* 2:75.)

This text stems from a lecture entitled "Powers of the Mind," delivered by Emerson originally in Boston in 1858 and used later as material for his lectures on the *Natural History of the Intellect* at Harvard University (1870–71). These lectures come close to Peirce not only historically but also philosophically, insofar as they insistently stress, first, the generalizing faculty of the human mind; second, the reciprocal attunement between mind and nature; and third, the importance of attitudinal obedience to the study of nature. In a lecture, "Natural Religion," Emerson called for "a certain realism, to judge not after appearances, for things are the reverse of that which they appear," in a word, an "attitude of obedience" (*LL* 2:191). Like Peircean science, Emerson's later philosophy is constantly *on the way* towards larger generalizations, the ultimate nature of which, however, is presently unknowable: "Every generalization shows the way to a larger; and every reform is only a cover, under which a larger reform, that dares not yet name itself, advances" (*LL* 2:92). In "Powers of the Mind," one finds a text where Emerson used the Peircean metaphor of *cherishing general ideas,* or more precisely, our *sensibility* to them, inasmuch as it is "not propositions that are of the first need, but to watch and tenderly cherish the intellectual and moral sensibilities,—those fountains of right thoughts, and woo them to stay and make their home with us" (*LL* 2:69). Emerson's realism is thus concerned with loyalty to truth and rightness: tending our intellectual and moral sensibilities helps us think rightly and truthfully.

In Emerson's earlier work (up to *Essays: Second Series,* 1844), one essay above others exemplifying his commitment to ontological realism is "Nominalist and Realist." The aptly titled piece concludes *Essays: Second Series,* and in particular, the book's overarching theme of Being *versus* Appearance, though Emerson's figurative way of treating his subject might blind the inadvertent reader to the depth of its theme.[15] Highly pertinent to a central philosophical problem yet woefully understudied, the essay's importance is heightened, in the American context of philosophy, by the problem of nominalism and realism having occupied not only Emerson but also Peirce and James. Historically, the nominalism-realism debate dates back to ancient Greek philosophy, and it was famously taken up again and discussed in great detail in the Middle Ages. "In the famous dispute with the Nominalists, the Realists had a good deal of reason. General ideas are essences. They are our gods: they round and ennoble the most partial and sordid way of living" (*CW* 3:136). In the context of contemporary philosophy, nominalism and realism are ambiguous and complicated terms, and in Emerson's time the significations of these terms were less technical than they are today. To the classical definitions of 'nominalism' and 'realism',[16] Emerson added his own twist by finding the thrust of nominalism in the view that we may know only parts of reality, never reality in its entirety, in its oneness, which in turn would be the realist view. To mention one difference with regard to the standard interpretation of the terms, universals or general ideas, for Emerson, have important practical relevance insofar as they guide and dictate the conduct of our lives. Thus Emerson appears to move beyond a scholastic-technical discussion of the problem.

A somewhat unexpected analogy between Emerson and Peirce opens up here: In an early review article of Fraser's edition of *The Works of George Berkeley,* Peirce (1871, 38) noted at the end that "though the question of realism and nominalism has its roots in the technicalities of logic, its branches reach out about our life." As an example he used the consideration whether the genus *homo* "has any existence except as individuals," noting this to be—perhaps with a degree of youthful zeal—"the most fundamental practical question in regard to every public institution the constitution of which we have it in our power to influence" (*ibid.*). We note in passing that the representativeness of persons is also addressed in Emerson's "Nominalist and Realist," and in any event, underscoring the practical relevance of the dichotomy heightens its Emersonian echoes.

As the essay "Nominalist and Realist" shows, Emerson, like Peirce, was repeatedly occupied by the alternate manifestations of the one and the many, a perspicuous overview of reality and its versatile appearances. In his essay on Plato in the *Representative Men,* Emerson stressed "the two cardinal facts" of philosophy: *Unity* and *Variety.* On the one end, we "unite all things by perceiving the law which pervades them; by perceiving the superficial differences and the profound resemblances." On the other end, however, "every mental act,—this very perception of identity or oneness, recognizes the difference of things" (*CW* 4:27–28). This is analogous to how Peirce, in his 1903 lecture on phenomenology—which for him, consisted in delineating the categories operating in all experience—drew attention to three faculties that one ought to observe particularly carefully, in order to bring about phenomenological perceptiveness. The first one we have already touched upon above: the *rare faculty of seeing.* The second faculty was characterized by Peirce as *resolute discrimination,* fastening itself upon "the particular feature" being studied and recognizing its concrete instances amidst various appearances. The third faculty was the *generalizing power* comprehending "the very essence of the feature" independent of all contingent appearances. This Peircean distinction, not to mention the first-named faculty of seeing *what stares one in the face,* is virtually identical to Emerson's Platonic insistence on the interplay of Unity and Variety in the surrounding world.

In the secondary literature on American pragmatism, Emerson has been more frequently compared with James and Dewey, though Peirce is of most help in making sense of Emerson's realism. This is because Peirce's version of realism, more successfully than James's, gives an extensive elaboration of the ways in which the quasi-Kantian dichotomy between our experience and the reality allegedly unknowable to us underlying those experiences can be broken down.[17] Peirce stressed the ways in which there are no truths in the world categorically unknowable to us: we may understand laws and general principles operating in the world for the very reason that the world itself is humanly understandable; this aspect of Peircean philosophy is anticipated by certain realist tendencies in Emerson's philosophy. As he puts the point in his first book, *Nature:* "Parts of speech are metaphors, because the whole of nature is a metaphor of the human mind" (*CW* 1:21). Whereas James (1907, 37) did admit, in his famous slogan, that "the trail of the human serpent is over everything," he did not go as far as Emerson and Peirce in spelling out the full metaphysical implications of such a correspondence between human mind and the world. Here it behooves us to remind ourselves of the fact that in a later work titled "The Natural History of the Intellect," Emerson (*CW* 12:17) went as far as to say what Peirce (*CP* 6:158) virtually reiterated some thirty years later, that "at last Matter is dead Mind."

The intriguing thing about Emerson's discussion of nominalism and realism is that he showed high sympathies for both, arriving at a kind of synthetical solution to the problem, according to which both positions are in their own ways defensible. On the one side, human inquiry is characterized by the incessant striving for universal truths, "our native love of reality," longing to transcend the particularity of our experiences (*CW* 3:134). "Let us go for universals; for the magnetism, not for the needles" (*CW* 3:135). But our yearning for universality is never fully satisfied, and we must thus content ourselves with observing the parts accurately. "If we cannot make voluntary and conscious steps in the admirable science of universals, let us see the parts wisely, and infer the genius of nature from the best particulars with a becoming charity" (*CW* 3:143). Note that Emerson is not claiming that we could make no steps in the science of universals whatsoever, but only that we cannot take such steps *willfully:* in other words, we need self-aversive obedience to impersonal truth, just as Peirce insisted. It follows from Emerson's pragmatic revisionism that since we can never attain an all-embracing view of the universe, we must grant the imperfectness of our knowledge—or in the Emersonian hyperbole, "giv[e] ourselves the lie" (*CW* 3:144). In other words, all alleged truths are only approximations of truths, to be superseded by better theories. As Peirce wrote to James regarding Royce and the Hegelians: "It is simply absurd to attempt, as they do, to attain absolute and perfect truth" (MS, L, WJ, 8–1905). This is what Peirce (*CP* 1:310) appears to mean, too, with his sympathetic allusions to Emerson's "The Sphinx": "Always it asketh, asketh; / And each answer is a lie" (*CW* 9:24).

4 MORAL SENTIMENT: ATTENTIVENESS AND OBEDIENCE

With these preliminaries in place, let us now turn to Emerson's peculiar stripe of ethico-ontological realism. With Emerson's revisionism, to begin with, is associated a sharp critique of excessive individualism: "So much as we can shove aside our egotism, our prejudice and will, and bring the omniscience of reason upon the subject before us, so perfect is the work" (*CW* 7:24). Anti-individualism, indeed, unveils itself as one of the chief tenets of his later essays, such as in the incisive criticisms of "Culture": "The man runs round a ring formed by his own talent, falls into an admiration of it, and loses relation to the world" (*CW* 6:70). With phenomenological echoes, Emerson stressed how the ideal student of philosophy "must have a catholicity, a power to see with a free and disengaged look every object," in other words, to see objects "for their own sake, and without affection or self-reference" (*CW* 6:70–1). Thus Emersonian realism is shown to have ethical underpinnings inasmuch as the inquirer must work hard to resist excessive egotism. This is similar to how Peirce admired artistically sensible scientists for their "craving to know," overbalancing "all prejudice, all vanity, and all passion" (*CP* 1:34).

In his later philosophy, Emerson identified his moral realism increasingly with what he terms the *moral sentiment* (*e.g., CW* 6:115). This is a complicated term denoting primarily (though not exclusively) the human ability to see and judge and make ethical decisions in the particular situations in life. Historically, the term derives from the Scottish common sense philosophers, above others Thomas Reid and Dugald Stewart, whom Emerson read and was influenced by since his student days at Harvard. For these philosophers, the key to ethics is the so-called moral sentiment or the *moral sense,* an inherent ability underlying our ethical actions of intuiting the right course of action in each particular case. Reid and Stewart, in turn, were inspired by the original theory of "moral sentiment" by Francis Hutcheson, propagated, among other philosophers, via Adam Smith.[18] It should be added, in passing, that Peirce was also influenced by Dugald Stewart, and Levi Hedge taught Thomas Brown's—a Scottish metaphysician, and student of Stewart's—work at Harvard in early to mid-nineteenth century.

In the context of Emerson's realism, however, his understanding of the *moral sentiment* was most proximally indebted to his Romantic—or more specifically, Coleridgean—appropriation of Kant's terminology of *Reason* and *Understanding.*[19] Emerson's distinction was loosely based on the two central operative faculties of Kant's transcendental idealism, the faculty of understanding (*Verstand*), whose domain is confined to the operations of the transcendental forms of intuition and categories, and that of reason (*Vernunft*), allowing the postulation of transcendent ideas.[20] The radical difference between Kant and Coleridge, however, lies in the latter taking Reason not so much as a faculty of enabling transcendent postulates, *per* Kant, but as the direct capability of intuiting truths that for Kant transcend the scope of intuition, to the transcendent realm above sense, incognizable through our Understanding. This is where Emerson's ontological realism and his realism about universals begin to merge.

For Coleridge (1825, 215–16), understanding meant "faculty of forming judgments through the senses." This accords, roughly speaking, with Kant's usage in the *Critique of Pure Reason* (B137). However, in contrast to Kant, reason consisted in inner and immediate perception or intuition of truth.[21] "Reason is the Power of universal and necessary Convictions, the Source and Substance of Truths above sense, and having their evidence in themselves" (Coleridge 1825, 216). That Emerson had in mind such an idiosyncratic interpretation of the two terms is evidenced, among other texts, by a letter from 1834 to his brother Edward, whom Emerson asked whether he was acquainted with the mentioned distinction, attributed in the letter to Milton, Coleridge, and the Germans. He went on to define Reason as

> the highest faculty of the soul—what we mean by the soul itself: it never *reasons,* never proves, it simply perceives; it is vision . . . revelations of Reason [include] the love of the beautiful & of Goodness as the highest beauty the belief in the absolute & universal superiority of the Right & the True. (Emerson 1939, 1:412–13)

It must be observed that Emerson here voices not only his idiosyncratic appropriation of Kant, but also—as may be seen from the referenced trinity of goodness, beauty, and truth—his interpretation of Plato. We allow ourselves to add the observation that such a notion of Reason is analogous to how Peirce regards Reason as the ultimate aesthetic ideal, manifested both through scientific inquiry and our ethical pursuits of increasing reasonableness in the world. Moreover, Peirce argued that we can perceive ideals as general ideas and that these become the bases for the conduct of our lives.

Emerson's moral sentiment bears relevance to different stripes of realism, then, because its significance extends from ontology to ethics. In his essays early and late, indeed, there are essentially two ways for Emerson to speak about the moral sentiment: the *ontological,* having primarily to do with the ways in which the moral sentiment is a non-human force operating in the world, and the *ethical,* having rather to do with the ways we ourselves may exercise it. While these two senses are intertwined, it will be useful to say a few words about them separately. In the ethical sense, first, *moral sentiment* refers to the human ability of obeying moral perceptions in each particular situation. In a concise formulation from the later essay "The Preacher," moral sentiment "is a commandment at every moment and in every condition of life to do the duty of that moment and to abstain from doing the wrong" (*CW* 10:225). This quote gives expression to both the lawlikeness as well as the particularity of the moral sentiment; it is a *commandment at every moment* . . . to do the *duty of that moment.* An appropriate illustration of what this means in concrete ethical situations may be found in a later journal entry, where appealing to the moral sentiment is contrasted with fictional thought experiments often employed in ethics:

> There are cases frequently put by the curious, moral problems like those problems in arithmetic which in long winter evening the rustics try the hardness of their heads in ciphering out. And it is said: What will you do when wife & children are assaulted, are murdered in your sight? Will your non-resistance go to this extent? . . . We are not careful to answer in this matter. A wise man will never impawn his future being & action & decide beforehand what he shall do in a given extreme crisis. Nature & God will instruct him in that hour. (*JMN* 12:219–20)

The similarities with Peirce's thought here are striking. Peirce routinely laid religious belief and reason—both guides to conduct—at the door of sentiment, instinct, and common sense. Like Emerson, he was influenced by the Scottish tradition of common sensism. In his review of Henry James Sr.'s *Substance and Shadow* Peirce (W 2:436) argued, "we do not in the least oppose the Scotch philosophy which makes all knowledge finally to repose on what are sometimes called ultimate beliefs; because these beliefs are the common sense of mankind which belongs to all men and which no man can resist." After the turn of the century Peirce systematically developed his "critical common sensism" which maintained this same relation of conduct to sentiment and common sense, but which also allowed for cultures to modify common sense beliefs as evolution and history revealed their successes and failures.

In the ontological sense, secondly, *moral sentiment* alludes to non-human lawfulness manifesting itself in the world. Emerson wrote in the later essay "Fate": "If thought makes free, so does the moral sentiment. The mixtures of spiritual chemistry refuse to be analyzed. Yet we can see that with the perception of truth is joined the desire that it shall prevail" (*CW* 6:15). This text may be used to supplement Emersonian *moral realism:* once we unveil a truth, we instantaneously will that it *ought* to prevail. It is the *ontological reality* of the moral sentiment, indeed, that leaves us no alternative but to obey its force: "the pure sympathy with universal ends is an infinite force, and cannot be bribed or bent. Whoever has had experience of the moral sentiment cannot choose but believe in unlimited power. Each pulse from that heart is an oath from the Most High" (*CW* 6:15–16). Observe what follows the quotation: "I know not what the word sublime means, if it be not the intimations in this infant of a terrific force" (*CW* 6:16). Here Emerson's understanding of the moral sentiment as an ontologically real force places him at a crossroads of Plato and Peirce.[22]

To make Emerson's *moral sentiment* intelligible in a contemporary context, permit us a comparative interlude on the so-called "Wittgensteinian moral philosophers," in particular, Cora Diamond and Iris Murdoch. These philosophers have a great deal to say about the importance of carefully perceiving and fine-tuning ourselves to the particular ethical situations confronting us, both as philosophers and as moral subjects. One central idea surfacing particularly in the work of Diamond is that of "attention" and "attending"; as Diamond (1991) argues, what matters in moral philosophy is not so much arguments, but our ability to attend to the joys and sufferings of other people. To heighten such attentiveness, we ought to engage a wide variety of different resources, including philosophy, art, music, and literature. As Iris Murdoch (1970, 41, 33, 82) has argued, moreover, ethical attentiveness goes crucially together with obedience: in the footsteps of Plato, her greatest philosophical teacher, she calls for "obedience to reality as an exercise of love," "a just and loving gaze directed upon an individual reality," in a word, *unselfing*. Emerson may be taken to share all the three different ideas here briefly outlined: those of attentiveness and obedience, the corollary caution with arguments, and the broadening of the scope of philosophy to encompass religion and art. Moreover, his emphasis on the ontological character of the moral sentiment brings him close to Murdoch's Platonic moral realism, and his criticism of individualism aligns with Murdoch's ideal of *unselfing*.

We are finally in a position to orchestrate an explicit juxtaposition of Emerson's and Peirce's ethico-ontological realism. What we suggest is that Emerson's understanding of the *moral sentiment* is analogous to Peirce's conception of *reality*. To persuade the reader of the textual evidence lending initial credibility to such a comparison, we may begin by citing Emerson's first book *Nature*, where we bring to our particular experiences "the innate universal laws," which "exist in the mind as ideas, yet stand around in nature forever embodied" (*CW* 3:113). "[T]he mind is a part of the nature of things; the world is a divine dream, from which we may presently awake to the glories and certainties of day" (*CW* 1:37). Emerson anticipates the first pragmatist in holding that a non-human mind pervades the fabric of the universe, knowable by virtue of science recognizing this very congruity between human beings and the world. "The very being of law, general truth, reason," in Peirce's (*CP* 8:136) words, "consists in its expressing itself in a cosmos and in intellects which reflect it, and in doing so progressively; and that which makes progressive creation worth doing—so the researcher comes to feel—is precisely the reason, the law, the general truth for the sake of which it takes place." Or, to cite another formulation: "The man of science has received a deep impression of the majesty of truth, as that to which, sooner or later, every knee must bow" (*ibid.*).[23] Now set these texts alongside the previously cited Emersonian (*CW* 6:15–16) reflection on the moral sentiment: "the pure sympathy with universal ends is an infinite force, and cannot be bribed or bent. Whoever has had experience of the moral sentiment cannot choose but believe in unlimited power. Each pulse from that heart is an oath from the Most High."

Indeed, one might go as far as to suggest that Emerson's *moral sentiment* is ultimately synonymous not only with Peirce's *reality* but also with Peirce's—and Emerson's—*God*. Their ways of expressing themselves are different, yet Emerson's *God* and Peirce's *reality* share the following three features: they are both ultimately mind-independent powers operating in the world, yet our minds cannot but obey them, once we attend to them submissively enough, and thus our knowledge approaches and ideally aims at their real being. According to both Emerson and Peirce, a non-human reality inevitably affects and even normatively guides the course of our actions, regardless of whether we momentarily grasp this reality. In other words, the very being of mind-independent reality consists in its expressing itself, and that is the basis on which we make—or are constantly *in the process of* making—impersonal truth the aim of our inquiry. In Emerson's (*EL* 2:346) words, the moral sentiment *is the most impersonal of all things*, hence, the ultimate power beyond the self, yet something with which our self can gain contact.[24]

5 CRITICAL SUMMARY REFLECTIONS

Emerson's and Peirce's discussions of realism as well as their original takes on the nominalism-realism dispute are multifaceted, and hence no comprehensive presentation can be attempted in a single essay. We have defended in this chapter a realist reading of both philosophers: though they occasionally entertain certain idealist notions, this is usually done in an attempt to problematize the dichotomy rather than to argue for metaphysical idealism as such. Both Emerson and Peirce, in the end, believe that the empirical world is what is ultimately real, though we—the imperfect inquirers—might never be able to fully fathom the world's character. Secondly, whereas Peirce's realism has been traditionally treated mostly in the contexts of epistemology and ontology, we have shown how his realism carries important ethical and even aesthetic implications, as can be seen from his very classification of the normative sciences. One simple sense in which Peircean realism is ethical lies in the reciprocal relationship between realism and moral perfectionism[25]: in order for our views to broach nonhuman reality, we must constantly be ready to perfect them, and hence ourselves also. Moreover, as we have shown, Peirce's take on realism is *not* independent of his discussion of religion, in particular, Christian ethics.

To draw our conclusions pertaining to the different stripes of realism operating in their work, for both Emerson and Peirce, to begin with, ontological realism (in contrast with ontological *idealism*) can be reconciled with realism about universals. Peirce, in his own vocabulary, considers himself a *scholastic realist* inasmuch as he understands universals or *generals* as *real, viz.* really operating in the natural world. Even such a real general as God—indeed, perhaps the ultimate real general—is eventually rooted in experience. "Where would such an idea, say as that of God, come from, if not from direct experience?" Peirce (*CP* 6:493) asks, and answers without hesitation: "as to God, open your eyes—and your heart, which is also a perceptive organ—and you see him." Here 'direct experience' must be understood broadly; Peirce's so-called Neglected Argument for the Reality of God, indeed, sets forth from observing order and meta-order in the empirical world, and arriving thence, via free play of the imagination, at the hypothesis of God's reality. Emerson's position on the nominalism-realism issue, in turn, can be summarized with help from ancient philosophy to suggest that he believed in the existence of Platonic "forms" *only in an Aristotelian sense:* Universals are always instantiated in the concrete, empirical particulars, and do not thus exist in an other-worldly realm. In this respect, he would agree with Iris Murdoch's Platonically inspired moral realism: The fountain of ethics is a loving gaze of knowledge directed at an individual reality, facing us each moment of our lives.

While Peirce's discussion of realism involves speculative as well as empirical aspects, Emerson was somewhat less interested in discussing technical aspects of the nominalism-realism debate, and he was historically less informed about the problem than Peirce. Instead, he was driven by the desire to find examples, from the realm of practical morality, of the ways in which universals manifest themselves in our lives. This creates a difference of emphasis in Emersonian and Peircean realism: whereas the ethical roots of Emerson's realism are mostly 'existential' and thus centered on interpersonal relationships, Peirce's realism is more emphatic in underscoring the role of community. Peirce went as far, indeed, as to suggest that the very conception of reality "involves the notion of a community, without definite limits, and capable of a definite increase of knowledge" (*CP* 5:311). To be sure, *some* of Emerson's essays, such as "Nominalist and Realist," do stress the sociality of knowledge, but they do not venture as far as Peirce's philosophy in explicating the importance of a scientific community.

Emerson's philosophy, in sum, is useful for fleshing out some of the concrete senses in which American pragmatism, including Peircean realism, has ethical repercussions. Emerson's version of moral realism, encapsulated in his original appropriation of *moral sentiment,* shows how the distinction between moral and ontological realism becomes ultimately blurred. Emerson hearkened back to Plato in thinking that we can, through perfect obedience, come in touch with the Platonic idea of the good, which the essayist termed *moral sentiment.* It is precisely the ontological reality of the moral sentiment that gives our moral actions their ethical power. Attentiveness to particulars is important, ethically speaking, yet there are moments in our lives—glimpses of inevitability—that are based on an experience of lawfulness transcending the particularity of experience. To take an everyday example, suppose one commits an unexpected and unconventional action, say, resigning from well-paid work because one feels one's calling to be elsewhere. In such a situation, it will not suffice to say that one is acting on the spur of the moment; one must feel, practically speaking, to be doing what is *right,* lest one's head might be turned by criticism.

Whereas Peirce usually avoided formulating his version of realism with explicit ethical overtones, the Emersonian comparison makes it evident how Peirce's very notion of real generals operating in the world has strong ethical implications. Perhaps this tends to be overlooked because Peirce often touched upon this dimension in the context of religion. The reality of God, to cite an example, is not to be understood "as a proposition of metaphysical theology, but in a form directly applicable to the conduct of life, and full of nutrition for man's highest growth" (*CP* 6:457). For Peirce, moreover, Ideals are what he often called "living powers." And for him they are constantly drawing attentive inquirers toward them though there is never a final interpreting moment. As he put it, "there never will be a necessarily ultimate opinion as a psychological phenomenon, but identical logically and metaphysically" (MS L, WJ 7–23–05). Throughout this essay, we have sought to explicate the ways in which Peirce's realism is grounded in ethics. *Without* constantly striving to make the object of our inquiries nonhuman and impersonal, we not only distance ourselves from a genuinely ethical life but also from genuine knowledge of reality. Thus, for Peirce, "science goes unswervingly its own gait. What is to be its goal is precisely what it must not seek to determine for itself, but let itself be guided by nature's strong hand." (*CP* 6:434). Since our inquiry will never be complete, our ideal—epistemic and ethical—will be endless perfectibility.[26]

NOTES

1. We use 'Christianity' interchangeably with 'religion' because of Christianity's focus on love, although Peirce occasionally remarked that Buddhism better exemplified the force of love in practice. Peirce's view on *aesthetics* will not be separately discussed in this essay, since in his classification of the normative sciences ethics is given more weight, and in a sense, aesthetics (*qua* inquiry as to which state of things reasonably recommends itself) is incorporated into ethics.
2. For example, the editors of the sixth volume of his *Collected Papers*, Charles Hartsthorne and Paul Weiss, remark that "The second book of the volume, devoted to religion or 'psychical metaphysics', has rather tenuous connections with the rest of the system, offering, apart from scattered flashes of insight, views which have a sociological or biographical, rather than a fundamental systematic interest" (*CP* 6:*v*). These claims are misleading and false in two important ways. First, Peirce's discussion of religion, and particularly of Christianity, is antecedent to and logically independent of his discussion of what he terms *psychics*, an umbrella term bringing together such pseudo-sciences as telepathy or spirituality. Second, it is not correct to say that Peirce's discussion of religion and particularly Christian ethics of charity have only "rather tenuous connections" with the rest of his system.
3. This part of the argument is developed in another essay by the present authors, tentatively titled "Peirce's Scientific Method of Inquiry and the Christian Ethics of Anti-Egotism," to be submitted for publication in 2013.
4. Quoted after Kruse (2010, 398).
5. Peirce quoted the poem in *CP* 1:310, 2:302; 3:404, 7:425. Among these, the first quotes eight lines from the poem; the others only quoted the line "Of thine eye I am eyebeam."
6. For this lecture by Emerson, however, only notes remain; see Bosco (1997).
7. This textual similarity is neglected by Kruse (2010), whose otherwise accurate analysis traces Peirce's *effete mind* passage back to Schelling.
8. Emerson's acquaintance with Fichte was limited, while with Schelling's writings he was somewhat familiar; see Wellek (1943), 49–53.
9. Peirce also noted how the term 'real' was originally used in legal contexts, as in 'real estate', before Albertus Magnus and Duns Scotus imported it into philosophy proper in the thirteenth century (*CP* 6:495). Peirce summed up the medieval meaning thus: "'Real' is a word invented in the thirteenth century to signify having Properties, *i.e.* characters sufficing to identify their subject, and possessing these whether they be anywise attributed to it by any single man or group of men, or not" (*CP* 6:435).
10. David Hume had asserted the contrary: "It is not contrary to reason to prefer the destruction of the whole world to the scratching of my finger" (*Treatise*, 2.3.3.6). When Peirce was accused by Paul Carus of being a modern Hume, he vehemently resisted.
11. Elsewhere Peirce made similar claims: in "the power of knowing one's precise sensations unaffected by any interpretation of them, the scientific men of most branches are decidedly inferior to the artist" (*CP* 7:607).
12. See, for example, *Contingency, Irony, and Solidarity* (1989, 6).
13. Here an excellent example is provided by the sixth chapter of Emerson's first book *Nature*, titled "Idealism." Whereas Emerson in this chapter explicitly advocated a version of ontological idealism, it is of foremost importance to realize that he was at the same time diagnosing cultural factors giving rise to the doctrine in the first place. On a meta-level, then, it might be suggested that he prioritized factors *in the empirical reality* that induce us to believe in the idealistic hypothesis in the first place; see Kovalainen (2010), 143–162.

14. In Peirce's wording: "Only one must not take a nominalistic view of Thought as if it were something that a man had in his consciousness. . . . Thought is more without us than within. It is we that are in it, rather than it in any of us" (Ms L, WJ, 11–25–02).
15. In reading "Nominalist and Realist" in conjunction with "Experience," where the former's "perspectival metaphysics" is taken to complement the "epistemology of moods" adumbrated in the latter, we are indebted to Goodman (1998); also see Goodman (2008).
16. As might be expected, Peirce discussed in greater detail than Emerson the classical varieties of nominalism and realism, but that aspect of Peircean thought is not our concern here. According to the traditional interpretation of the problem, 'nominalism' refers to the view that universals are mere names and do not refer to anything really existing, and consequently all there is are particulars. We are not discussing here the second variety of nominalism, according to which there are no abstract objects. 'Realism', in turn, affirms the reality of the universals, although there are different possibilities as to how the universals are taken to be manifested (*cf.* Rodriguez-Pereyra 2010).
17. James (1902, 384), to begin with, shared the Emersonian emphasis on the particularity of our experience: "We must frankly recognize the fact that we live in partial systems." While Jamesian radical empiricism strove to depict human experience in all its concrete richness, it perhaps fell short of making sense of how our radical experiences, even allowing for the inclusion of versatile relations and connections between them, would hook onto something other than just those experiences—and James's notion of "pure experience" is a notorious example of him perhaps somewhat arbitrarily transgressing the boundaries of his own pragmatistic radical empiricism, in an attempt to make sense of the primordial unity underlying all our experiences.
18. For a good albeit somewhat outdated general discussion, consult Hirst (1917). In addition, Emerson's use of the moral sentiment is obviously related to its Christian underpinnings, particularly to the Quaker idea of God within each one of us, the divine light of conscience we may attend to in our hearts. For an excellent article on Emerson's debts to the Quakers, consult Tolles (1939).
19. Among the various mediators of Kant read by Emerson near the coming of age of his own philosophy, Samuel Taylor Coleridge was the most influential, and his quasi-mystical reading of Kant provided great spiritual encouragement for the religiously inclined Emerson. On Emerson and Coleridge, see Lentricchia (1973), Liebman (1973), Bégout (2002), and Keane (2005).
20. In order to grasp this distinction in the full sense, it is necessary to study all the three *Critiques* of Kant, in particular, *Critique of Pure Reason* and *Critique of Practical Reason.* For discussions of Reason and Understanding in the secondary literature on Emerson and Kant, see in particular Wellek (1943), Davis (1944), Keane (2005), and Flynn (2009).
21. See Coleridge (1825, 223–224, 233–234); Beer (1993, *lxxix–lxxxiv*).
22. One quasi-paradox about the moral sentiment interpreted as obedience to lawfulness is that a law would customarily be understood as something holding above and beyond particular instances or exceptions of it, while the moral sentiment seems, in its ethical sense, to refer to a certain moral perceptivity, as opposed to rigid lawlikeness. Perhaps Emerson is thinking that what is lawful in all the particular moral situations is that we attend to the same principle of attending to the particularities of the situation in each case.
23. Or in a better-known formulation explicitly talking about *reality:* "The real, then, is that which, sooner or later, information and reasoning would finally result in, and which is therefore independent of the vagaries of me and you" (*CP* 5:311).

24. For erudite readings of Emerson's early essays, with decisive accents on his notion of the *impersonal,* consult Cameron (1998) and Lekan (2007).
25. Stanley Cavell (1990) has famously drawn attention to the moral perfectionist element running through Emerson's work. He has nothing to say, however, about the perfectionist elements figuring in Peirce's work.
26. This essay is based, in addition to the Helsinki Realism conference, partly on papers delivered in conferences organized by the Nordic Pragmatism Network in Copenhagen in August 2011 and by a project on the Possibility of Metaphysics (led by Leila Haaparanta) at the University of Tampere in August 2010. Of the various persons to have kindly commented on or assisted with these papers, we would like to thank, in particular, Michael Devitt, David Pfeifer, John Kaag, Bjørn Ramberg, Ulf Zackarisasson, Erkki Kilpinen, and Sami Pihlström.

REFERENCES

Primary

Emerson, R. W., 1960–1982. W. H. Gilman, *et. al.,* eds., *The Journals and Miscellaneous Notebooks of Ralph Waldo Emerson,* 16 vols. Cambridge, Mass., The Belknap Press of Harvard University Press; cited as '*JMN*', by volume:page numbers.

———, 1971–2003. R. E. Spiller, *et. al.,* eds., *The Collected Works of Ralph Waldo Emerson,* 6 vols. (as of 2004). Cambridge, Mass., The Belknap Press of Harvard University Press.

CW 1	(1836–44)	*Nature, Addresses, and Lectures.*
CW 3	(1844)	*Essays: Second Series.*
CW 4	(1850)	*Representative Men: Seven Lectures.*
CW 6	(1860)	*The Conduct of Life.*

———, 1903–1904. E. W. Emerson, ed., *The Complete Works of Ralph Waldo Emerson,* Centenary Edition, 12 vols. Boston and New York, Houghton Mifflin.

CW 7	(1870)	*Society and Solitude.*
CW 9	(1841/46–47/63)	*Poems.*
CW 10	(1848–1884)	*Lectures and Biographical Sketches.*
CW 12	(1840–1893)	*Natural History of Intellect and Other Papers.*

———, 1939. R. L. Rusk, ed. *The Letters of Ralph Waldo Emerson,* 6 vols. New York, Columbia University Press; rpt. with corrections, 1966.

———, 2001. R. E. Bosco and J. Myerson, eds., *The Later Lectures of Ralph Waldo Emerson,* 2 vols. Athens, GA, University of Georgia Press; cited as '*LL*', by volume:page numbers.

Peirce, C. S., 1931–1958. P. Hartshorne, P. Weiss, and A. Burks, eds., *The Collected papers of Charles Sanders Peirce,* 8 vols. Cambridge, MA: Harvard University Press; cited as '*CP*' by volume:paragraph numbers.

———, 1963–1966. *The Charles S. Peirce Papers.* The Houghton Library. Cambridge, Mass., Harvard University Library Microreproduction Service; cited as 'MS' by number and page number; letters cited as 'L', with date of correspondence.

———, 1992, 1998. N. Houser and C. Kloesel, eds., *The Essential Peirce,* 2 vols. Bloomington and Indianapolis, Indiana University Press; cited as '*EP*' by volume: page numbers.

———, 2000–. N. Houser, *et al.*, eds., *Writings of Charles S. Peirce*. Bloomington and Indianapolis, Indiana University Press; cited as 'W' by volume:page numbers.

Secondary

Anderson, D. R., 1990. "Three Appeals in Peirce's Neglected Argument." *Transactions of the Charles S. Peirce Society* 26.3: 349–362.

———, 1995. "Peirce's God of Theory and Practice." *Revista Portuguesa de Filosofia* 51: 167–178.

Beer, J., 1993. "Editor's introduction." In: J. Beer, J., ed., *Aids to reflection* (Princeton, Princeton University Press), *xli–cxlix*.

Bégout, B., 2002. "Le transcendant et le transcendental: une lecture croisée de Coleridge et d'Emerson." *Revue Française d'Études Américaines* 91.1: 61–79.

Bosco, Ronald A., 1997. *'His Lectures were Poetry, his Teaching the Music of the Spheres': Annie Adams Fields and Francis Greenwood Peabody on Emerson's 'Natural History of the Intellect' University Lectures at Harvard in 1870. Harvard Library Bulletin*, n.s., 8.2; http://pds.lib.harvard.edu/pds/view/2573358?n=20025&s=6&printThumbnails=no

Cavell, S., 1990. *Conditions Handsome and Unhandsome: The Constitution of Emersonian Perfectionism, The Carus lectures 1988*. Chicago and London, University of Chicago Press.

Coleridge, S. T., 1825. *Aids to reflection*. Princeton: Princeton University Press, 1993.

Davis, M. R., 1944. "Emerson's 'Reason' and the Scottish Philosophers." *The New England Quarterly* 17.2: 209–228.

Dewey, J., 1934. *A Common Faith*. In: J. A. Boydston, ed., 1969–1991, *The Later works of John Dewey, 1925–1953*, vol 9: *Essays, Reviews, Miscellany, and A Common Faith*. Carbondale and Edwardsville, Southern Illinois University Press.

Diamond, C., 1991. "Anything but Argument?" In: *idem.*, *The Realistic Spirit: Wittgenstein, Philosophy, and the Mind* (Cambridge, Mass., MIT Press), 291–308).

Flynn, E. E., 2009. "Intellectual Intuition in Emerson and the Early German Romantics." *Philosophical Forum* 40.3: 367–389.

Goodman, R. B., 1998. "Ralph Waldo Emerson, 1803–82." In: E. Craig, ed., *Routledge Encyclopedia of Philosophy* (London, Routledge), 3:269–272.

———, 2010. "Paths of Coherence Through Emerson's Philosophy: The Case of 'Nominalist and Realist'." In: B. Arsi and C. Wolfe, eds., *The Other Emerson* (Minneapolis, University of Minnesota Press), 41–58.

Hausmann, C. R., 1974. "Eros and Agape in Creative Evolution: A Peircean Insight." *Process Studies* 4.1: 11–25.

Hirst, E. W., 1917. "Moral Sense, Moral Reason, and Moral Sentiment." *Mind* 26.102: 146–161.

James, W., 1902. *The Varieties of Religious Experience: A Study in Human Nature*. In: F. H. Burkhardt, F. Bowers and I. K. Skrupskelis, eds., *The Works of William James* (Cambridge and London: Harvard University Press, 1975–1985), vol. 13.

Kant, Immanuel, 1871, 1876. *Kritik der reinen Vernunft*. In: *Kants gesammelte Schriften* (Berlin: Walter de Grutyer, 1902–), vols 3, 4; cited as '*KrV*', by 'A' and 'B' editions.

Keane, P. J., 2005. *Emerson, Romanticism, and Intuitive Reason: The Transatlantic 'Light of All Day'*. Columbia, MO, University of Missouri Press.

Kruse, F. E., 2010. "Peirce, God, and the 'Transcendentalist Virus'." *Transactions of the Charles S. Peirce Society* 46.3: 386–400.

Lentricchia, F., 1973. "Coleridge and Emerson: Prophets of Silence, Prophets of Language." *The Journal of Aesthetics and Art Criticism* 32.1: 37–46.

Liebman, S. W., 1973. "Emerson's discovery of the English Romantics, 1818–1836." In: L. N. Neufeldt, ed., *Ralph Waldo Emerson: New Appraisals. A Symposium* (Hartford, Transcendental Books), 36–45.

Orange, D., 1984. *Peirce's Conception of God: A Developmental Study.* Bloomington, Indiana University Press.
Pfeifer, D., 1981. "Charles Peirce's Contribution to Religious Thought." In: K. L. Ketner, *et al.*, eds., *Proceedings of the C.S. Peirce Bicentennial International Congress* (Graduate Studies 23; Lubbock, Texas Tech University Press), 367–373.
Raposa, M., 1989. *Peirce's Philosophy of Religion.* Bloomington, Indiana University Press.
Rodriguez-Pereyra, G., 2011. "Nominalism in Metaphysics." In: E. N. Zalta, ed., *The Stanford Encyclopedia of Philosophy;* http://plato.stanford.edu/entries/nominalism-metaphysics/, accessed: 08.08.12.
Rorty, R., 1989. *Contingency, Irony, and Solidarity.* Cambridge, Cambridge University Press.
Short, T. L., 2000. "Peirce on the Aim of Inquiry: Another Reading of 'Fixation'." *Transactions of the Charles S. Peirce Society* 36.1: 1–23.
Smyth, R. A., 1997. *Reading Peirce Reading.* Lanham, MD, Rowman & Littlefield.
Thompson, F. T., 1926. "Emerson's Indebtedness to Coleridge." *Studies in Philology* 23.1: 55–76.
Tolles, F. B., 1939. "Emerson and Quakerism." Rpt. in: E. H. Cady and L. J. Budd, eds., *On Emerson* (Durham, N.C.: Duke University Press, 1988), 19–42.
Wellek, R., 1943. "Emerson and German Philosophy." *The New England Quarterly* 16.1: 41–62.

12 Concepts and the Real in C. I. Lewis' Epistemology

Lauri Järvilehto

1 INTRODUCTION

The Harvard pragmatist Clarence Irving Lewis did not write many explicitly metaphysical papers. In fact, he shunned metaphysical theses in the manner of a philosopher drawing heavily from both the traditions of pragmatism and of analytic philosophy. He did not, however, subscribe to the positivistic tenet of eschewing metaphysical discourse all together. Indeed, an intriguing position on the reality of what is experienced can be drawn from his epistemology.

In the center of Lewis' epistemology are two ideas: what is given in experience independently of the experiencing subject, and the concepts the experiencing subject uses to differentiate and categorize that given. In an interesting twist, Lewis argued that what is *real* in fact depends on how experience is interpreted. Dreams may be given a solid reality in one conceptual scheme, whereas they may be thought of as ephemeral illusions in another. At first glance, this position seems to put Lewis in an awkwardly relativistic position: the real is whatever anyone happens to think is the real. Lewis' take on reality is, however, more intricate. He does not subscribe to a complete conceptual relativism, but instead seeks to maintain a very robust realism at the core of his thinking. Confusion arises, however, from an ambiguity in Lewis' use of the term 'real'. Once this ambiguity is rectified, Lewis' position gives rise to a thoroughly pragmatic realism where experience can be interpreted one way or another, but is still rooted to a metaphysical reality of how things stand with respect to our interpretation of them.

2 CONCEPTS IN LEWIS' EPISTEMOLOGY

According to Lewis, the given element in experience is what we must accept, come what may. The given element in experience is what is independent of the activity of thought (Lewis 1929, 37). It is the brute fact of experience that we cannot affect by our conceptualizing activity (Lewis 1952, 326). It is what is left once we strip out of an experience everything we ourselves bring to it (Lewis 1946, 182–183). Concepts, in turn, guide our attention to what is given in experience. As Rosenthal (1976, 21) notes, our "conceptual schemes do not limit or determine the given, but they determine our attention to the given, as well as the attitude we take toward that to which we do attend."

A concept allows us to differentiate and classify experience into distinct categories. Only by having concepts can we make sense of the "buzzing blooming confusion" (Lewis 1926, 250) that is given in experience. In *Mind and the World Order* (1929; hereafter: '*MWO*') Lewis states: "Experience does not categorize itself" (14). In order to make sense of what is given to us by our senses, we need first to anticipate some distinctive regularities by which we may direct our attention so as to tell certain experiences apart from some others, and to classify certain experiences as akin to some others. Lewis (1926, 245) further notes: "until we have certain definite concepts or meanings in mind, we cannot even approach the problem of acquiring knowledge of any sorts of things to which such concepts might apply. We have no handle to take hold of them by."

When rendered explicit, a concept is a schema: an aggregate of anticipatory criteria on the basis of which we can determine whether an experienced object falls into one class of objects or another (Lewis 1946, 134; *cf. MWO*, 86–88).[1] These criteria can be expressed as judgments that concern only immediate experience. They may be phrased as counterfactual conditionals of the form "S being given, if I were to do A, E would ensue" (Lewis 1946, 182; *cf. MWO*, 80, 192). In other words, they express the anticipation of some consistent effects that should be producible with whatever the concept is applicable to. Concepts are those criteria in mind on the basis of which we can guide our attention to what we experience so as to be able to produce consistent results. A concept is a disposition or habit on the basis of which we may classify experience. The concept allows us to guide our attention to some parts of experience to the neglect of others.

For example, the concept of duck allows us to pay attention to the relevant features of the famous duck-rabbit image of Joseph Jastrow (1901) so as to perceive a duck, whereas the concept of rabbit allows us to pay attention to the same image in a different way, and thus to perceive a rabbit. Or, to view the issue from another point of view, the duck-rabbit given in experience satisfies the anticipatory criteria of the concept of duck one way, and the anticipatory criteria of the concept of rabbit another way. But it does not, for example, satisfy the anticipatory criteria of the concept of monkey wrench. Concepts allow us to anticipate results that follow consistently from a certain kind of experience. By being able to anticipate given results consistently, we can classify which experiences are alike and which different. Therefore concepts function as the basis of differentiating and classifying experience into distinct categories (*MWO*, 99–100).

Concepts are also needed for sharing experiences. We cannot share what we directly experience (*MWO*, 91). I have no way of determining whether what I experience as red is the same as what someone else experiences as red. It could well be that the sensation I experience when I perceive strawberries, tomatoes, and fire trucks is the sensation another person experiences when she perceives plums, orchids, and violets. If I were to point at a tomato and say, "I see red," she would agree: she would have learned to attribute the term 'red' to the sensation she has when seeing a tomato or a strawberry by our conventions of language use. We cannot, therefore, share what is given to us in experience. But we can share what we anticipate from an experience.

An experience functions as a sign of other possible experience (*MWO*, 192). Furthermore, an experience signifies potential future activity. If I see a paper, it signifies, among countless other things, the possibility to write. If I see a door, it signifies the possibility to open it and pass through it. Insofar as we act together in these predetermined ways, we assume that we conceive our experiences in a similar manner. But if someone were to scream at the top of her lungs when seeing a paper, or stand on his head when seeing a door, we would soon come to believe that these people were conceiving of something altogether different from us. We cannot share the ideas in our minds. But by observing the ways we behave, we can infer whether we have the same, or similar, concepts guiding our action: "Congruity of behavior is the ultimate practical test of common undertaking" (*MWO*, 90).

Concepts are anticipatory patterns of potential future action: anticipatory schemata. By designating concepts by terms, we can share and coordinate our actions by the use of language. In this way, concepts allow us to differentiate experience by delineating different possible activities a given experience allows, and to share this experience by pointing to potential future action and behavior by using language.

3 CONCEPTUAL PRINCIPLES AND THE REAL

Conceptual principles consist in relationships of concepts (Järvilehto 2011, 98). They are such patterns of anticipation that can be characterized as categorial laws or universal principles. Conceptual principles are legislative with respect to experience because, like simple concepts, they direct our attention in what we expect to experience. They are protected from experience because if we were to experience something at odds with our conceptual principles, we would simply count that experience as non-veridical. If something failed to conform to such expectations we place on experience—such as: something cannot be in two places at the same time, or that anything must be some one thing and not another—we would simply judge such experience as invalid.

Consequently, the very reality attributed to experience depends on the concepts we employ. Experience that does not conform to our conceptual principles is thrown out of court as unreal: "We know that any experience which does not conform to our categorial principle will not be veridical because the principle states the criteria of reality of that categorial type" (*MWO*, 225).

Even hallucinations are, in a sense, real: they are real hallucinations. The fact that we classify some experience as hallucinatory, but others not, depends upon the criteria on the basis of which we identify hallucinations. In a famous quote Lewis quips that a "mouse which disappears where there is not a hole, is no real mouse" (*MWO*, 261). We assume, antecedently to experience, that mice do not vanish on their own accord. If our experience would stand contrary to such an assumption, we would no doubt classify the experience itself as faulty, or look for hidden factors. Conceptual principles function as categorial laws, and in this sense they cannot be overthrown by experience:

> In the case of an empirical law, a mere generalization from experience, if the particular experience does not fit it, so much the worse for the 'law'. But in the case of the categorial principle, if experience does not fit it, then so much the worse for the experience. (*MWO*, 224)

We will not give up the conceptual principles to which we are committed in the face of any particular experience. They are the legislative principles we adopt in order to make sense of what we experience. In this way, they also legislate our attributions of reality to what we experience, and our withholding or denying such attributions.

Once we have adopted a set of conceptual principles—a conceptual scheme—those conceptual principles hold come what may. They are not vulnerable to experience because any experience at odds with the conceptual principles would be thrown out of court as unreal. Consequently, what experience we count as real, and what not, depends at least in part on the concepts we use.

4 THE REAL AND THE METAPHYSICALLY REAL

Up to now, it appears that we have adopted a type of relativism: that is real which appears real in terms of a given conceptual scheme. However, this apparent relativism arises from the ambiguity in the way Lewis uses the term 'real'. Lewis' epistemology does, indeed, invite a more robust realism that dissolves this apparent relativism once the terminological ambiguity is rectified.

Lewis employs the term 'real' to denote both what is considered real in terms of the currently employed conceptual scheme and also what is real independently of the experiencing subject. In *Mind and the World Order*, Lewis writes that "decisions of reality and unreality are themselves interpretations involving principles of the same order as scientific law" (*MWO*, 261) and "whatever is denominated 'real' must be something discriminated in experience by criteria which are antecedently determined" (*MWO*, *x*), thus restricting the term 'real' to mean what is judged to be real in terms of the accepted conceptual scheme. In *MWO*, Lewis also writes, however, about the relationship of the given to reality (*MWO*, 157) and of the presentation to reality (*MWO*, 159), referring by 'reality' to that which is independent of the conceptual interpretation put upon it. Elsewhere, Lewis speaks also explicitly of 'independent reality' and 'ultimate reality' (Lewis 1955, 339, 340, respectively).

To avoid this ambiguity, I introduce the following terminological division of two ways of employing the term 'real':

1. *The real* means whatever is construed as real on the basis of the conceptual principles accepted within a conceptual scheme.
2. *The metaphysically real* means whatever exists independently of the observer.

As I argued above, what is considered *real* depends upon the conceptual principles employed within a conceptual scheme. However, what is considered *real* does not only depend upon those conceptual principles, it also depends upon whatever we encounter and classify. Though the term can be troublesome, some of Lewis' characterizations permit calling this the *metaphysically real.*[2] From whatever we experience, we can infer that whatever is given to us in that experience has, independently of our constructive activity in making sense of it, at least the metaphysically real property of its being conceivable in the present fashion (Lewis 1955, 344).

In other words, while we cannot infer from experience the entire nature of any object of experience—we might be dreaming or be brains in vats—we can still infer that regardless of whether that object is made of atoms, dream images, or a stream of bits, it allows for being interpreted in the way we presently do. While the object in its entirety may remain unknowable, we can know from experience some aspects of its metaphysical reality: "We never know or can know all the properties of any individual thing, but what we do or may know is metaphysically veridical; these properties are in the things themselves as in our knowledge of them" (Lewis 1955, 345).

If we explicate the structure of our conceptual principles, we render explicit some such relationships that also bear a reference to what actually exists. The truth of the statement "all cats are animals" is not a merely conceptual truth: It also bears relevance to what there is in the metaphysically real sense. The concept 'cat' is a criterion of classification. One part of this criterion is animality. For whatever there is or could be that we classify as cats, we must also classify as animals. Furthermore, whatever it is that is experienced is certainly structured in some way and not another to allow the applicability of such concepts as 'cat' and 'animal'. Not anything whatsoever can be conceived of as a cat on a whim. Rosenthal frames the point as follows:

> if the epistemic process is such that knowledge arises by the application of concepts to an independent element, then certain conditions must

> hold of this independent element. In short, the universe must be one that allows for the knowledge situation as Lewis's pragmatic epistemology interprets it. (Rosenthal 2007, 98)

For the attribution of animality to cats to work, their metaphysical reality need not be made of atoms; it could just as well be made of dream images or bits. But however their metaphysical reality is configured, it must be so configured that the attribution of animality to cats makes sense. Otherwise such a classification would be utterly useless to us. What is metaphysically real must, therefore, be structured in such a way as to render such an attribution useful for our purposes. This is well characterized by a passage in *MWO*:

> If relative to *R*, *A* is *X*, and relative to *S*, *A* is *Y*, neither *X* nor *Y* is an absolute predicate of *A*. But "*A* is *X* relative to *R*" and "*A* is *Y* relative to *S*," are absolute truths. Moreover, they may be truths about the independent nature of *A*. Generally speaking, if *A* had no independent character, it would not be *X* relative to *R* or *Y* relative to *S*. These relative (or relational) characters, *X* and *Y*, are partial but absolutely valid revelations of the nature of *A*. (Lewis 1929, 168)

Let us say that Jack possesses the concept of duck but not that of rabbit, and Jill possesses the concept of rabbit but not that of duck. When shown Jastrow's duck-rabbit (*DR*)—a metaphysically real image—Jack can only see a duck, and Jill can only see a rabbit. Nonetheless the image allows both interpretations: it supports both kinds of realities: a duck-reality and a rabbit-reality. Both 'duck' and 'rabbit' are valid interpretations of the image. "*DR* is a duck" and "*DR* is a rabbit" are both thus provisionally true statements, respectively to the conceptual principles employed by Jack and Jill. There exists something, *DR*, that satisfies both Jack's duck-interpretation and Jill's rabbit-interpretation.

While it follows from Lewis' position that what we hold to be real reflects our conceptual commitments, these commitments themselves also bear a link to the metaphysically real by virtue of their being commitments that work. Those regularities that we anticipate in terms of such commitments must somehow be genuine: "Our categories are guides to action. Those attitudes which survive the test of practice will reflect not only the nature of the active creature but the general character of the experience he confronts" (*MWO*, 21).

While there is no necessity that we should, for example, lump cats and dogs into the class of animals, making such a classification works: there is something about the metaphysical reality of cats and dogs such that it makes sense to place them in a shared class, whereas there is something about the metaphysical reality of train engines and lamp posts such that it makes sense to exclude them from such a class.

For everything that is classified as real within a conceptual scheme, there is something metaphysically real that allows such a classification. Whatever classifications we make are also grounded in the metaphysically real insofar as such classifications work—more or less well—at all.

The reason we classify some things we experience as real rather than others is that such classification facilitates our needs and purposes. Therefore, what is attributed reality within a conceptual scheme must also be attributed an aspectual metaphysical reality. An experienced object is not simply the metaphysically real object itself, but an aspect or aspects of it.

5 CONCLUSION

According to Lewis, concepts guide our attention in what we experience. Concepts combine to form conceptual principles, which in turn function as the categorial laws which we do not give up in the face of any random experience. If an experience does not conform to our conceptual expectations, we will classify that experience as non-veridical. Consequently, what is attributed reality and what not depends in part on the conceptual principles we employ. This seems to produce a very strong relativism: what is real depends on the concepts we use. The apparent relativism arises from a terminological ambiguity in Lewis' writing. He employs the term 'real' both to designate what is attributed reality in terms of a given conceptual scheme, as well as what actually exists, independently of the observer. Once this ambiguity is clarified, the apparent relativism of Lewis' position dissolves.

In order for something to be conceived as real in the first place, there must be some such conceptual principles that it makes sense to apply. The applicability of some concepts rather than others makes what is experienced as real also metaphysically real, although only in part. The fact that a classification works for our purposes, and is thus attributed reality, goes to show that it also reflects, albeit aspectually, the metaphysically real. As Lewis (*MWO* 225) puts it, "What is not reality of one sort is reality of another; what we do not understand in one way, we shall understand in another."

In the end, Lewis' position is a perspectivalist or an aspectualist realism rather than relativism. We cannot know things simply as such, but neither are things determined solely by our conceptions of them because, as Lewis (1929, 187) stressed, "'Thing as known' is a function of two variables; it depends on the mind, but also it depends on the thing."

NOTES

1. While Lewis hints at an explicit analysis of concepts in *MWO* (*e.g.*, 82*ff.*), the method of analysis employed in *An Analysis of Knowledge and Valuation* (1946; hereafter '*AKV*') can be used somewhat more powerfully to clarify how concepts work in his epistemology. Because it might be argued that Lewis' position regarding concepts changed between the two works, I provide references for similar arguments in *MWO*. Any change in his views is irrelevant to

the issues considered here. For a more thorough analysis of concepts in *AKV*, see Järvilehto (2011), 62–77, 91–100.

2. It may be questioned whether calling this deeper dimension of the real 'metaphysical' is warranted or wise; one might refer to it as well as the naturally or the physically real, or employ Lewis' occasional 'independent' and 'ultimate' notions of the real. All such terms are, of course, one or another way to classify reality itself, aspects of which may ultimately lie beyond the grasp of any one's, or any one, set of concepts. In keeping with good philosophical tradition, I have, however, chosen to employ the term 'metaphysical' here to designate what there is as such, which we classify more or less effectively.

REFERENCES

Järvilehto, L., 2011. *Pragmatic A Priori Knowledge. A Pragmatic Approach to the Nature and Object of What Can Be Known Independently of Experience.* Dissertation, University of Jyväskylä.

Jastrow, J., 1901. "The Mind's Eye." In: *idem.*, *Fact and Fable in Psychology* (Cambridge, Mass., Houghton, Mifflin and Co.), 275–295.

Lewis, C. I., 1926. "The Pragmatic Element in Knowledge." Rpt. in: Lewis (1970), 240–257.

———, 1929. *Mind and the World Order.* Rpt., with corrections, New York: Dover, 1956.

———, 1952. "The Given Element in Empirical Knowledge." Rpt. in: Lewis (1970), 324–330.

———, 1955. "Realism or Phenomenalism?" Rpt. in: Lewis (1970), 335–347.

———, 1946. *An Analysis of Knowledge and Valuation.* La Salle, IL, Open Court.

———, 1970. J. D. Goheen and J. L. Mothershead, Jr., eds., *Collected Papers of Clarence Irving Lewis.* Stanford, Stanford University Press.

Rosenthal, S. B., 1976. *The Pragmatic A Priori. A Study in the Epistemology of C. I. Lewis.* St. Louis, MO, Warren H. Green, Inc.

———, 2007. *C. I. Lewis in Focus. The Pulse of Pragmatism.* Indianapolis, Indiana University Press.

13 Pragmatic Realism

Sami Pihlström

1 INTRODUCTION

This essay synthesizes my reflections on the relations between pragmatism and realism, and the possibility of pragmatic realism, from my *Structuring the World* (1996) to my recent work. Here I summarize my version of pragmatic realism, contrast it with contemporary debates on pragmatic realism, especially Joseph Margolis (2012), and compare my view critically to the more realistically oriented views of Rein Vihalemm and Kenneth Westphal. From its inception, my work has been characterized by philosophizing historically.[1] I begin by reflecting on the development of the realism issue in classical pragmatism and in recent neopragmatism. Only against this historical background can any systematic proposal to reintegrate pragmatism and Kantian (or quasi-Kantian) transcendental idealism make sense. To develop my systematic position within this debate, I investigate the history of that debate and place my view within it. These two investigations are inseparable; what I say about the history of the realism issue is central to my understanding of the nature of that issue, and *vice versa*. In particular, there is no credible denial of the Kantian aspects of this history, although pragmatists—early and late—have been reluctant to admit their Kantian assumptions and commitments.

2 THE REALISM *VS.* IDEALISM TENSION IN CLASSICAL PRAGMATISM

2.1 Pragmatism, Classically Speaking

Charles Peirce, William James, and John Dewey all defended views that are, arguably, *to some extent realistic but to some extent nonrealist*—if not straightforwardly antirealist, then at least in some sense idealist or constructivist. Yet, none of these classical pragmatists was a *relativist* in any clear sense, though James's (1907, chap. 6) notorious views on truth as something that is good or satisfactory to believe have often been (mis)read as relativist or subjectivist. The tensions we can find in their reflections on realism and its various alternatives illuminate the ways in which the realism issue has been and still is at the heart of the pragmatist tradition.[2] In this section I briefly consider the classical pragmatists' contributions to the realism debate by commenting on some recent scholarly controversies, as advancing the conjointly historical and systematic aims and character of the present analysis.

To address the classical pragmatists' views on realism and nonrealism, we must first be a bit more specific about these concepts, though this paper proposes no novel definitions of them. Let us start from the obvious idea that there are various aspects of realism and, correspondingly, of the realism debate: ontological, semantic, epistemological, methodological, axiological, and so forth.[3]

The *ontological realist* affirms the existence (or reality) of a largely mind-, experience-, language-, concept-, theory-, and practice-independent world, while the ontological nonrealist in one or another way rejects this thesis.[4] The *semantic realist* maintains that language (or, more generally, semantic or semiotic systems) can be used to refer to elements of such an ontologically independent reality, whereas the semantic nonrealist rejects such a referential picture of language and the conception of truth (typically, the correspondence theory of truth) that tends to go together with it. The *epistemological realist,* furthermore, believes that the real, independent world (in the sense of ontological realism) is at least to some extent knowable, that is, that there are facts or states of affairs that can be known by human beings (or, possibly, other cognitive agents). In contrast, the epistemological nonrealist can either be an ontological nonrealist denying the existence of the independent world claimed by the realist to be knowable, or a skeptic accepting its existence but denying its knowability.[5] Furthermore, *methodological realism* in a way applies the general considerations of epistemological realism to the philosophy of science, maintaining that there are methods that enable us to know (elements of) the independent world, in contrast to versions of methodological nonrealism that deny the idea that truth or knowledge is the (or even a) goal of scientific inquiry. Finally, *axiological realists* believe that there are values or normative truths (or both) independently of our experiences and opinions, while nonrealists again deny this. Methodological and axiological realism can be closely connected insofar as the key value whose objective status is debated is truth.

Although several versions of nonrealism are available—I deliberately use an inclusive term to cover both antirealism(s) and various forms of relativism, constructivism, *etc.*—it will be helpful to examine the classical pragmatists' ideas primarily in terms of the traditional opposition between *realism and idealism.* This, after all, is the opposition in terms of which the realism debate has been primarily approached since Kant, and one of the tasks of my inquiry is to reconnect pragmatism with its Kantian origins. This task is nowhere as vital as in the case of the realism issue.

2.2 Peirce

Peirce is a major classic of scientific realism—in particular, of the idea that science progresses toward the truth—but he seems to be ambivalent between realism and idealism. Indeed, largely as a result of his work, the problem of realism *vs.* idealism seems to be unavoidable in the pragmatist tradition even today. It is, of course, possible to object that this contrast is not appropriate in a discussion of pragmatism, as pragmatists, early and late, have attempted to transcend the oscillation between realism and idealism instead of defining their views in terms of it. Such an objection is legitimate; yet, it is equally legitimate to employ this traditional opposition to uncover the tensions that remain in pragmatists' particular combinations of realism and idealism (even if we may in the end agree that the contrast has been transcended). What makes this tradition philosophically interesting is, I believe, its tendency to result in fruitful albeit not easily resolvable tensions between realism and idealism. Peirce's, James's, and Dewey's (or their recent followers') views are as complex as those of Kant, who attempted to combine transcendental idealism with empirical realism.

In a short essay on Peirce's pragmatism, Dewey concluded:

> Do not a large part of our epistemological difficulties arise from an attempt to define the 'real' as something given prior to reflective inquiry instead of as that which reflective inquiry is forced to reach and to which when it is reached belief can stably cling? (Dewey 1923, 308)

This suggestion, that the "real" should *not* be defined as "something given prior to reflective inquiry," leads to the elusiveness of the contrast between realism and idealism that can be found throughout the pragmatist tradition. Does inquiry produce the real by being forced to reach it? How independently does the real exist already before inquiry, if it is not 'given' prior to it? There is no clear answer to these worries to be found in the writings of any of the pragmatists, even Peirce, as these philosophers tried to undercut the realism *vs.* idealism distinction instead of relying on it. However, if philosophical traditions are at least as much constituted by open questions as by definite answers, the tension between realism and idealism we find in Peirce and in virtually all subsequent pragmatists should by no means be considered unfortunate. On the contrary, it is a sign of the philosophical strength of pragmatism. Troubles with realism make the tradition vital. Those troubles to a great extent obviously have their origin in Peirce's views.[6]

It seems to me that Peirce and other pragmatists were fully aware of their difficulties in reconciling the at least partly conflicting demands of realism and idealism. On the one hand, Peirce defined, as is well known, truth as "[t]he opinion which is fated to be ultimately agreed to by all who investigate" and reality, or "the real," as "the object represented in this opinion" (1878, *CP* 5:407/*W* 3:273). On the other hand, the real, as he always emphasized, must be thought of as something that is "independent of the vagaries of me and you."[7] Traditional realists usually require that the nature of reality itself be absolutely independent of our—even our most considered, rationally acceptable, or 'final'—opinions. Peirce was prepared to claim that reality "depends on the ultimate decision of the community" (1868, *CP* 5.316/*W* 2:241). Claims like this do make his pragmatism (including, arguably, his later pragmaticism) somewhat ambiguous between realism (connected with something like a correspondence analysis of truth, according to which the final opinion of the community of inquirers corresponds to the way things are) and idealism (connected with a coherence or perhaps rather consensus account of truth). Peirce attempted to face this tension by reminding us that reality, while being independent of "what you or I or any finite number of men may think about it," may not be independent "of thought in general" (1878, *CP* 5.408/W 3:274, *cf. CP* 7.336). This reference to the role played by "thought in general" in the constitution of reality does in some sense make him an idealist. Realizing the instability of his position, Peirce remarked that the claim that "[t]he object of final belief which exists only in consequence of the belief, should itself produce the belief" sounds paradoxical, but this is not to say that the object of the belief "begins to exist first when the belief begins to exist" (1873, *CP* 7.340). Even though Peirce characterized inquiry nonpsychologically when defining the real and truth in terms of the final outcome of inquiry, we may claim that it is not easy for him (or his followers) to make sense of the idea of inquiry as a process of genuine discovery, if inquiry, which is fated to lead to a consensus of opinion in the long run, somehow *constitutes* the way the world is (*cf.* Hookway 1985, 37–39).

The secondary literature is full of attempts to reconcile the tension between realism and idealism in Peirce. For example, Carl Hausman endorses the idea that Peirce was a "metaphysical realist" (although preferably to be called an "evolutionary realist"), and sets out to defend this view against nonmetaphysical realists like Hilary Putnam. As Hausman (1993, 144) notes, Peirce is *both* a scholastic realist about generality (*i.e.*, an antinominalist) *and* an anti-idealist, postulating an extra-mental and extra-linguistic or extra-semiotic "constraining condition" for human thought, language, and *semeiosis*. Yet, the Peircean realist's world does not consist of brute facts or any given metaphysical "furniture"; the independent external world is, rather, a "dynamical object" (Hausman 1993, 199, 214–215, resp.). While Peirce rejected what Dewey later called the "spectator theory of knowledge" (as all pragmatists did), he insisted that there are conditions of inquiry that were never made by us, *i.e.*, that there is "resistance" in experience (Hausman 1993, 224–225). However, one may worry whether it is possible to reject the spectator theory, denying that the object of knowledge is 'given' to us, and yet claim that there is an external, independent world that is the object of knowledge. If the object of knowledge is constituted as the final outcome of inquiry, if truth is definable as belief that cannot be improved upon through further inquiry, it is hard to see how the world can be totally independent of us in the sense in which realist interpreters like Hausman claim it to be.

Certainly it would be too simple to say that (according to Peirce or indeed any of the pragmatists) the progressive dynamics of the scientific enterprise, the collective belief-fixation rational inquirers engage in, decides or determines (let alone constructs) the nature of the world. But it would also be too simple to say that reality exists in a ready-made form, as a Kantian thing in itself, entirely independently of the inquirers' habits of action. Peirce's pragmatism, and the countless versions of pragmatism to be found in the literature after him, all the way up to and including controversial contemporary figures like Putnam and Rorty, can be interpreted as attempts to deal with this opposition between realism and idealism. Perhaps the question, "Is Peirce assuming an external, objective world independently of inquiry, or is the world constituted through the process of inquiry?," is somehow a bad question, but it still remains an open issue to determine exactly what sort of questions should replace it.

The recent volume on Peirce by Hausman and his colleague Douglas Anderson, *Conversations on Peirce* (2012), offers one of the most up-to-date accounts of Peirce's relation to the problem of realism *vs.* idealism. The fact that there is a tension between these authors' interpretations of Peirce beautifully illustrates the complexity of this issue in Peirce scholarship (and pragmatism scholarship more broadly). Anderson emphasizes Peirce's idealist background, whereas Hausman focuses on Peirce's realism, not only in the sense of scholastic realism about "real generals" but also in the sense of nonidealism (2012, *x*). Peirce, indeed, was a unique philosopher in the sense of integrating metaphysical realism with objective idealism (2012, 95). In Hausman's (2012, 80) words, what this metaphysical realism basically means is that "reality is not construed as wholly constituted by mind or as isomorphic with mind or mentality as such." Peirce's concept of the dynamical object is fundamental for this form of realism.

Despite their different emphases, Anderson and Hausman find a common enemy in neopragmatist views threatening to deflate Peirce's distinctive realism; they (2012, chaps. 3, 4) examine Peirce's differences from neopragmatism through two examples, Joseph Margolis and Richard Rorty. In the only jointly authored chapter, Anderson and Hausman argue against Margolis's constructivist interpretation of "real generals," suggesting (2012, 45) that Margolis "ignores Peirce's conception of conditionality, or the 'would-be' character of generality." The Peircean realist need not follow Margolis into full-blown constructivism, they (2012, 52) argue, "because he hypothesizes a condition of constraints on the part of something that is not internal to the constructions," namely, the "dynamical object." Nevertheless, they (2012, 52) contend, Peirce's realism is *not* committed to any "independent, pre-established structured reality," and the Peircean convergence of the scientific opinion toward the final opinion is not a "convergence toward a pre-fixed structure to be matched, but convergence toward a reality that is dynamic" and constrains inquiry. Nor are Peircean "real objects" Kantian "things-in-themselves" (2012, 88). Reality is, crucially, an evolutionary process (2012, 77); this, in a way, makes Peirce a process philosopher (2012, chap. 5). Hausman (2012, 76) also notes, *contra* Margolis, that "what is real, although it is necessarily related to thought, maintains an independence insofar as it functions as a constraint on what anyone and every intelligent being thinks."

While Anderson claims that pragmatism "degenerated" on the way from Peirce to Rorty, Dewey's views on realism were still close to Peirce's, or at least bear a family resemblance to the latter: according to both (2012, 67), "real generals are neither merely discovered nor simply constructed." The pragmatist position is thus a genuinely third way between an implausibly strong metaphysical realism, according to which inquiry seeks to 'mirror' a pre-structured reality that remains independent of inquiry, and an equally implausible radical constructivism, according to which no independent reality constrains our inquiries at all.

While we should warmly welcome these careful comparisons of Peirce to other pragmatists, I believe Hausman (2012, 79) exaggerates when he says that Peirce differed from other pragmatists by not equating reality with mental action. It would require much more evidence to show that *any* of the other pragmatists actually endorsed this equation. Moreover, the argument seems to lose important distinctions within pragmatism, as it makes Margolis's and Rorty's quite different positions seem identical: from Hausman's (2012, 97) Peircean (realist) perspective, these neopragmatists maintain that there are only "immediate objects," "the products of theories and interpretations." Insofar as the views of Margolis and Rorty merge into one another, one may wonder whether something is missing from the perspective from which this merging seems inevitable. Margolis has argued for decades (at least since Margolis 1986) that Rorty's neopragmatism loses normative ("second-order") questions such as the issue of realism itself. As we will see below (§3), Margolis has offered one of the most interesting recent contributions to pragmatic realism.

2.3 James

In contrast to Peirce, James has often been read as an antirealist—after all, he does say that our interests and what we find satisfactory play a role in determining what is true and what is real—yet he also affirms that there is a reality out there that our truths are in agreement with. So again there seems to be a tension between realism and a version of idealism in Jamesian pragmatism, though in a manner somewhat different from the other classical formulations of pragmatism.

In his essay, "Pragmatism, Realism, and Religion," Michael R. Slater (2008) defends James against a number of widespread misunderstandings (see also Slater 2009). As Slater focuses on the issue of realism in James's pragmatism, it is worthwhile to explore his interpretation in order to consider in more detail what kind of a realist James—or the contemporary Jamesian philosopher—is, or might be.

Slater seeks to show that James's pragmatism not only is compatible with realism but is even a species of metaphysical realism. James, we are told (Slater

2008, 655), "combines a 'humanist' account of truth with a commitment to metaphysical realism," while also defending "a pragmatic version of the correspondence theory of truth." I agree about the latter point, provided that the correspondence theory is not understood as being necessarily connected with metaphysical realism. That is, far from being a naïve subjectivist or relativist, James clearly starts from the commonsense idea that truth involves some sort of agreement between our "ideas" (beliefs, statements, theories, *etc.*) and reality, seeking to specify this notion of agreement pragmatically, in concrete experiential terms. This is clear in James's responses to his critics in *The Meaning of Truth* (1909), in particular. Slater offers sufficient textual evidence for this reading. However, the claim that James is a metaphysical realist, even in a general sense, is inaccurate, or at least needs qualification.

Much of the issue here depends on what exactly we should mean by 'metaphysical realism'. Slater's (2008, 658 n5) definition of this notion follows that given in Robert Audi's *Cambridge Dictionary of Philosophy* (1999, 33): according to the metaphysical realist, "there is a world of mind-independent objects." Slater thus rejects Putnam's threefold characterization of metaphysical realism as comprising these three theses:

1 The world consists of a definite set of mind- and scheme-independent objects and properties.

2 The world can in principle be described by means of a single complete true theory, *i.e.*, there is an "absolute conception of the world," formulated from an ideal "God's-Eye View."

3 Truth is a nonepistemic correspondence relation between our statements (or whatever the ultimate truth-bearers are) and the mind-independent entities and facts of the world. (Putnam 1981, 1990; *cf.* Pihlström 1996).

Now, while there certainly is a sense in which James does accept the realist view that there is a mind-independent world, and the related account of truth as agreement with such a world, he can hardly be taken to endorse even the relatively broad notion of metaphysical realism understood as the commitment to there being "a world of mind-independent *objects*" (my emphasis). This is because James finds the very notion of an object dependent upon human purposes. This is especially clear in the seventh lecture of *Pragmatism* (1907, 121–123), where he argues that reality is something we (help to) create, to which we inevitably make an "addition"—it is something "still in the making"—and anything we may call a *thing* is "carved out" by us in and through our purposive, interest-driven practices.

Moreover, even James's theory of truth, albeit starting from the idea of "agreement" (as Slater correctly shows), is not a version of the correspondence theory, insofar as the latter is viewed (as it is by most metaphysical realists) as an essentially nonepistemic account of truth. That is, metaphysical realists and correspondence theorists would clearly not accept the Jamesian epistemic characterization of correspondence (or, in James's own terms, agreement) as something based on confirmation and experiential satisfaction (Slater 2008, 661) as characterizing the proper sort of correspondence required by a realist theory of truth.

A much more helpful description of the Jamesian conception of truth could be achieved by emphasizing the fact that James, unlike metaphysical realists, refuses to draw a principled distinction between the *meaning* of truth and the *criteria* of truth. Even strong metaphysical realists may regard the latter as epistemic, but for James (1907, chap. 6) the very meaning of truth is epistemic all the way down, and inseparable from our pragmatic criteria of arriving at whatever truths our inquiries may reach. Metaphysical realists can hardly accept this as a correspondence theory of truth proper, but James deliberately—and in my view with good reason—blurs the distinction between meaning and criteria, pragmatically proposing that the meaning of truth largely lies in the epistemic criteria we employ for arriving at truths. There is no pragmatic meaning of truth available *in abstracto*; what the concept of truth means for us is inseparable from the plurality of epistemic practices we engage in in order to reach truths.

Slater (2008, 663–664 n16) is one of the very few commentators on James who perceives that there are "transcendental arguments that are basic to his pragmatism," one of them being the realistic batch of arguments seeking to show that "if beliefs or statements about reality are to be either meaningful or true, there must first be a reality to which they refer," and that "in order for a belief or statement about some reality to be true it must also *agree* with that reality." The existence of a mind-independent reality is argued to be a necessary condition for the possibility of truth (Slater 2008, 665).

Slater (2008, 665) points out that James has difficulties in balancing his realist and "humanist" ideas,[8] even at times "undercutting his commitment to realism" by suggesting that "reality is somehow altered or reconstructed through our cognitive interactions with it." This formulation is revealing, because right here James arrives at what is essential (in my view) to his "realism," namely, its subordination to (what I have elsewhere proposed to call) his *transcendental pragmatism*. We need not read James as claiming that we alter or reconstruct reality through our cognitive interactions with it in any causal, factual, or empirical sense; these alterations and reconstructions—or the dependence of reality upon us and our cognitive powers—is, rather, in Kantian terms 'transcendental'. It is only in the sense in which the spatio-temporal framework and the categories, such as causality, are established by us, or imposed by us upon the world we are able to experience, according to Kant, that reality is, for James, dependent upon us or altered by us in cognition. Our 'worldmaking', to borrow a term from Nelson Goodman's (1978, 1984) more recent neopragmatism which is in some ways analogical to James's pragmatism, is transcendental, not factual—although neither James nor Goodman is willing to put the matter in precisely these terms.

Nevertheless, just as Kant was able to defend empirical realism within, and indeed held that it is made possible by, his transcendental idealism, James is able to endorse a form of empirical (though not metaphysical) realism within his transcendental pragmatism.

I thus agree with Slater (2008, 666) when he argues that we need both realism and humanism about truth. But it is important to see that the kind of realism that we may maintain compatibly with humanism is empirical realism, and the kind of humanism we may maintain compatibly with realism is transcendental humanism (or transcendental pragmatism). This Jamesian suggestion will be developed as the argument of this essay unfolds.

2.4 Dewey

Dewey's basic epistemic and metaphysical position, as articulated in his major works such as *Experience and Nature* (1929a), *The Quest for Certainty* (1929b), *Logic: The Theory of Inquiry* (1938), and *Reconstruction in Philosophy* (2nd ed., 1948)—a view variously labeled 'instrumentalism', 'experimentalism', or 'operational thinking'—also accommodates an intriguing tension between several standard realisms and antirealisms regarding the status of the objects postulated in inquiry. Dewey's position has clearly *instrumentalist* and more generally *empiricist* elements, roughly in the sense in which these terms are used in the philosophy of science, although his talk about instrumentalism should *not* be simply equated with later philosophers' of science narrower treatment of the empiricist doctrine carrying the same title.[9]

Dewey (1929b, 79) emphasizes the "overtly executed operations of interaction" that are needed to obtain scientific knowledge, thereby rejecting the gap traditionally thought to lie between knowledge and action. This seems to lead to a form of antirealism:

> There is something both ridiculous and disconcerting in the way in which men have let themselves be imposed upon, so as to infer that scientific ways of thinking of objects give the inner reality of things, and that they put a mark of spuriousness upon all other ways of thinking of them, and of perceiving and enjoying them. It is ludicrous because these scientific conceptions, like other instruments, are hand-made by man in pursuit of realization of a certain interest—that of the maximum convertibility of every object of thought into any and every other . . . [W]hen the physical sciences describe objects and the world as being such and such, it is thought that the description is of reality as it exists in itself. [. . . However, the] business of thought is not to conform to or reproduce the characters already possessed by objects but to judge them as potentialities of what they become through an indicated operation. . . . [T]o think of the world in terms of mathematical formulae of space, time and motion is not to have a picture of the independent and fixed essence of the

> universe. It is to describe experienceable objects as material upon which certain operations are performed. (Dewey 1929b, 135–137)

Science, accordingly, has no more privileged relation to "the real" than (some) other human practices. "There are as many conceptions of knowledge as there are distinctive operations by which problematic situations are resolved," Dewey (1929b, 221) says. In brief, "scientific conceptions" are not "revelations of antecedent properties of real Being and existence" but "instrumentalities which direct operations of experimental observations" (1929b, 192). Natural laws, similarly, are "intellectual instrumentalities," "*formulae for the prediction of the probability of an observable occurrence,*" rather than statements about "ultimate and rigid uniformities of being" (1929b, 205–206; original emphasis). A law supposed to "govern phenomena" ought to be understood as "a way of transacting business effectively with concrete existences" (1929b, 207).[10] Scientific concept(ion)s and theories are tools, "open to development through use" (Dewey 1948, 145).

Similarly, in his naturalist *magnum opus, Experience and Nature,* Dewey (1929a, 115) equates "the proper objects of science" with "nature in its instrumental characters." The key move for a pragmatic naturalist is to stop treating the objects of science as "complete and self-sufficient"; this only results in an insoluble problem, *viz.*, the question concerning the relation between the "perceptual order" and the order of "inferred and logically constructed real objects," with two incompatible kinds of knowledge and kinds of objects of knowledge (1929a, 116). It seems, indeed, that mainstream philosophers of science struggled with such a problem throughout the twentieth century. Dewey, in contrast, seems to avoid the problem of scientific realism altogether. In this context, he also reminds his potential misunderstanders that his "instrumentalism" is not a theory about personal satisfaction in knowing, but about "the proper objects of science" (1929a, 126).[11] Precisely as such, it might, however, be regarded as straightforwardly antirealistic.

Dewey's commitment to instrumentalism is relatively obvious when he (1929b, 240) argues that the problem of Eddington's (1928, *ix–xi*) "two tables" is illusory, because the table we perceive and use is "the only table," "for it alone has both individuality of form . . . and also includes within itself a continuum of relations or interactions brought to a focus." Thus, there is hardly any significant problem of reality (or realism) in Dewey. Reality is simply what is given—or, better, actively "taken"—in experience.[12] In *The Quest for Certainty,* he writes:

> The world as we experience it is a real world. But it is not in its primary phases a world that is known, a world that is understood, and is intellectually coherent and secure. Knowing consists of operations that give experienced objects a form in which the relations, upon which the onward course of events depends, are securely experienced. It marks a

> transitional redirection and rearrangement of the real. It is intermediate and instrumental. . . . (Dewey 1929b, 295)[13]

Is Dewey saying that the scientific table—or the "scientific image" according to which the table is not, in the final analysis, a concrete, perceived, practically used object, after all, but a collection of microphysical particles—is unreal or illusory? If so, he seems to subscribe to an instrumentalist antirealism, which several philosophers of science, following Sellars (1963), have powerfully called into question (*cf.* Tuomela 1985, Niiniluoto 1999). Or is he instead subscribing to idealism, or to what would now be called constructivism, when maintaining that scientific objects are not independent of inquiry but pragmatic constructions arising out of an intelligent use of the methods of inquiry (*cf.* Shook 2000)?

In an earlier work, Dewey (1916, 30–31, *cf.* 60) admits that he may sound like an idealist when he holds that thinking, which "is instrumental to a control of environment,"[14] in a way constructs its objects: "[P]rocesses of reflective inquiry play a part in shaping the objects—namely, terms and propositions—which constitute the bodies of scientific knowledge. . . . Insofar as it is idealistic to hold that objects of knowledge *in their capacity of distinctive objects of knowledge* are determined by intelligence, [this view] is idealistic. It believes that faith in the constructive, the creative, competency of intelligence was the redeeming element in historic idealisms." Dewey's Hegelian background is visible here, though he (1916, 31) immediately adds that his instrumentalism (or idealism) does not postulate an entity or substance which constitutes the world, but rather "defines thought or intelligence by function, by work done, by consequences effected."

The charges of instrumentalism and idealism have their obvious justification, but we are already able to see that it would be overhasty to judge Dewey to be simply an antirealist in his philosophy of science. For one thing, he—like James—uses the terminology of 'objects', 'concepts', 'conceptions', 'theories', 'hypotheses', *etc.*, more broadly than is customary in more recent analytic philosophy of science, which easily leads to confusions. For another thing, Dewey was, primarily, a *naturalist,* always arguing that experience and knowledge, including the production of scientific knowledge (and its objects), are natural phenomena in a natural world in which we try to cope, to settle the problematic situations we arrive at. But this naturalism, obviously, goes well together with humanism or culturalism. Whatever is natural to *our* practices of inquiry, our human, culturally developed habits of settling indeterminate situations by critically using the intelligence that itself naturally arises as a human capacity of reacting to problematic situations, is to be accepted, as well as both empirically and conceptually understood in actual detail, rather than treated with philosophical suspicion. A philosophical skepticism about, say, the existence of unobservable entities referred to in our most advanced and most successful scientific theories and practices would be an utterly *unpragmatist* and *unnatural* attitude from Dewey's point of view. Rather, we should take seriously the natural practices of inquiry we engage in. It was always the actual practice of inquiry itself that Dewey appealed to when, for instance, attacking the ancient ideal of certainty and the "spectator theory of knowledge."

Deweyan pragmatism is definitely *not* scientific realism, *if* the latter is defined as the thesis that scientific theories provide us with the only true (or truthlike) picture of reality, or that the "scientific image" offered by theorization is ontologically prior to the "manifest image" we are more directly acquainted with in our ordinary experience.[15] But a Deweyan may be, and Dewey himself may also be regarded as having been, a scientific realist in the sense of rejecting unnatural skeptical doubts about the reality of theoretical entities postulated in the actual course of scientific theorization, as long as such theorization is firmly rooted in its humanly natural practical context, that is, in the natural processes of settling problematic situations that arise in the course of our experience (*cf.* French 1989). The key pragmatist move is to liberate not only science but also scientific realism—and all other philosophical interpretations of science we might practically *need*—from foundationalist and noncontextualist pursuits of certainty, essences, and other remnants of "first philosophy." Pragmatism offers, then, a middle path (or, rather, several such paths) between scientific realism and instrumentalist antirealism, and between realism and constructivism, by contextualizing the reality of scientific postulations to specific aims and purposes of inquiry.

So, finally, there *is* a realism issue in Dewey, just as there is such an issue in Peirce and James—indeed several. These problems of realism require contextualization; they are not reducible to any essential, global formulation but must, rather, be investigated in a piecemeal fashion, case by case, respecting the pragmatic plurality of the cases themselves and the quite different needs served by our practices of inquiry in the relevant contexts. For example, we need to be sensitive to the different aspects of the realism dispute as it arises in relation to science, art, religion, or morality, to name just a few humanly central contexts in which it may arise.

2.5 Some Provisional Conclusions

It should now be evident that the classical pragmatists were involved in the realism discussion not only in terms of realism *vs.* idealism, they also made very important contributions to the discussion of realism about universals, that is, the debate between realism and nominalism. Accordingly, the pragmatists contributed to virtually all dimensions of the realism debate: realism *vs.* idealism, realism *vs.* constructivism, realism *vs.* relativism, realism *vs.* nominalism, realism *vs.* empiricism, realism *vs.* instrumentalism—you name it. In each case, while the debate may have received more specific formulations recently, the pragmatists were there already. They hardly ever resolved these problems, but they shaped our understanding of them in ways that still haven't been adequately investigated in the philosophical community exploring realism.

In particular, we may argue, on the basis of the pragmatists' ideas, that the debates between realism and idealism on the one hand and realism and nominalism on the other are more closely related than has sometimes been supposed. This link between these two debates can be particularly appreciated when we take Peirce's approach to realism seriously (*cf.* §2.2 above).

3 THE REALISM *VS.* IDEALISM TENSION IN NEOPRAGMATISM: TOWARD PRAGMATIC TRANSCENDENTAL IDEALISM

3.1 Putnam and Rorty

Moving from classical to more recent pragmatism, we may begin by noting that Hilary Putnam's and Richard Rorty's neopragmatist positions continue to exhibit the kinds of tensions familiar from classical pragmatism. Putnam's internal realism, in particular, is characterized by a very similar tension, or even oscillation, between realism and idealism. His *pragmatic pluralism* (close, though not identical, to what he has called "conceptual relativity") also contains elements of relativism, even though both Putnam and Rorty explicitly deny any association with relativism. Putnam is also a key figure in the emergence of Kantian pragmatism, because his internal realism was influenced both by Kant and by the classical pragmatists—and Putnam (1987, 17) himself suggested that internal realism should have been labeled "pragmatic realism." In this sense, Putnam's influence on the kind of position I develop here is crucial, though I neither simply follow him nor claim to get his views exactly right.[16]

However, I shall not detail Putnam's and Rorty's widely discussed views here. Here it suffices to say that one way of making sense of the realism *vs.* idealism tension they share with the classical pragmatists is the Kantian reading (or, better, rearticulation) of pragmatism as a version of transcendental idealism that I have defended on a number of occasions (*cf.* Pihlström 1998, 2003, 2008, 2009, 2011a, 2012) and continue to defend here: the world is practice- or scheme-dependent at a transcendental level while being practice- and scheme-independent at the empirical level. Despite the pragmatists' generally critical reception of Kant,[17] they can be said to synthesize pragmatic realism and what may be called 'transcendental pragmatism' (albeit a label most pragmatists—including Putnam—tend to avoid) roughly in the way in which Kant integrated empirical realism and transcendental idealism. Here we may see how this synthesis of realism and idealism emerges by taking a critical look at a recent neopragmatist effort to advance such a synthesis *without* Kantian transcendental idealism.[18]

3.2 Margolis

Joseph Margolis has explored the problem of realism extensively over several decades. His latest book, *Pragmatism Ascendent* (2012), discusses a wide spectrum of issues, including the transformations of transcendental philosophy today in terms of Hegelian historicity and "flux," as well as the question, "what it is to be a human self" (2012, x), yet the realism debate is one of the central themes covered, largely in relation to these other complex debates. In this section, I examine Margolis's arguments for the special kind of integration of realism and idealism (or "Idealism," as he prefers to write)[19] in relation to his attempt to develop a viable version of pragmatism conscious of its Kantian and especially Hegelian roots.

One of Margolis's (2012, 4) characterizations of his view is this: "a descendent strategy argumentatively (or genealogically) derived from the transcendental turn turned pragmatist by refusing to concede any strong disjunction between broadly 'empirical' first-order inquiries and broadly 'rational' second-order speculations about the legitimacy of both the first *and* the second." In this context of inquiry, the realism issue is never the primary topic; it is commented upon repeatedly as the genealogical and quasi-transcendental examinations of Hegel's response to Kant, of Peirce's fallibilism, and of the inadequacies of contemporary philosophy of mind and social ontology unfold. These philosophical and metaphilosophical contexts turn out to be relevant to the unusual integration of realism and Idealism Margolis proposes. (Note the parallel to the Peircean integration of realism and idealism discussed in §2.2 above.)

A more substantial comment on the realism issue is this:

> The viability of the transcendent*al* question (apart from the fortunes of transcendental*ism*) makes no sense, unless we also concede that the viability of empirical realism cannot be separated from 'idealism' (the 'Idealism' already implicated in the transcendental question itself): *that* consideration already signals the importance of deciding whether the *human* version of 'reason' reflexively affects what we affirm to be possible regarding 'what there is' in the whole of reality independent of human cognition. . . . I take 'realism' and 'Idealism' to be inseparable within any 'constructivist' form of realism—it being the case that there is no other viable form of realism. I take that to be both Hegel's and Peirce's view. (Margolis 2012, 8)

So realism ought to be, and is, maintained by Margolis: there is something that can be called "reality independent of human cognition," and we can and do, quite legitimately, affirm things about "what there is" in that reality. Yet this is something that *we* affirm, or fail to affirm, and the transcendental question (that is, the second-order question concerning legitimation itself), as Margolis notes, reflexively addresses whether human reason—its structure, or perhaps rather its history—inevitably affects these affirmations. Thus, realism, when considered transcendentally, cannot be all-inclusive or full-blown. It must be restricted to a human perspective available within a pragmatic analysis.

This is what it means to take the issue of realism not just metaphysically but also epistemologically seriously: we need to construe a form of realism that we are able to "live with" within our historically situated and finite inquiries, processes that are themselves continuously in flux. There is no return to what Kant labeled "transcendental realism" (which, notoriously, conflates appearances with things in themselves) or to what Putnam later called "metaphysical realism" (which postulates a "God's-Eye View" on the world). These appeals to an imagined super-human perspective on what there really is go beyond the more minimal realism that Margolis favors, a realism that accepts the idea that there is a reality independent of us but insists on there being only human views or perspectives—no divine ones—on that reality. "We construct our 'pictures' of reality, but not reality itself," even when the inseparability of realism and Idealism is recognized (Margolis 2012, 30).

Constructivism, for Margolis (2012, 39), is fully compatible with realism and does not entail any "ontic construction of the whole of 'reality' itself." Accordingly, while what we, within Idealism, may find "determinately real" presupposes "the ability of a cognitive agent to discern the fact," this by no means requires the real world to be constructed by such agents; again, "what *is* constructed is one or another picture of the world" (2012, 59)—presumably including, reflexively, this picture of realism itself. Peirce's realism, in particular, is a "constructivist posit supported in terms of what we rationally Hope holds true at the end of infinite inquiry" (2012, 60). This is—and here Peirce's special kind of fallibilism truly comes into the picture—because realism cannot be a "free-standing epistemological option," if inquiries "must be infinitely extended"; rather, realism requires (possibly a naturalized version of) an "Idealist supplement" (2012, 73).

Given his Peircean elaborations, it is easy to see that Margolis's realism is not just the minimal affirmation that there is *something* we never constructed. It is a significantly richer position, accommodating an acknowledgment of realism itself being a distinctively *human* conception of reality. Our realism itself is a "human face" that the world, as seen from our human perspective, has.[20] Realism, as Margolis has put it in several previous works, is a human "posit" rather than the world's "own" picture of itself. It is not Nature's image of itself but *our* image of the world, natural and historic-cultural.[21]

In what sense is this view on realism and Idealism pragmatic, or related to the historical developments that led to the emergence (and the current re-emergence) of pragmatism? Margolis links these processes of development with Peirce's fallibilism, which along with "the inseparability of realism and Idealism" is one of the defining features of "the sense in which pragmatism . . . cannot fail to be construed as an ingenious and especially promising spare variant of Hegel's own undertaking, now naturalized . . ." (2012, 10). Hegel, then, is Margolis's historical hero—not Kant, even though Kant was the first to insist upon the compatibility of realism and idealism (*viz.*, empirical realism and transcendental idealism). Indeed, Kant's account of empirical realism is claimed by Margolis (2012, 11) to be "completely subjectivist" and incoherent by Hegelian lights; Kant cannot "recover any robust form of empirical realism" (2012, 20). The proper recovery of realism then eventually takes place, after Hegel, in Peirce; indeed, Peirce and Hegel form a pair of philosophical heroes for Margolis.

It is understandable that Margolis emphasizes Hegel's role as a background figure of pragmatism and pragmatic realism in contrast to Kant's. The latter has been emphasized by other pragmatism scholars (including the present author), and Hegel has often been unduly neglected. Margolis's criticism of Kant's arguments for transcendental idealism could be compared to Kenneth Westphal's, who also suggests that Hegel was the first "pragmatic realist."[22] Westphal would, I believe, happily join Margolis in acknowledging that "objectivity becomes historicized and constructivist" in a pragmatist reinterpretation (or, as Margolis says, "fragmentation") of Hegel's "recovery of realism" (2012, 21) and that Hegel's critique of Kant leads to a robust realism about the objects of experience, so long as "historicized" and "constructivist" objectivity does not preclude realism. But few other contributors to the realism debate would acknowledge Hegel's role in this way.[23]

Margolis (2012, 52–3) argues, thus, that Peirce develops further the Hegelian unity of realism and Idealism—to the extent that it would be wrong to say that Peirce was a realist *and* an Idealist. He is both but *not* in the disjunctive sense these doctrines are traditionally understood in relation to each other. On Margolis's (2012, 54) reading, Peirce "believed that it was *only* within the terms of his Idealism that the realist thesis would prove compelling at all" and therefore "saw no viable disjunction in pressing the realism of science and the need for an Idealist metaphysics." Or, more precisely, Peirce was "not a realist *and* also, independently, an Idealist" (2012, 55); these two doctrines could not coherently be maintained independently according to him.[24] What Peirce uniquely developed, in a post-Hegelian environment, is a constructivist (Idealist) version of scientific realism.

By emphasizing the inseparability of realism and Idealism, or the inevitable embeddedness of realism in constructivism, within Peirce's views, Margolis insightfully criticizes some of the leading more strongly realist interpreters of Peirce, all the way from Max Fisch to Cheryl Misak and others.[25] He also suggests—as he has done earlier—that Peirce's scholastic realism about the reality of "generals" should be replaced by a constructivist account of "predicable 'generals'" (2012, 76).[26] Such generals, in a way, are not independent of human thought (2012, 78)—though again I would rather emphasize their *transcendental* dependence and, correspondingly, *empirical* independence.

Margolis recognizes very important similarities in Peirce's and Dewey's fallibilisms and their versions of the union of realism and Idealism, which, in Dewey, is more explicitly naturalized and evolutionarily laid out. Where Margolis in my view goes wrong is in his neglect of James. He claims that James was "little more than a secondary figure" in comparison to the two other great classical pragmatists (2012, 86) and speaks about James' "disastrous" conception of truth (2012, 92). Though he may be right to point out that the realization that the Peircean notion of what is "independent of the vagaries of you and me" is itself constructed by pragmatist means does not validate James's account of truth (2012, 106), he unfortunately seems to ignore James's independent role in the development of the pragmatic method—and in the pragmatist articulation of the realism issue in terms of that method, which in effect makes ontology dependent upon, or entangled with, ethics.[27]

Occasionally—to go backwards in the historical story Margolis is telling us—it also seems that Margolis fails to do full justice to Kant's transcendental considerations. For example, though I very much appreciate Margolis's Hegelian and Peircean project of "pragmatizing" and historicizing Kant, I remain unconvinced by the criticism that Kant does not introduce "a working distinction between appearances and the objects they are appearances *of*" (2012, 19). A "one world" Kantian response to this charge is obviously that appearances *are* appearances of things in themselves; these are not two different classes of objects (as more traditional "two worlds" interpretations maintain) but, rather, the "same" objects considered from two different perspectives, or articulated through two different considerations.[28] Moreover, Margolis does not pay due attention to the distinction between the quite different empirical and transcendental ways in which, say, space and time can be said to be "in us" (2012, 29). He partly relies on P. F. Strawson's (1966) relatively conventional interpretation which has been heavily criticized by "one world" Kantians. Margolis thus claims repeatedly that Kant's transcendental question is incoherent from the very start, but he never (as far as I can see) explains in any great detail, or in full communication with relevant scholarship, why this is so. This is a serious setback in his otherwise admirable treatment of the realism issue.

The story of the development of pragmatic realism can be told by starting from Kant—and partly skipping Hegel—just as it can be told (and generally is compellingly told by Margolis) by beginning from Hegel's historicization of Kant. Such a story, even when it remains more Kantian than Hegelian, may also join Margolis in rejecting any "principled disjunction between the empirical and the transcendental" (Margolis 2012, 30).[29] In brief, I am not entirely convinced by Margolis about the idea that it is *only* Hegel, not Kant, who offers a compelling version of the inseparability of realism and Idealism. Kant rejects such an exclusive disjunction as firmly as Hegel (though I cannot justify these historical claims here).

Margolis (2012, 39) is relatively modest when insisting upon the possibility of preserving the distinction between metaphysical and epistemological questions within the "constructivist" character of his realism. He could have gone further by saying that inasmuch as realism itself is inevitably constructivist (pragmatic), all metaphysical questions about the way the world is are inevitably also epistemological, or invoke epistemological positioning. This would yield a more radical pragmatism—presumably closer to James and Dewey than Peirce. In fact such an entanglement of the metaphysical and the epistemological would also have been a more Kantian position and would come closer to a Kantian-cum-pragmatist (instead of Hegelian-*cum*-Peircean) naturalization of transcendental philosophy.

Consider this claim by Margolis:

> Kant defeats the realist metaphysics of the rationalists all right; but, then, he also obliges the 'objectivity' of science and metaphysics to depend on transcendental (subjective) sources and, in doing that, he makes 'empirical realism' no more than an artefact of those same *subjective* sources: accordingly, he *cannot* separate, as Hegel can, epistemological and metaphysical constructivism. (Margolis 2012, 39.)

Now isn't Margolis in his own way doing more or less the same when making realism constructivist? Epistemological and metaphysical versions of constructivism may themselves be claimed to be entangled, necessarily, in a transcendental sense, while of course being disentangled at the empirical level. We (mostly) do not construct the world in any empirical, factual, causal, or concrete sense; here I of course agree with Margolis. But it does not follow that the concept of construction fails to do important transcendental work that is not just epistemological but also, albeit in a qualified sense, metaphysical. Similarly, the claim that Kant's objects (*Gegenstände*) are, as objects of possible experience, mere "internal accusative of *subjective* experience itself" (2012, 40) is misleading, because, again, we can endorse a one world reading (*à la* Allison and others) and view empirical objects and things in themselves as (ontologically) identical and only methodologically distinguishable.

Therefore, when Margolis (2012, 84) writes that Peircean "Idealism" is "construed 'epistemologically' (in the constructivist way) rather than 'metaphysically' (disjunctively)" and is thus restricted to "our constructed picture" of reality rather than the "actual 'constitution' of reality itself," one might ask whether he isn't himself resorting to new versions of dichotomies or disjunctions he wants to set aside. Instead of the realism *vs.* Idealism dichotomy, we now (still) have the dichotomy between epistemology and metaphysics, and also the corresponding dichotomy between our picture of reality and reality in itself. These dichotomies—or, to be fair, more absolute versions of them—are habitually used in the kind of mainstream analytic philosophy that Margolis wisely wants to leave behind. In my view, *all* these dualisms should be critically examined in terms of the pragmatic method and thereby *aufgehoben* as versions of the age-old subjective *vs.* objective disjunction to be given up (at least in its conventional versions) in any viable post-Kantian (and post-Hegelian) pragmatism.

Yet, my proposed re-entanglement of the metaphysical and the epistemological at the transcendental level—the level at which a kind of constructivism provides, I here agree with Margolis, a framework for any viable realism—must somehow also accommodate the (re-)entanglement of the transcendental and the empirical. Here I see the real challenge for the contemporary pragmatist who seeks to develop further the insights of naturalized transcendental philosophy and apply them to the realism debate by arguing for the compatibility of pragmatic realism with transcendental pragmatism. However that challenge can be met, the pragmatist can certainly agree with Margolis's *préçis*:

> [W]e must, as realists, replace representationalism with some form of constructivism; . . . we must, again as realists, avoid characterizing reality as itself constructed . . . and hold instead that what we construct are only conceptual 'pictures' of what we take the real world to be . . . and . . . we must acknowledge that the realism thus achieved is itself cognitively dependent on, and embedded in, our constructivist interventions. (Margolis 2012, 55)

This can, I think, be offered as *a* useful characterization of the program of pragmatic realism, insofar as we are able to give up Margolis's—in my view too sharp—distinction between (the construction of) reality itself and our pictures of it. When developed in Margolis's way, pragmatic (constructivist) realism is reflexively conscious of its own status as a human pragmatic posit, rather than an imagined God's-Eye View picture of how things absolutely are. Ironically, Margolis (2012, 85) notes, "Peirce's most strenuous insistence on reality's being independent of belief . . . *is, contrary to what he actually says, not* 'independent of what anybody may think them to be'." For the reflexively sophisticated pragmatist, the real is indeed independent of what anyone of us thinks, but this independence itself is affirmed within and on the basis of human thought.

Even if I mildly disagree, in the manner explained above, with Margolis about Kant's place in the story of pragmatism, and in the ways in which the story could continue into the future, I warmly agree with him on the deep integration, or even inseparability, of realism and Idealism—and even Idealism and naturalism, because these need not be any more incompatible than realism and Idealism in Margolis's (2012, 41–2) Hegelian pragmatism. I would again just prefer to rephrase this in more explicitly Kantian transcendental terms. I also agree that this transcendental approach itself, though naturalized within pragmatism, precludes any reductive naturalism or eliminativism, which in a sense presuppose the God's-Eye View that only stronger—metaphysical—forms of realism (avoiding *any* link to *any* form of idealism) hopelessly try to presume.[30]

4 PRAGMATIC REALISM *WITHOUT* TRANSCENDENTAL IDEALISM AND METAPHYSICAL REALISM?

4.1 Two Recent Developments

While Margolis's project provides us with a highly interesting case study of the dialectics of realism and idealism in contemporary pragmatism, we should not fail to notice some other recent—quite different—attempts to defend pragmatic forms of realism without *either* any kind of (transcendental) idealism *or* any substantially realistic commitments that could be criticized as 'metaphysical'.

A very important criticism of Kantian transcendental idealism has been presented in a series of contributions by Kenneth R. Westphal, who argues—transcendentally—for realism "*sans phrase.*" The view I am defending through my readings of the pragmatists gets support from Westphal in the sense that he plausibly suggests that pragmatists or pragmatic realists can also engage in transcendental argumentation. However, such arguments should—and can—in his view avoid any commitment to transcendental idealism. In order to complete our discussions of the prospects of "transcendentalized" pragmatic realism, we must examine Westphal's argument for realism *sans phrase*. I suggest that it at least remains an open issue whether his—in my view highly plausible—transcendental argument for realism, based on the concept of "transcendental affinity," yields anything more than empirical realism (in Kant's terms) or pragmatic realism (in mine). It therefore remains an option (albeit not a necessity) to contextualize this realism within a more fundamental transcendental pragmatism analogous to transcendental idealism. I cannot here settle this issue, but I will summarize my earlier critical discussions of Westphal's views (*cf.* Pihlström 2003, 2012).

4.2 Westphal

Westphal (*e.g.*, 2003, 2004, 2005, 2006) argues transcendentally that realism (*sans phrase*) about ordinary objects and events in our environment is a necessary condition for the possibility of human self-consciousness, and that such an argument can be based on Kant's own transcendental philosophy, insofar as we give up the dubious transcendental idealism Kant himself associated with it. Westphal (2005) claims to find such an argument for realism even in Wittgenstein's later work. In a world of "transcendental chaos," a world lacking a certain, by us recognizable, degree of regularity and variety, we could not be self-conscious, nor *a fortiori* entertain skeptical doubts.

Contra some other contemporary Wittgensteinians, I agree with Westphal that there *are* elements of transcendental philosophy, or even transcendentally supported metaphysical conclusions, to be found in Wittgenstein, even the later Wittgenstein.[31] However, I am concerned about the strength of the kind of realism transcendentally established through such arguments.

A mere empirical realism, such as Kant's, is, arguably, sufficient for the possibility of self-consciousness, if we have a rich enough transcendental (yet pragmatic) account of subjectivity as our background theory.[32] A transcendentally acceptable yet pragmatic realism must be carefully distinguished from what Kant called transcendental realism and from what we today, following Putnam, may call metaphysical realism.

A slightly closer look at Westphal's argument may illuminate the matter. Westphal (2004, chap. 3) maintains that Kant's transcendental idealism is unable to account for one of the crucial transcendental conditions for the possibility of self-conscious experience recognized by Kant himself, namely, the "transcendental affinity of the sensory manifold." In order for human beings (that is, beings with the kind of cognitive capacities and incapacities that we can, with careful epistemic reflection, identify in our own case) to be self-conscious, the world in which we live must contain a certain minimal degree of regularity and variety in its contents. Below such a minimum degree, a transcendental chaos would obtain; such a world would be utterly cognitively incapacitating for us. We could not form causal judgments in such a world, and thus we could not identify external objects; hence, we could not be conscious of ourselves as subjects of temporally sequenced experiential states at all—*per* Kant's argument in the "Refutation of Idealism" (*KdrV* B274–9) to the effect that outer experience of spatiotemporal objects in our environment is necessary for us to be able to experience our (temporally ordered) internal states. On these anti-Cartesian grounds,[33] Westphal offers a truly transcendental argument for an "unqualified" realism, "realism *sans phrase*," which in his view should *not* be conflated with Kant's merely empirical realism, nor with Kant's sense of "transcendental realism."[34]

Moreover, Kant's own conception of transcendental affinity as a transcendental material condition for the possibility of self-conscious cognitive experience—a condition that is, in fact, both material and formal, yet neither intuitive nor conceptual (*i.e.,* it is not reducible either to Kant's *Anschauungsformen,* space and time, or to the categories)—provides a strong case against his transcendental idealism, Westphal claims. The objects of the world we live in (a world in which we can so much as be self-conscious) must themselves be regular and variable enough; their characteristics (including their spatiotemporality) cannot, *contra* transcendental idealism, be derived from our cognitive capacities, if we are to explain the very possibility of self-conscious experience.[35]

Westphal's impressive argument, as far as it goes, seems to me to be sound, and I happily leave a number of intricacies of Kant scholarship involved to more dedicated Kant specialists.[36] However, it may still be claimed that in the end it only yields empirical or pragmatic realism, a moderate form of realism compatible with (though presumably not requiring) a pragmatically reinterpreted transcendental idealism ("transcendental pragmatism"), or (if you prefer) with a Wittgensteinian variant of this Kantian doctrine—if not exactly with Kant's own idealism. The conception of transcendental philosophy as an "ontology of the human world" (see Pihlström 2009, 2012) may be argued to be committed to such an idealism, which in my view is *not* ruled out by the argument from transcendental affinity. At least so it seems, insofar as that conception entails that our human world is, transcendentally speaking, thoroughly linguistically or conceptually shaped (or both), *i.e.*, categorized by us, and that there is, for us humans, no primary metaphysical, underlying, conceptualization-independent reality grounding our language and, more broadly, underlying our conceptual categorization—which itself is world-involving and embedded in our natural practices—except in a trivial causal or ontic, instead of genuinely ontological and transcendental, sense.

The world- and practice-embeddedness of language, linguistic meanings, and conceptual categorization ought to be taken seriously. As language-users, we are inevitably in the natural world, instead of being mere spectators of that world placed in an imagined point of view outside it, or hermetically sealed by language from it. It is a fundamental mistake—as the classical pragmatists, the later Wittgenstein, as well as some (though not all!) contemporary neopragmatists have argued—to think of language or meanings as something that can be found exclusively, or even primarily, "in the head" of the speakers, or to think of language and reality as separable and exclusive. (Self-styled "post-linguistic-turn" pragmatists may have partly caused such confusions.) A kind of semantic externalism is so profoundly built into the genuinely pragmatist picture of human beings as active creatures involved in their habits of action in the natural world that pragmatists rarely even care to explicitly argue for externalism. Rather, the idea that language, meaning, and conceptualization are world-involving, and the related idea that language-users and conceptualizers—all those beings, specifically humans, that can mean anything by their marks and noises—are actors embedded in nature (and society) instead of being mere spectators are among the key premises of virtually any serious pragmatist argument, including the arguments pragmatists may develop in favor of the kind of pragmatic realism developed here.

Largely for these reasons, the pragmatic version of transcendental idealism that I have claimed to be compatible with my pragmatic realism is no standard idealism, and it seems to be quite far even from Kant's original transcendental idealism, after all (let alone more metaphysical idealisms). The transcendental pragmatist, I agree with Westphal (and Margolis), need not claim that the world, or its categorization, or its spatiotemporal structure, simply depends upon us, let alone upon our mental capacities.[37] Rather, any world inhabitable and cognizable by us must, for transcendental reasons, arise out of an interplay between our representational and cognitive capacities, on the one hand, and what *we* can recognize as (what Peirce called) an "external permanency," on the other. Crucially, as we have learned from Margolis's arguments (§3.2 above), this recognition of an external, independent reality is, inevitably, *our* acknowledgment of there being something empirically and factually mind- and conceptualization-independent, not an acknowledgment from an imagined "God's-Eye View" that there is, absolutely, a metaphysically independent fundamental reality *an sich*.

The relevant "dependence relation" is, then, neither simply from our concepts to the world nor from the world to our concepts, but holds in both directions: the structure of the world, or any world we are able to experience, depends, transcendentally, on our conceptualizing activities (on the categories we provide through our categorizing practices), while those activities themselves are causal (ontic, factual, natural) products of the natural world itself, in which we engage in those activities as fully natural beings, amidst material and social practices.[38] The latter, naturalist, part of this thesis can be easily defended by referring to the results of the natural and the human sciences, while the former, properly transcendental part requires a transcendental argument focusing on the very possibility of experience (*e.g.*, along Wittgensteinian lines, with special emphasis on language-games or forms of life as the background against which alone meaningful identification of objects is possible for us). There is, in this heavily qualified idealistic picture, no firm transcendental ground for our activity of (linguistic) categorization itself, because that on-going process is itself the transcendental ground for whatever structure our world may have; yet, it is undeniable that no such process would obtain, if we did not inhabit a natural world whose objects and processes possess a sufficient degree of regularity and variety (as argued by Westphal)—although, again, it is only through our actual employment of the conceptual framework in which these notions have their habitually established meanings that we can reach such a transcendental conclusion.

Insofar as Westphal emphasizes the importance of (fallible) transcendental reflection on human cognitive capacities and the need to embrace an essentially pragmatic form of realism about the objects of our environment, a pragmatic yet transcendental metaphysics of "the human world" is compatible with his view (*cf.* Pihlström 2012). The transcendental argument for (not unqualified but pragmatic) realism provided by Westphal offers a splendid example of a truly metaphysical result that transcendental reflections may produce. This kind of realism *is* a part—a crucial part—of our pragmatically justified picture of the human world. Yet (and this may not be entirely congenial to Westphal), precisely for this reason pragmatic realism must be subordinated to a transcendental account of our way(s) of categorizing reality—that is, transcendental pragmatism (rather than transcendental idealism). Realism, just like any other ontological commitment, is a human way of structuring the world.[39]

In any event, Westphal's struggle with the realism issue in relation to transcendental arguments must be applauded: he succeeds in his transcendental demonstration of our pragmatic need to commit ourselves to pragmatic realism about our human reality. That is, even if he fails to secure realism *sans phrase,* he succeeds in his transcendental defense of a human, pragmatic realism.

4.3 Vihalemm

Having thus at least tentatively settled my debate with Westphal, I will try to briefly respond to Rein Vihalemm's (2011, 2012) challenge to compare pragmatic realism with what he calls "practical realism." I do not want to quarrel about words, and I warmly welcome Vihalemm's proposal to examine more closely Karl Marx's concept of practice in relation to the realism issue and to compare it with the analogous pragmatist notions; however, I still maintain that my view, which unlike Vihalemm's incorporates a pragmatically rearticulated idea of the construction of the empirical (knowable, experienceable) world adapted from Kantian transcendental idealism, offers a more plausible approach to the realism debate. Indeed, Vihalemm's tendency to soften the boundary between the concepts of construction and identification (of objects) supports my view, which integrates realism and transcendental pragmatism. Insofar as by identifying scientific objects we in a way construct them within our scientific practices (or more generally our practices of inquiry), those objects—and practices—cannot be accounted for in terms of metaphysical realism (*cf.* above, §2.4, on Dewey's views on the construction of the objects of inquiry). The objects of inquiry do not possess any 'ready-made' identity prior to practices of inquiry—here I definitely agree with Vihalemm. In this respect, I would also be happy to join Vihalemm's criticism of Ilkka Niiniluoto's (1999) stronger form of realism, "critical scientific realism," which postulates *unidentified* objects (that is, objects existing, with full ontological identity, prior to and independently of inquiry) and incorporates a version of truth as correspondence. As neither Vihalemm nor I endorse that strong a form of realism, we are, I think, more united than divided here, whether the preferred position is labeled pragmatic or practical realism.

Scientific realism in the philosophy of science today must take seriously not only the practicality of science—that is, the idea, strongly emphasized by Vihalemm among others, that scientific knowledge is produced and scientific objects and facts identified (or even constructed) within practices of inquiry—but also the *plurality* of such practices and the resulting need to understand the *interdisciplinary* character of inquiry. A pragmatic, local, and contextualized approach to the realism issue encourages us to relativize the opposition between realism and antirealism in its different dimensions. Our interpretations of scientific theories and their postulations—that is, theoretical concepts and entities—must be contextualized not only within areas of discourse but also within local, practice-laden, and often interdisciplinary inquiries.[40] Moreover, not only is the identity of the postulated entities a contextual, practice-embedded matter (as Vihalemm argues), the very distinction between the context- or practice-independence of such postulations, on the one side, and their dependence on contexts or practices must itself be contextualized (*cf.* Pihlström 2011a).

At this point, the following problem arises. Can we simultaneously interpret (say) a theory *T* realistically within inquiry (practice) *I* and interpret it antirealistically within another, possibly overlapping inquiry *I*′, assuming that *T* is employed—for instance, either presupposed or tested—within both *I* and *I*′? If inquirers are allowed, or required, to truly 'mix' their disciplinary identities when engaging in interdisciplinary inquiries, such problems may have to be faced.

Broadening our scope from philosophy of science to cover ethical inquiries, we may ask whether it is possible to understand moral values and norms realistically when engaging in moral philosophy (including the ethics of science) and, at the same time or by the same inquirer, antirealistically (that is, as reducible to, say, natural interests of survival, *etc.*) when engaging in an interdisciplinary inquiry into the evolutionary origins of moral behavior, that is, an inquiry incorporating biological, psychological, social-scientific, and other dimensions? Can one and the same person—or group, or research project, or Peircean community of rational inquirers, that is, whatever the 'subject' of inquiry may be—do both?

In such a situation, in which *T* is in the "common area" of two (or more) overlapping inquiries constituting an interdisciplinary set of practices of inquiry, we cannot interpret *T* either realistically or antirealistically from a global perspective. We can only interpret it contextually, that is, in the context of either *I* or *I*′ (or some other inquiry or practice). This may require a fundamental reconsideration of our own identities as inquiring subjects. It may be a perspectival and context-dependent matter whether we are committed to a realistic or an antirealistic account of *T* and its theoretical postulations. This contextuality of the subject of inquiry is something that a "practical realism" in the philosophy of science today ought to recognize, along with emphasizing the practice-embeddedness of the identities of the objects of inquiry. The kind of transcendentally established pragmatic realism I have sketched here may, I submit, be better equipped to explore this contextuality than Vihalemm's practical realism.

5 MINIMAL REALISM?

Westphal's realism *sans phrase* can be understood as a version of what might be labeled "minimal realism," that is, realism avoiding *both* idealism (in any, including transcendental, form) *and* any strong, substantial commitments to, say, there being states of affairs that correspond to true propositions, or something similar.[41] Many other versions of minimal realism can also be distinguished, though they cannot be examined here. The brief discussion of Westphal's position (above, §4.2) may serve to illustrate how an allegedly modest and minimal realism may—if it is not too minimal to serve any proper philosophical purpose—be accounted for by transcendental pragmatism. I propose that we use *the pragmatic method* to examine the 'core meaning' of various minimal realisms, and whether or to what extent they are fully compatible with transcendental pragmatism.[42] I submit that they are: Using the pragmatic method reveals no genuine difference between the minimal realist who affirms the existence of a mind-independent world while avoiding idealistic commitments, and the empirical or pragmatic realist who also affirms the existence of a mind-independent world while proposing to ground this affirmation—understood as a pragmatic commitment enabling inquiry (not an imagined God's-Eye View on the world independent of any actual or possible inquiry)—in a more fundamental pragmatist orientation.

By applying the pragmatic method, we may note that there is no difference in the *practices of inquiry* engaged in by a minimal realist and a pragmatic realist (who is also a transcendental pragmatist). Their expected practical results and experiences are the same. We thus come close to Wittgenstein's famous observation (in *Zettel,* §§413–414) that the realist and the idealist teach their children exactly the same ways of using words: they talk about tables and chairs, and use these objects in the same ways in their lives; hence the differences between them are mere differences in "battle cries." There is no difference here that "makes a difference," and so none that has pragmatic significance. Accordingly, we may specify the core meaning of (minimal) realism by means of the pragmatic method, observing that it plays the role of a pragmatic commitment of inquiry, roughly along the lines proposed by Margolis (above, §3.2): We must actively make a commitment to there being a world 'out there' in order for us to be able to engage in practices of inquiry at all. (This could, again, be reformulated as a transcendental argument.) But this, then, is indeed the true pragmatic meaning of our realism. Realism, for us, means little beyond this practical commitment compatible, as Margolis argues, with idealism. Here I see no major difference between, say, Vihalemm's practical realism and my pragmatic realism. Thus conceived, transcendental pragmatism makes realism much more pragmatically relevant than "purely theoretical" defenses of realism that claim to be independent of both pragmatism and idealism. The further questions to consider, which may only be posed here, are these: To what extent or in what ways does our understanding of minimal realism (pragmatic or otherwise) make a difference to our understanding of our practices of inquiry, including our inquiries into the character of human inquiry and knowledge, and to what extent or in what ways do those (potential) differences in our understanding of our practices of inquiry make a difference to the character, conduct, or outcomes of any or all of our inquiries?

6 CONCLUSION

Fruitful tensions keep philosophical ideas and entire philosophical traditions alive. In this sense, the unavoidability and irresolvability of issues about realism within pragmatism—classical and recent—is a positive feature of this tradition and part of its current (and future) vitality. I join Margolis in reaffirming that realism and idealism must be intimately joined in any plausible pragmatism we can hope to defend today. This tension is one of the many that pragmatists should live with—critically! The general pragmatist (and Kantian) tendency to think beyond the received dichotomies that have shaped the issue of realism up to the present day is enormously significant.[43]

NOTES

1. For my pragmatist approach to realism in ethics, philosophy of science, metaphysics, and philosophy of religion, see (respectively) Pihlström (2005), (2008), (2009), and (2013).
2. A species of ontological realism is *scientific realism;* see below, §2.4; for detailed pragmatist discussion, see Pihlström (2008).
3. For more detailed discussion of these versions of realism, see Pihlström (1996), chap. 2, and Niiniluoto (1999). Here I use 'nonrealism' generically. Some nonrealists seek to transcend the realism *vs.* antirealism debate; antirealism and nonrealism cannot simply be equated.
4. On the distinction between existence and reality, see below, §2.2, on Peirce's realism.
5. In principle a skeptic about the possibility of *human* knowledge of reality can postulate a superhuman cognitive subject, *e.g.*, a divine one.
6. In Pihlström (2003) I characterized Peirce's scholastic realism as a species of metaphysical realism, *i.e.*, an essentially unpragmatist doctrine which assumes we can discover Nature's own "joints," at least in the long run, a doctrine whose rejection some other pragmatists (*e.g.*, Jamesians and Deweyans) regard as a virtue rather than a vice.
7. 1868, *CP* 5:311/*W* 2:239; *cf. CP* 5:405/*W* 3:271, *CP* 5:430.
8. James's occasional references to "humanism" must be set in the context of his exchange with F. C. S. Schiller, the somewhat more radical British pragmatist who called his view "humanism"; *cf.*, *e.g.*, Pihlström (1998).
9. When acknowledging Peirce as a precursor of his operational thinking, Dewey (1936) endorsed, with qualification, a Peircean view of universals ("generals"), yet avoided hypostatizing them as real entities. For standard instrumentalists and operationalists, a qualified acknowledgment of anything like the medieval universals is unacceptable. According to Dewey (1936, 532), such Peircean generals perform a vital function as "formulae of operations," guiding us in our inferences; thus, their status is logical, not metaphysical (in Dewey's broad sense of "logical").
10. For Dewey's instrumentalism about laws, *cf.* Dewey (1929a), 121–123.
11. The entire chapter 4, "Nature, Means and Knowledge," of *Experience and Nature* (Dewey 1929a) is essential to his views on the status of scientific objects.
12. Here Dewey disagreed with the "new realists" of his time; see Morgenbesser (1977).
13. On the importance of distinguishing between knowledge and experience in Dewey's naturalism, see Hildebrand (2003).
14. See Dewey (1916), 331, for his characterization of instrumentalism as a "logical version of pragmatism," according to which "knowing is literally something which we do." On this intimate relation between knowledge and action, see esp. Dewey (1929b).
15. For such a strong conception of scientific realism, see Sellars (1963) and Tuomela (1985); for some criticism, see Pihlström (1996).

16. The development of Putnam's views on realism in the 2000s deserve separate study. While Putnam completed his journey from realism "back to realism" in the 1990s, he seems now to have even more clearly returned to realism, though a form still distinguishable from the metaphysical realism he held prior to his internal realism of the 1970s. Thus, Putnam's major changes of view seem to have occurred after mid-1970s, *ca.* 1990, and again *ca.* 2010. (Hence at least "four Putnams" are to be distinguished in discussions of realism.) In his most recent essays, Putnam (2012) returns to his *nonverificationist* metaphysical realism, stressing that he never held any relativism or idealism and that internal realism was never supposed to conflict with scientific realism; *cf.* Putnam (2006), in which he responds to several critics, including Pihlström (2006).
17. Putnam is an exception; his internal realism (Putnam 1981, 1990) was explicitly indebted to Kant, and he also referred to the Kantian roots of pragmatism (*e.g.*, Putnam 1995). My surmise is that Putnam (esp. his 2012) resists assimilating his views—even his internal realism—to anything like transcendental idealism due to his deflationary reading of Wittgenstein, including the remarks on solipsism in the *Tractatus:* Putnam thinks transcendental idealism *makes no sense,* not that it is false. However, he never replies, *e.g.*, to Allison's (2004) claim that it does make sense (*cf.* Pihlström 2003, 2009).
18. This transcendentally enriched pragmatic realism can be applied to several special instances of the realism issue, including the debates over moral realism and realism in philosophy of religion; *cf.* Pihlström (2005), (2012).
19. I will write "Idealism" when referring to the kind of idealism Margolis subscribes to (and finds compatible with realism). This must be distinguished from some more traditional idealisms that contrast with realism. The difference between the two doctrines (or sets of doctrines) is that "idealism" is "either independent of or neutral with regard to 'realism' or disjunctively opposed to 'realism'," while "Idealism" with a capital 'I' is "hospitable to incorporating some forms of constructive 'realism'" (Margolis 2012, 91).
20. *Cf.* Putnam's (1990) notion of realism with a human face.
21. Margolis (1986), (1995) also discuss realism in relation to pragmatism; *cf.* his (2010).
22. Westphal (2003), (2004). I return briefly to Westphal's realism below (§4.2).
23. Since this Kantian-Hegelian narrative is so central to Margolis's argument, it might have been appropriate for him to acknowledge Westphal's work (especially Westphal 2004), both his careful historical examinations of both Kant's and Hegel's arguments and his more systematic efforts to show both that Kantian transcendental idealism is not a viable option—neither for Kant himself, given his anti-Cartesian and fallibilist commitments, nor for us—and that the realist can nevertheless argue transcendentally (yet fallibly). The reader familiar with Westphal's ideas would find a comparison highly valuable.
24. Margolis (2012, 67) also points out that "Peirce's fallibilism is meant to explain just why we *cannot,* post-Kant and post-Hegel, fall back again to any separate realism or Idealism"—and that Dewey's version of fallibilism in a way inherits this feature from Peirce's, although (2012, 67) the realism/Idealism issue is, by comparison, only marginally present in Dewey.
25. Margolis (2012, 170 n26) also remarks that "most of those who have followed Max Fisch's reading of Peirce are also committed to what Putnam opposes as the 'God's-Eye view'." This relation between the two slightly different realism disputes deserves more historical scrutiny.
26. For my own earlier reflections on Peircean realism about generality, already inspired by Margolis's constructivist and historicist account of predicable generals, see Pihlström (2003) and (2009); *cf.* Anderson's and Hausman's (2012) critique (above, §2.2).

27. I argue for this in Pihlström (2009). See above, §2.3, for a qualified defense of James and for Slater's (2008, 2009) recent realistic interpretation.
28. See Allison (2004). In Pihlström (2003) and elsewhere, I have suggested that Allison's one-world reading of transcendental idealism is the proper picture of Kant at least for the purposes of the pragmatist, though I do wish to take any fixed stand on interpretive issues concerning Kant.
29. I attempt this in Pihlström (2003); Margolis (2010, 110–111) comments on my effort.
30. Pihlström (2003) examines the proper sense of "naturalism" available to the transcendental pragmatist.
31. This paper does not claim to interpret Wittgenstein, though it is interesting to connect his views with the realism debate.
32. For some pragmatist reflections on subjectivity, see Pihlström (2011b), 54–62.
33. Westphal (2004, 2007) is emphatic about Kant's radically anti-Cartesian and (in contemporary terms) externalist (instead of internalist) conception of mental content.
34. *KdrV* A369–70, *cf.* A491–3/B519–21; Westphal (2004), 34, 71, 125–6, 223.
35. Versions of this transcendental (and pragmatic) argument for realism and against idealism, an argument which essentially defends the dependence of the very possibility of thought and experience upon objects in our environment, have, Westphal argues, been provided by Hegel and by Frederick Will (see Pihlström 2003, chap. 5). Westphal (2003, 2004) emphasizes the *fallibility* of transcendental proof and transcendental reflection.
36. These include the complex relation between *Kritik der reinen Vernunft* and *Die metaphysische Anfangsgründe der Naturwissenschaft*, detailed by Westphal (2004).
37. In Pihlström (2003), esp. chap. 5, I criticized Westphal's (earlier) formulations, defending a reinterpreted (and to some extent naturalized) transcendental idealism. I now see this dispute as more terminological than substantive, though disputes about words are important—even substantial—in philosophy. While I still believe that a picture resembling idealism (in the Kantian sense) is defensible, on pragmatist grounds, this picture may be quite far from what transcendental idealism meant for Kant himself, and may be compatible with what Westphal says about his "realism *sans phrase*." It is important to note that Westphal's (2004) unqualified realism is a minimal doctrine distinct from all kinds of "qualified" realisms debated recently, including both metaphysical and internal, as well as commonsense and scientific, realism.
38. It is not my view that human conceptualizing practices, language-use, *etc.*—our transcendental bases of world-constitution—have a privileged position interpreted (naively) realistically, while the world they are used to structure would (idealistically) be claimed to depend on them. Rather, the combination of transcendental idealism and pragmatic realism and naturalism I defend has a *circular* (though not viciously circular) structure: we transcendentally constitute the world through engaging in the worldly (and entirely natural) practices that are also themselves constituted through this very same on-going process. There simply is no Archimedean point or *fundamentum* for our transcendental world-constitution. (On the notions of ontological dependence and independence, pragmatically and transcendentally relativized, see Pihlström 2011a.)
39. See, *e.g.*, Pihlström (1996), (2003) for possible further developments of this thesis.
40. Interdisciplinarity extends beyond science insofar as we acknowledge, with Dewey, that there are also ethical, political, religious, aesthetic, and other inquiries.

41. In contrast, Armstrong's (2004) realism is not "minimal"; *cf.* Pihlström (2009).
42. In Peirce's famous formulation: "Consider what effects, which might conceivably have practical bearings, we conceive the object of our conception to have. Then, our conception of these effects is the whole of our conception of the object" (1878, *CP* 5:402/*W* 3:266); *cf.* James (1907), 28–29.
43. My views on realism have developed during more than two decades, that is, since I worked toward my doctorate under the supervision of Ilkka Niiniluoto—one of the great scientific realists today—in the early 1990s. In addition to Professor Niiniluoto, I should like to thank Professors Joseph Margolis, Kenneth R. Westphal, and Rein Vihalemm (among many others) for particularly stimulating exchanges of views related to the topic of this paper. I am also grateful to the audience commenting my paper in the symposium on "Realism in Its Multiple Forms" at the Helsinki Collegium for Advanced Studies in June 2011.

REFERENCES

Allison, H. E., 2004. *Kant's Transcendental Idealism: An Interpretation and Defense, Enlarged Edition.* New Haven, CT, Yale University Press. (1st ed. 1983.)

Anderson, D. R. and C. R. Hausman, 2012. *Conversations on Peirce: Reals and Ideals.* New York, Fordham University Press.

Armstrong, D. M., 2004. *Truth and Truthmakers.* Cambridge, Cambridge University Press.

Devitt, M., 1991. *Realism and Truth.* Oxford, Blackwell. (1st ed. 1984.)

Dewey, J., 1916. *Essays in Experimental Logic.* New York, Dover.

———, 1923. "The Pragmatism of Peirce." In: M. R. Cohen, ed., C. S. Peirce, *Chance, Love, and Logic* (rpt.: Lincoln, NB, and London: University of Nebraska Press, 1998), 301–308.

———, 1929a. *Experience and Nature.* Rpt. La Salle, IL: Open Court, 1986.

———, 1929b. *The Quest for Certainty: A Study on the Relation Between Knowledge and Action.* Rpt. Boston, G. P. Putnam's Sons, 1960.

———, 1936. "What are Universals?" *The Journal of Philosophy* 33: 253–261.

———, 1938. *Logic, Theory of Inquiry.* Boston, Holt, Rinehart, and Winston.

Eddington, A. S., Sir, 1928. *The Nature of the Physical World.* Cambridge, Cambridge University Press; New York, The Macmillan Company.

French, S., 1989. "A Peircean Response to the Realist–Empiricist Debate." *Transactions of the Charles S. Peirce Society* 25, 295–307.

Hausman, C. R., 1993. *Charles S. Peirce's Evolutionary Philosophy.* Cambridge, Cambridge University Press.

Hildebrand, D., 2003. *Beyond Realism and Anti-realism: Dewey and the Neopragmatists.* Nashville, TN, Vanderbilt University Press.

James, W., 1907. F. H. Burkhardt, F. Bowers, and I. K. Skrupskelis, eds., *Pragmatism: A New Name for Some Old Ways of Thinking.* Cambridge, Mass., Harvard University Press.

———, 1909. F. H. Burkhardt, F. Bowers, and I. K. Skrupskelis, eds., *The Meaning of Truth.* Cambridge, Mass., Harvard University Press.

Margolis, J., 1986. *The Foundations of Pragmatism.* Oxford, Blackwell.

———, 1995. *Historied Thought, Constructed World.* Berkeley, University of California Press.

———, 2010. *Pragmatism's Advantage.* Stanford, CA, Stanford University Press.

———, 2012. *Pragmatism Ascendent: A Yard of Narrative, a Touch of Prophecy.* Stanford, CA, Stanford University Press.

Niiniluoto, I., 1999. *Critical Scientific Realism.* Oxford, Oxford University Press.
Peirce, C. S., 1931–58. C. Hartshorne, P. Weiss, and A. Burks, eds., *The Collected Papers of Charles Sanders Peirce,* 8 vols. Cambridge, Mass., Harvard University Press; cited as '*CP*' by vol:paragraph numbers.
———, 1980–. N. Hauser, *et al,* eds., *The Writings of Charles Sanders Peirce,* 8 vols. to date. The Peirce Edition Project. Bloomington, Indiana University Press; cited as '*W*', by volume:page numbers.
———, 1992–98. *The Essential Peirce,* 2 vols. The Peirce Edition Project. Bloomington, Indiana University Press.
Pihlström, S., 1996. *Structuring the World: The Issue of Realism and the Nature of Ontological Problems in Classical and Contemporary Pragmatism.* Dissertation, Helsinki, The Philosophical Society of Finland.
———, 1998. *Pragmatism and Philosophical Anthropology: Understanding our Human Life in a Human World.* New York, Peter Lang.
———, 2003. *Naturalizing the Transcendental: A Pragmatic View.* Amherst, NY, Prometheus/Humanity Books.
———, 2005. *Pragmatic Moral Realism: A Transcendental Defense.* Amsterdam, Rodopi.
———, 2008. "How (not) to Write the History of Pragmatist Philosophy of Science?" *Perspectives on Science* 16: 26–69.
———, 2009. *Pragmatist Metaphysics: An Essay on the Ethical Grounds of Ontology.* London, Continuum.
———, 2011a. "The Problem of Realism, from a Pragmatic Point of View." In: R. Frega, ed., *Pragmatist Epistemologies* (Lanham, MD, Lexington Books), 103–126.
———, ed., 2011b. *The Continuum Companion to Pragmatism.* London, Continuum.
———, 2012. "Transcendental Philosophy as Ontology." In: L. Haaparanta and H. J. Koskinen, eds., *Categories of Being: Essays on Metaphysics and Logic* (Oxford, Oxford University Press), 453–477.
———, 2013. *Pragmatic Pluralism and the Problem of God.* New York, Fordham University Press.
Putnam, H., 1981. *Reason, Truth and History.* Cambridge, Cambridge University Press.
———, 1987. *The Many Faces of Realism.* La Salle, IL, Open Court.
———, 1990. J. Conant, ed., *Realism with a Human Face.* Cambridge, Mass., Harvard University Press.
———, 1995. *Pragmatism: An Open Question.* Oxford, Blackwell.
———, 2004. *Ethics without Ontology.* Cambridge, Mass., Harvard University Press.
———, 2012. M. De Caro and D. Macarthur, eds., *Philosophy in an Age of Science.* Cambridge, Mass., Harvard University Press.
Rorty, R., 1982. *Consequences of Pragmatism.* Brighton, The Harvester Press.
———, 1991. *Objectivity, Relativism, and Truth.* Cambridge, Cambridge University Press.
Sellars, W., 1963. *Science, Perception, and Reality.* London, Routledge and Kegan Paul.
Shook, J. R., 2000. *Dewey's Empirical Theory of Knowledge and Reality.* Nashville, TN, Vanderbilt University Press.
Slater, M. R., 2008. "Pragmatism, Realism, and Religion." *Journal of Religious Ethics* 36.4: 653–681.
———, 2009. *William James on Ethics and Faith.* Cambridge, Cambridge University Press.
Tuomela, R., 1985. *Science, Action and Reality.* Dordrecht, D. Reidel.
Vihalemm, R., 2011. "Towards a Practical Realist Philosophy of Science." *Baltic Journal of European Studies* 1.1(9): 46–60; www.ies.ee/iesp/No9/articles/

———, 2012. 'Practical Realism: Against Standard Scientic Realism and Anti-Realism'. *Studia Philosophica Estonica* 5.2: 200–280. www.spe.ut.ee

Westphal, K. R., 2003. "Can Pragmatic Realists Argue Transcendentally?" In: J. R. Shook, ed., *Pragmatic Realism and Naturalism* (Amherst, NY, Prometheus), 151–175.

———, 2004. *Kant's Transcendental Proof of Realism*. Cambridge, Cambridge University Press.

———, 2005. "Kant, Wittgenstein, and Transcendental Chaos." *Philosophical Investigations* 28: 303–323.

———, 2006. "How does Kant Prove that We Perceive, and not Merely Imagine, External Objects?" *The Review of Metaphysics* 59: 781–806.

———, 2007. "Consciousness and its Transcendental Conditions: Kant's Anti-Cartesian Revolt." In: S. Heinämaa, V. Lähteenmäki, and P. Remes, eds., *Consciousness: From Perception to Reflection in the History of Philosophy* (Dordrecht, Springer), 223–243.

Wittgenstein, L., 1970. G. E. M. Anscombe and G. H. von Wright, eds., *Zettel*. Oxford, Basil Blackwell.

14 McDowell's Pragmatist Anti-anti-realism

Eirik Julius Risberg

1 INTRODUCTION

A long-standing issue central to the discussions between Richard Rorty and some of his friendly critics, notably Donald Davidson and John McDowell, has been the place of a notion of *objectivity* in a philosophy cleansed of the dualism of conceptual scheme and empirical content.[1] While Rorty has commended Davidson for undermining the idea of a dualism between different conceptual schemes or languages and uninterpreted content—"the idea that something like 'mind' or 'language' can bear some such relation as 'fitting' or 'organizing' to the world" (Rorty 1986, 126)—he has increasingly come to criticize Davidson for what he takes to be a metaphysical residue in the latter's thinking. Though dismissing the scheme-content distinction, Davidson has nevertheless continued to insist on the irreducibility of the mental or intentional to the physical or non-intentional. "Davidson," Rorty writes,

> . . . thinks that even antirepresentationalists, people who believe (as he and I do) that our links with the world are merely causal, rather than representational or justificatory, ought to recognize the importance of the difference between the intentional and other idioms. (Rorty 1999, 577)

However, by insisting on the importance of differentiating between intentional and non-intentional idioms, Davidson's thinking, as Rorty (1999, 576) sees it, is marred by an underlying commitment to a Quinean ontology, to "the very idea of a distinction between the presence and absence of what Quine called a 'fact of the matter'."[2] Maintaining that the intentional is somehow special, distinct from, and irreducible to the non-intentional, Rorty argues, is nothing but

> . . . a hang-over from an early, pre-Darwinian epoch in the history of philosophy—the epoch of Kant. As I read the history of philosophy, Brentano distilled the essence of Kant's grandiose scheme-content distinction into his criterion of the psychical, and Quine and Davidson swallowed the resulting poisoned pill. (Rorty 1998a, 394)

More objectionable to Rorty than Davidson, however, is McDowell, who positively encourages those same elements in Davidson that Rorty finds misguided. Pleading with Davidson to shed his commitment to a distinction between vocabularies, which Rorty thinks can only be ontological, he chastises McDowell for emphasizing that very distinction, for reading Davidson as accommodating the Kantian distinction between the receptivity of the senses and the spontaneity of the understanding, and thus "leading us down a garden path at the entrance to which Sellars and Davidson have posted warnings" (Rorty 1998a, 390). McDowell's talk of thought's 'answerability to the world' or 'world-directedness', Rorty (1998b, 139) fears, "will provide yet another way to prolong the tedious controversies about realism *vs.* anti-realism." Thus, rather than alleviating the worry about how mind or thought can be about the world, Rorty thinks the search for objectivity or 'answerability' will leads us to an impasse where the anxiety McDowell seeks to exorcize instead becomes acute. Only by ousting the notion of objectivity can philosophy have the peace that McDowell seeks to give it.

Recently, however, this issue has been raised from within pragmatism proper, in the revisionary or *new pragmatism* of Brandom, Ramberg, *et al.* Seeking to re-appropriate the label 'pragmatism' from Richard Rorty's influential interpretation, the *new pragmatists* share the conviction that, contrary to Rorty's idiosyncratic interpretation, the notion of objectivity is not intrinsically at odds with pragmatism. Whereas Rorty (1991, 13) argues that we should "reinterpret objectivity as intersubjectivity, or as solidarity . . . [and thus] drop the question of how to get in touch with a 'mind-independent and language-independent reality'," the *new pragmatists* are "united in their efforts to articulate a position that tries to do justice to the objective dimension of *human* inquiry" (Misak 2007, 1). And so, whereas Rorty (1985, 22) identifies 'pragmatists' as those who "wish to reduce objectivity to solidarity," and contends (1985, 23) that, "[f]or pragmatists, the desire for objectivity is not the desire to escape the limitations of one's community, but simply the desire for as much intersubjective agreement as possible"; the *new pragmatists* challenge Rorty's conclusion that nothing of interest can be said about the notion of objectivity.

What can be meant by objectivity in this context? In denouncing talk of objectivity as illegitimate, Rorty seems to equate a commitment to objectivity with a commitment to some form of metaphysical realism, to a "metaphysics which has room for a special relation between beliefs and objects which will differentiate true from false beliefs" (Rorty 1985, 22). Thus there is no reason to wonder why, as one new pragmatist puts it (Stout 2007, 8), Rorty thinks "[t]he attempt to rehabilitate the rhetoric of objectivity within the context of a pragmatic philosophy actually gives away the store." If we equate talk of objectivity with a commitment to metaphysical realism, the only viable way to remain a pragmatist, it seems, is to embrace Rorty's redescription of objectivity in terms of solidarity. While wishing to maintain Rorty's conversationalism,[3] epitomized in the Jamesian slogan that "the trail of the human serpent is everywhere," the *new pragmatists* are not content to give up the notion of objectivity in favor of solidarity.[4] Thus the challenge appears to be to find a notion of objectivity that is neither compromised by, nor entails any metaphysical position; one that involves no ontology, and respects pragmatism's therapeutic or anti-metaphysical aspirations.

In attempting such a maneuver, some *new pragmatists*—most prominently, Bjørn Ramberg—have turned to Davidson for material for a revised pragmatism. Despite Rorty's objections, Davidson's distinction between the intentional and the non-intentional, Ramberg argues, does not commit him to an ontological dualism between our intentional thought and talk and any determining 'fact of the matter'. On the contrary, properly understood, Ramberg argues, Davidson's claim that the intentional is irreducible opens the prospect of a *post-ontological* distinction between the intentional and the non-intentional, providing pragmatism with what we may call a *post-ontological notion of objectivity.* The compatibility of pragmatism with a notion of objectivity thus turns on how to understand Davidson's insistence that the intentional is irreducible to the non-intentional. Criticizing Rorty for misunderstanding Davidson's claim, Ramberg places the distinction on a new footing, showing how it is not only compatible with, but conducive to pragmatism.

Rorty has acknowledged that Ramberg has made him see more clearly Davidson's claim about the irreducibility of the intentional, and thus how Davidson may provide a post-ontological distinction between the intentional and the non-intentional conducive to his own pragmatism. However, Rorty has denied that this requires him to revise his view of McDowell. As with his initial criticism of Davidson, Rorty's objections to McDowell seem to be based on the conviction that there is a metaphysical distinction operating behind the latter's talk of "answerability to the world," an ontological distinction, which Rorty used to believe was present in Davidson's work, but which he still takes to be present in McDowell's thinking. Like Ramberg, however, McDowell finds no ontological distinction in the *sui generis* character of the intentional, or, as he often puts it, using Sellars's phrase, in distinguishing 'the space of reasons' from the 'space of nature'. McDowell differs with Davidson about what can justify our beliefs. Whereas Davidson has argued that "only a belief can justify another belief," McDowell has argued that the dictum can be extended to read: "only something with a propositional structure can justify a belief." However, McDowell's rejection of metaphysics is entirely Davidsonian. In refusing to see the anti-metaphysical nature of Davidson's distinction and, McDowell argues, thus continuing to purvey the dualisms he seeks to abolish, Rorty's pragmatism remains unfinished. Turning the tables on Rorty, McDowell contends that, by exorcizing the last remnants of the dualism of scheme and content, his own view represents a pragmatism less "half-baked" than Rorty's.[5]

How should we adjudicate this standoff? Can McDowell be described as a pragmatist, representing a more complete pragmatism than Rorty's? Or is

Rorty right that McDowell's yearning for 'answerability' to the world traps him in the metaphysical quagmire from which he seeks to deliver us? Far from prolonging the "tedious realism–anti-realism debate," as Rorty claims, this chapter argues that McDowell's request for 'answerability' is provided within a non-ontological understanding of Davidson's distinction between intentional thought and the non-intentional world, which Rorty concedes is an improvement on his own pragmatism. In defending the Davidsonian distinction between the intentional and the non-intentional from allegations that it harbors a realist ontology, McDowell develops a view he labels *anti-anti-realism,* which, in being neither realist nor anti-realist, seeks in good pragmatist spirit to *end* rather than further that debate. In light of Rorty's concession to Davidson and Ramberg, McDowell's view deserves to be seen as a pragmatism of the *new* stripe. McDowell's claim, that thought is "answerable to the world," is therefore—Rorty's claim to the contrary notwithstanding—compatible with a revised pragmatism, and may be characterized as McDowell's *pragmatist anti-anti-realism.*

This chapter proceeds as follows. I first sketch Ramberg's arguments that Davidson's distinction between the intentional and the non-intentional should be seen as *post-ontological* rather than ontological (§2). On a proper understanding, Davidson's distinction thus provides the pragmatist with a *post-ontological notion of objectivity,* which, Ramberg argues, is not only compatible with, but also congenial to Rorty's own pragmatism. McDowell has consistently and repeatedly defended Davidson's distinction from objections—notably by Michael Dummett—that it rests on a notion of truth as verification-transcendent, on the idea of an in principle unknowable 'fact of the matter'. I then outline McDowell's view that a Davidsonian truth-conditional theory of meaning should rather be seen as *anti-anti-realist,* as upholding the irreducibility of the intentional without making an ontological claim (§3). I argue that the very different verdicts that Ramberg and McDowell on the one hand, and Rorty, on the other, pass on Davidson's philosophy—as held captive by an ontological dualism, or as potentially paving the way for a pragmatism compatible with a notion of objectivity—turn on the role played by the notion of *truth* in Davidson's philosophy, and consequently, the role truth can play in a renewed pragmatism. Finally I argue that in light of Rorty's recent concession to Davidson and Ramberg, we are justified in perceiving McDowell's anti-anti-realism as a pragmatism (§4).

2 DAVIDSON'S POST-ONTOLOGICAL DISTINCTION BETWEEN THE INTENTIONAL AND THE NON-INTENTIONAL

The tension Rorty detects in Davidson's thinking comes from his dismissal of the scheme-content distinction—the idea that our conceptual schemes represent or correspond to features of an independently existing reality—combined with his upholding Quine's distinction between the indeterminacy of translation and the underdetermination of empirical theories. Reflecting upon his long engagement with Davidson, Rorty (1999, 575–576) notes his surprise, when, having suggested "that all the links between mind and world are causal, and none representational . . . it turned out that [Davidson] heartily agreed with Quine about the indeterminacy of translation." While Rorty applauds rejecting the scheme-content distinction as a lasting contribution to philosophy and as completing what Quine's (1951) "Two Dogmas of Empiricism" and Sellars's (1956) "Empiricism and the Philosophy of Mind" had started, thus finally liberating philosophy from the throes of empiricism, Rorty views the intentional–non-intentional distinction as backtracking on the anti-dualistic nature of Davidson's thinking and suggesting an underlying commitment to a physicalist ontology. Distinguishing between the underdetermination of empirical theories—roughly, that there will "always be alternative descriptions of the world in physical terms . . . always more than one way of systematically accounting for any body of observations" (Ramberg 2000, 355)—and the indeterminacy of translation—the idea that having chosen our theory, and thus settled the *facts,* there is still an additional indeterminacy in trying to make sense of or interpret an agent—is tantamount, Rorty argues, to re-opening the ontological gap between scheme and content.

Quine explicitly committed himself to an ultimate 'fact of the matter' and openly embraced an ontological gap between the intentional and non-intentional: Distinct from the indeterminacy of translation, of ascribing intentional content to an agent, Quine stressed the additional underdetermination of our choice of scientific theory in settling the facts. Dismissing this talk of facts, however, Rorty (1999, 576) characterizes Quine's distinction as an "invidious distinction between the 'baselessness of intentional idioms' and the better 'based' idioms of physical science," which, he immediately adds, seems like "a residue of the unfortunate positivist idea that we can divide culture into the part in which there is an attempt to correspond to reality and the part in which there is not." Davidson's arguments against the scheme-content distinction, however, cut the other way, and should have led him, Rorty argues, to disassociate himself from this aspect of Quine's thinking. In the ensuing quarter century, Rorty persistently pressed Davidson to see that

> the anti-dualist line of thought developed in "Two Dogmas," a line of thought which led Davidson to reject the spontaneity-receptivity and scheme-content distinctions, should also have led him to reject the very idea of a distinction between the presence and absence of what Quine called a 'fact of the matter'. (Rorty 1999, 576)

However, by following Quine's distinction between indeterminacy and underdetermination, Rorty takes Davidson to exhibit a commitment to an ontological distinction between the intentional and the non-intentional at odds with pragmatist philosophy.

Ramberg (2000, 354–355) agrees with Rorty that Quine's distinction between underdetermination and indeterminacy involves an ontological ranking of vocabularies, but argues that it ". . . is not clear . . . that in the form that Davidson actually endorses the idea of the indeterminacy of interpretation[,] the Quinean commitment to ontology is retained." Affirming the indeterminacy-underdetermination distinction does not *in itself* commit Davidson to Quine's physicalist ontology, according to Ramberg. The intentional actions we ascribe to an agent may be indeterminate even when our physical theory is fixed—thus upholding the distinction—without this committing us to the further claim that "the latter mode of description is the fact-expressing one" (Ramberg 2000, 355). Only affirming the latter claim, Ramberg argues, would commit someone who maintains the indeterminacy-underdetermination distinction to the ontological distinction Rorty attributes to Davidson. Davidson, however, seems precisely to bar that conclusion by placing *facts* squarely within the intentional domain. As Ramberg argues:

> the ineliminable possibility of alternative ways of stating the facts does nothing to threaten the factuality of our statements. "Facts," evidently, does not refer to the truths captured by the vocabulary of some science, since those truths are not stated in the vocabulary of psychological ascription; facts, for Davidson, are just what true statements express, and the truths expressed by psychological and semantic ascriptions cannot, according to anomalous monism, be restated in terms of non-intentional theory. (Ramberg 2000, 356)

Contrary to what Rorty thinks therefore, upholding the Quinean distinction between the underdetermination of theory and the indeterminacy of translation does not entail that facts are to be found in our choice of physical theory, as it is according to "Quine's commitment to physics as ontology, as fixing the first-rate kinds" (Ramberg 2000, 356). The distinction between indeterminacy and underdetermination does not commit Davidson to an *ontological* distinction, to there being a 'fact of the matter' to which our intentional idioms must correspond. "It is," Ramberg (2000, 356) says, "*simply* the point that making a physical-theory choice will not settle agency-description." Davidson's distinction, he argues, should instead be seen as *post-ontological.*

However, if the distinction does not signal an ontological commitment, what, asks Rorty (1999, 577), is the rationale for upholding it and the claim that the intentional is distinct from and irreducible to the non-intentional? Why is the gap between psychology and biology supposed to be deeper than that between biology and chemistry? While Rorty (1998c, 55) is "happy to grant that attributions of intentional states are far more holistic than any other attributions we make," he sees the distinction between the intentional and the non-intentional as at best dispensable and at worst a backhand recognition of an underlying 'fact of the matter'. "Rorty's fear," McDowell notes,

> is that when Davidson thus singles out concepts of the mental as subject to a special irreducibility, that encourages a familiar sort of philosophical mind-boggling at how peculiar the mental is, and a familiar sort of project in which we take ourselves to have to tell supra-empirical stories to reestablish connections to ordinary reality for minds, conceived thus as peculiar and concomitantly as separated from ordinary reality. (McDowell 1998, 210)

Thus, even if it *can* be shown that the distinction isn't necessarily ontological, Rorty sees no reason for upholding it and argues that Davidson should drop it, lest it lead us to ponder the metaphysical relation between mind and world.

Far from dispensable, however, the distinction between the intentional and the non-intentional is central to Davidson's thinking. In a recent reply to Rorty, Davidson (1999, 599) makes clear that he "persists in thinking that the intentional idioms of psychology . . . differ from the rest." Moreover it should do so, according to Ramberg, within Rorty's own pragmatism. "Rorty," Ramberg (2000, 353) argues, "is insufficiently Rortyan in his negative assessment of the motivations for drawing a philosophical distinction between the vocabulary of agency and the vocabularies of scientific explanation." Rorty's dismissal of the distinction as philosophically sterile and uninteresting, Ramberg (2000, 353) argues, turns on a narrow and reactive conception of philosophical interest which "chains Rorty to an understanding of what philosophical theory is that he has spent most of his professional life trying to undermine."

What interest can a pragmatist see in reserving a special place for the intentional idiom? While it may be somewhat difficult to see how pragmatism could benefit from maintaining the irreducibility of the intentional, Ramberg (2000, 366) makes the case that Davidson's post-ontological distinction between the intentional and the non-intentional allows us a vocabulary that is not exhausted by the predictive explanations it can provide of agents, but one that "allows us to *recognize ourselves* in what we are talking about, and to bring to bear all those complicated considerations that we gesture at with the moral notions of a person." This, Ramberg (2007, §4.2) suggests, would serve Rorty's view of "philosophy as the hand-maiden of a humanist politics."

Ramberg's restoration of Davidson's distinction within a pragmatic philosophy starts by correcting Rorty's tracing the irreducibility of the intentional to the indeterminacy of translation. While Davidson made the connection in "Mental Events," that claim is dropped in later writings. Indeterminacy of translation is not peculiar to intentional descriptions; indeterminacy infects all attempts to translate between idioms, irrespective of whether they are mental or non-mental. As Davidson (1999, 596) says, "indeterminacy . . . is endemic in all disciplines." What is important, as Davidson (1970, 322) makes clear in an early paper, "is the *methods* we invoke in construing theories of belief and meaning that ensures the irreducibility of the concepts essential to those theories." The distinctness of the intentional from the non-intentional idioms, Ramberg suggests, lies rather in

> the particular ways in which the application-conditions of predicates of agency are holistically constrained; Davidson invokes the norms embodied in the principle of charity to venture a proposal about the *kinds* of similarities we seek when we describe creatures as agents. (Ramberg 2000, 360)

Contrary to Rorty's account therefore—which, in deflating the intentional–non-intentional distinction makes no principled difference between a non-intentional and an intentional characterization of action—for an interpreter governed by the principle of charity, Ramberg (2000, 360) contends, "not *any* complex pattern of resemblance traced in behavior will be identifiable as a pattern of *linguistic* behavior." In invoking the principle of charity, what counts as salient features marking off intentional action from non-intentional action such as speech production, is *already* visible to the interpreter through the availability of "a normativity we invoke when we try to explain to ourselves what it is that makes communication possible," Ramberg (2000, 360) argues. While we may find it convenient occasionally to describe intentional action in purely non-intentional terms, the principle of charity provides a "background [which] is *inescapable* for language users" (Ramberg 2000, 362), and which, in being inescapable, is what makes us intentional agents. The norms involved in characterizing noises as speech production or behavior as intentional behavior are not something against which we can hold up new samples to be classified, but instead are a basis from which all classification starts.

Now since describing something as a 'kind of thing'—to be able to classify it—is to tell under which conditions the classification is appropriately attributed, Ramberg (2000, 362) argues, the "theory of behavior of the sort that the ideal interpreter produces is a theory of truth." While this suggestion may set off pragmatist alarm bells, Ramberg argues that the idea not only can cohere with, but can strengthen Rorty's central idea that different vocabularies, and our way of moving between these vocabularies and redescribing them, are ways of bringing salience to different causal patterns in the world. Conceived as the ability to redistribute salience across the world, language, Ramberg (2000, 363) points out, "is the ability to reprogram our causal dispositions [to respond to the world] through salience-alteration." However, in

> changing our causal dispositions, redistributing causal significance across kinds, we affect how we engage with the world, and thus also

> the world [itself]. . . . On the pragmatist view I impute to Rorty, changing descriptions matters, just because it makes a causal difference in the world. This it does because it changes *us*, our dispositions. (Ramberg 2000, 363)

Against the inescapable background of the principle of charity, our linguistic utterances—our descriptions, as opposed to mere noises—will have appropriateness-conditions, and may therefore, Ramberg argues, be seen as a theory of truth. In describing or redescribing the world around us we expose ourselves to the possibility of *error*, which "takes the form in (assertively-used language) of the possibility of a failure of the claim to be in accord with what it says something about" (Ramberg 2000, 363). To say that 'snow is white' is to describe snow according to the norms, compliance with which makes us language users. Or, as Ramberg (2000, 362) argues: "*That* we generally conform to [these norms] is what makes us language-users, and so thinkers and knowers." In making the case that the irreducibility of the intentional is constituted by the inescapable norms of the principle of charity, rather than by any lingering ontological commitment to a 'fact of the matter', Ramberg shows us how Davidson's post-ontological distinction between the intentional and the non-intentional provides what we may call a *post-ontological notion of objectivity:* a notion of correctness or of objectivity that is neither dependent upon a 'fact of the matter', nor reducible to intersubjective agreement, but prior to and constitutive of the use of language.

3 McDOWELL'S ANTI-ANTI-REALIST NOTION OF OBJECTIVITY

Rorty (1999, 371) has acknowledged Ramberg's criticism, and credits him with showing us "how Davidson offers what [Ramberg] nicely describes as a 'post-ontological philosophy of mind'." Nevertheless, Rorty continues to deny that this concession requires him to adjust his position regarding McDowell. Rhetorically raising the question of what consequences his change of mind has for the disputes with his "'realist' opponents," he denies that the change comes to much. While sharing a deep admiration for Davidson, Rorty takes McDowell to emphasize exactly those passages in "Mental Events" about which he, Rorty, has misgivings. Hence restoring Davidson to pragmatism does not encompass McDowell, whose cultivation of a few ill-conceived comments places him at odds with Davidson's core insights. "*Pace* McDowell," Rorty (1999, 375) writes, "there is no second norm given us by the facts, in addition to the norms given us by our peers." Apparently, Rorty is convinced that McDowell's claim that thought is to be answerable to the world, is a nod to metaphysical realism and hence prolongs the realism–anti-realism debate. Is Rorty's refusal to accommodate McDowell's claim about answerability sustainable in light of his recent concession to Davidson and Ramberg? Does acknowledging the possibility of a post-ontological interpretation of the distinction between intentional and non-intentional idioms provide a non- or post-ontological interpretation of McDowell's claim about thought's answerability to the world?

Long a champion of Davidson's truth-conditional theory of meaning, McDowell has consistently defended that theory from the charge that it is inextricably tied to the idea of a fact of the matter to which our intentional talk must correspond: the very idea Rorty imputes to McDowell. McDowell's most sustained defense of Davidson's program against that objection can be found in his extended exchange with Michael Dummett, who explicitly takes truth-conditional theories of meaning to be committed to realism; on Dummett's construal, to the idea, as McDowell (1989, 355) puts it, that "we can contemplate in thought a reality that is determinate *beyond our access to it*" and to which our thoughts must correspond. McDowell objects to this alleged commitment:

> There is a familiar philosophical temptation to suppose that the notion of a fact can be sharply separated from the notion of the content of a thought or assertion that would express it; and it seems that Dummett's assumptions saddle adherents of the truth-conditional conception of meaning with the idea that truth consists in correspondence to such items . . . this is plainly wrong about some prominent adherents of the truth-conditional conception. (McDowell 1987, 92)

On Dummett's semantic construal of realism, the idea of a determinate reality amounts to the view that a sentence *S* has a truth-condition, irrespective of our ability to discover it; in short, that truth is verification-transcendent. Conceived as dependent upon a determinate reality potentially beyond our access, the truth or falsity of some sentences is not only in practice but also in principle unverifiable, such as certain mathematical statements or statements about the past. The sentence, 'James the II suffered a migraine in 1665, on the afternoon of his 32nd birthday' has a determinate truth-condition according to the realist, but one which, due to the loss of any evidence that could settle the matter, is in principle unverifiable.[6] Regarding these undecidable sentences, McDowell (1981, 317) quotes Dummett (1978, 224–225) saying: "the condition which must, in general, obtain for it to be true is not one which we are capable of recognizing whenever it obtains, or of getting ourselves in a position to do so."

This seems to raise a particular problem for the truth-conditional theory of meaning, which takes the meaning of a sentence to be given by knowledge of the conditions that must hold for the sentence to be true, conditions which, according to realism, may be beyond our access. The upshot seems to be that not only the truth-conditions but also the meaning of these sentences is beyond our access; we cannot accept this view without making unintelligible how there can be meaning or understanding at all. McDowell (1981,

314) concurs, noting that "we cannot be content with a conception of meaning that makes it a mystery how those states can be knowledge, and consequently how there can be such a thing as understanding." Now why, exactly, should the truth-conditional theory of meaning be wedded to realism?

The line of thought by which Dummett attributes realism to the truth-conditional theory of meaning starts with this constraint on a theory of meaning: A credible theory of meaning must maintain that "the significance of utterances in a language must, in general, lie open to the view, in publically available facts about linguistic behavior in its circumstances" (McDowell 1981, 314). The alternative, to take meaning to reside in an inner, private realm, would make communication impossible, and so to avoid falling to psychologism, the knowledge ascribed to anyone who is said to understand the sentence, is knowledge which cannot therefore transcend the capacity to use the sentence. However, Dummett's construal of the truth-conditional theory as necessarily committed to realism, McDowell argues,

> turns on the assumption that, when we seek to ground a theory of truth-conditions in linguistic practice, the behavior to which we may look is restricted to (behavior construable as) acknowledgements of the truth or falsity of sentences, in response to the recognizable obtaining, or not, of those truth-conditions. (McDowell 1981, 317)

On this assumption, McDowell concedes, the truth-conditional theory of meaning seems constrained by realism and thus utterly inadequate as a theory of understanding, unable, as it is, to prompt behavior—in the form of acknowledgements of truth or falsity—in the face of sentences whose truth-conditions are verification-transcendent. "There simply is not enough," McDowell (1981, 317) notes, "in the place where we are instructed to look, to manifest knowledge of what it would be for some sentence to be true." And since the alternative, Dummett argues, to allow more into our theory would be to revert to psychologism, making that assumption seems unavoidable.

However, while accepting the slogan that meaning cannot transcend use, and so the requirement on a theory of meaning that the meaning of our sentences must be publically available to view in our behavioral responses, McDowell questions whether this requirement really is incompatible with realism, as Dummett claims. "What I want to do," McDowell (1981, 316) says, "is to accept that the slogan expresses an insight, but question whether it undermines the possibility of realism." This characterization, however, is somewhat misleading as a characterization of McDowell's project. Rather than defending the view that realism is compatible with the requirement that meaning not transcend use, McDowell seeks, more radically, to alter the concepts under discussion—to draw up a new vocabulary—so that the 'realism' he seeks to defend is more appropriately described, he suggests, as an *anti-anti-realism*:

> A realist (in the relevant sense) need not go beyond rejecting the distinctive thesis of anti-realism—the thesis that an account of how understanding connects with the world must restrict itself to decidable circumstances. . . . [Thus,] it may help to conceive the position I am talking about as anti-anti-realism rather than realism. (McDowell 1989, 356)

As a contrastive thesis, anti-anti-realism is therefore primarily a rejection of Dummett's anti-realist challenge to realism, rather than a straightforward defense of realism. "My main purpose," as McDowell (1981, 321) once said, "is not to speak in favor of realism, but to question the cogency of Dummett's arguments against it." In particular, McDowell questions the generalization of Dummett's anti-realist argument against realism in mathematics—mathematical platonism—to realism across the board.[7] Taking his cue from the case of mathematics, Dummett makes his case against realism on the specific anti-realist objection against realism in mathematics, the view that

> the meaning of [Goldbach's] conjecture is such as to associate with it a specific possible configuration in arithmetical reality . . . [a] configuration [which] must determinately either obtain or not obtain, regardless of our lack of an effective procedure for deciding which of these alternatives is the case. (McDowell 1989, 345)

Since we do not know how to prove or falsify the conjecture, the anti-realist claim goes, the conjecture is verification-transcendent and the realist or Platonist can give no explanation of the "understanding of the relevant sentences in such a way that it is . . . fully displayable in the use that is made of them" (McDowell 1989, 347). The mathematical Platonist's error, Dummett argues, is not that he construes mathematical reality on the model of empirical reality, but that he construes the meaning of a sentence in such a way that it is not fully displayable in use. This, he argues, is a mistake the realist imports into non-mathematical discourse as well, and which, he contends, causes insuperable difficulties for the truth-conditional theory of meaning.

The assumption underlying Dummett's claim that the difficulty is transposed to realism across the board is that only behavior construable as acknowledgements of the truth or falsity of sentences in response to recognizably obtaining truth-conditions is publically available to view in communication. In thus delimiting what is publicly available to view, McDowell finds Dummett to be overly restrictive. While the restriction is motivated by the attempt to avoid psychologism, Dummett, McDowell argues, errs by modeling ordinary reality on mathematical reality. This places an undue burden on truth-conditional theories of meaning, McDowell (1981, 318) argues; indeed, "to force the truth-conditional conception of meaning into this mould—or rather, to try to do so, and then take failure as discrediting it—is actually not even to take note of the best version of the truth-conditional proposal." Instead of taking the truth-conditional theory of meaning to rely on the idea of a determinate reality providing the truth-conditions, and thus content or meaning of our sentences, we should see meaning as already visible *in* the speech of a fellow speaker of a language—in the specification of the conditions under which the sentence can be truly asserted. According to McDowell,

> We understand what the norms of disquotability come to, potential utterance by potential utterance, from the midst of a current practice of claim-making; we understand it by the lights constituted by being a (more or less) competent party to the practice. (McDowell 1987, 118)

A theory of meaning of this kind would be 'modest' in providing "no account of the concepts expressed by the primitive terms of the language" (Dummett 1975, 5).[8] Dummett (1975, 20) takes this to be defeatist, because such a theory of meaning "accomplishes no more than a translation manual, and hence fails to explain what, in general, someone knows when he knows a language." In contrast, McDowell believes that modesty is all we can and should hope for from a theory of meaning.[9]

While we do not need here to adjudicate the intricate debate between Dummett and McDowell, the above sketch of McDowell's anti-anti-realism should make clear that it does not return to any verification-transcendent conception of truth or objectivity. Thus Rorty is mistaken about the nature of the realism of his 'realist opponents', as he calls them. McDowell's anti-anti-realism is not subject to Dummett's anti-realist objections, nor to Rorty's objections to the idea of an underlying fact of the matter. However, neither does McDowell embrace anti-realism, but attempts instead to carve out a third alternative to Dummett's dichotomy—to deny an image of the mind, as McDowell famously argues in *Mind and World,* which first creates the problem of mind's relation to the world. Thus, while agreeing with Rorty in dismissing the idea of the world as a standard over and above our conceptual schemes, McDowell (2000, 110) simultaneously denies that we need to follow Rorty and "discard the very idea of being answerable to something other than ourselves." The problem, McDowell urges, is *not* the thought that we are answerable to something other than ourselves, but the idea of a standard which is *inaccessible,* and the idea of philosophy (drawing a metaphor from Dewey) with 'priestly intentions', capable of explaining how we can have *knowledge* of an inaccessible realm. It is the idea of the world as inaccessible, of our lack of *knowledge* of a realm of verification-transcendent truth-conditions, that Dummett takes to be the problem realism poses to a truth-conditional theory of meaning. But the idea of answerability to the world, McDowell argues, is distinct from that claim once we dismiss the idea of a reality beyond our access.

Attempting to tweak the issues apart, in the Introduction to *Mind and World* McDowell states his key question in these terms:

> Surely . . . the relevant question should rather be this: 'How is it possible for there to be empirical *knowledge?*' What that comes to, in terms of Quine's juridical image, is something on these lines: 'How can experience, standing in judgement over, say, a belief, return a verdict sufficiently favorable for the belief to count as a case of knowledge?' . . . [Now] suppose that we find it hard to see how experience can function as a tribunal, standing in a judgement over our belief. This would be a difficulty about how experience can return *any* verdict on our thinking at all, and that is surely more fundamental than a difficulty about how experience can return a particular kind of verdict, one that reaches some high level of favourableness. It is true that modern philosophy is pervaded by apparent problems about knowledge in particular. But I think it is helpful to see those apparent problems as more or less inept expressions of a deeper anxiety—an inchoately felt threat that a way of thinking we find ourselves falling into leaves minds simply out of touch with the rest of reality, not just questionably capable of getting to know about it. (McDowell 1996, *xii-xiv*)

Once we give up the idea of a reality beyond our access, it no longer seems to make sense to ask how we can come to know that inaccessible realm, how we can straddle the gap which separates us from a verification-transcendent realm of truth-conditions. In denouncing this kind of epistemological project, McDowell follows Davidson loyally. Thus, both Davidson and McDowell agree with Rorty that the idea of a correspondence relation between our thought and facts is an empty idea, since no one, Davidson (2000, 66) says, has succeeded in showing "which fact or slice of reality it is that makes a particular sentence true." However, Davidson (2000, 73) argues, we do not need to give up *any* correspondence relation: "Correspondence, while it is empty as a definition [of truth] does capture the thought that truth depends on how the world is." Davidson argues that we must dismiss the idea that we can stand in an epistemic correspondence relation with the world, though not the idea that what we hold to be true depends upon how the world is—a distinction and an idea which McDowell urges upon Rorty. As Ramberg (1989, 44) puts Davidson's point: "Correspondence, while epistemologically useless is semantically necessary, and amounts to a notion of correspondence which Davidson [1986, 307] . . . has called a 'correspondence without confrontation'."

In denying the existence of a fact of the matter, McDowell and Davidson both place the whole burden of explaining thought's answerability to the world on another notion, which Rorty finds equally troublesome: truth. The anti-anti-realist advocates the negative view that we need not perceive truth as the anti-realist does. What conception of *truth* can take us beyond Rorty's idea that the true is simply what it is good for us to believe, and take us to a world? In what sense can truth provide a notion of objectivity or answerability compatible with pragmatism?

Rorty has relentlessly questioned the central role of truth in Davidson's theory of meaning. Finding no good reason for Davidson to retain a notion of truth, which only prompts suspicion of an underlying realism, Rorty has sought to prove that Davidson should give up talk of truth and join the pragmatist ranks. When we give up the idea of a correspondence relation between thought and talk, on the one hand, and the world on the other, Rorty insists, there is no interesting role left for a notion of truth. However, while Rorty infers, in Davidson's (2000, 67) words, "[f]rom the fact that we will never be able to tell which of our beliefs are true . . . we may as well identify our best researched, most successful beliefs with the true ones, and give up the idea of objectivity," Davidson contends that there is another option, another role for truth to play, and another conception of objectivity. A correspondence theory of truth is out of the question; pragmatic or verificationist conceptions of truth are equally inadequate for a truth-conditional theory of meaning. And despite its popularity, the disquotational conception of truth, Davidson (2000, 69) categorically states, "does not give the entire concept of truth." None of these conceptions of truth, Davidson (2000, 70) contends, "come close to capturing what is important or interesting about the concept" of truth: its connection to meaning. "Truth," according to Davidson (2000, 72), "is important, . . . not because it is especially valuable or useful . . . but because without the idea of truth we would not be thinking creatures." Truth provides the inescapable normative background against which all talk becomes meaningful. Having previously criticized Davidson's emphasis on truth, Rorty finds himself persuaded by Ramberg to lay down arms against Davidson. Accepting Davison's point that 'true' is a relation holding between 'Snow is white' and snow—a relation that holds, not between "that sentence and what philosophers like to call 'reality as it is in itself', but only between those expressions and snow" (Rorty 2000, 374)—Rorty finally concedes the key role of truth Davidson (2000, 65) has consistently stressed "in our understanding of the world and of the minds of agents." Echoing Davidson's claim that without truth we would not be thinking creatures, Rorty (2000, 375) compliments Ramberg for leading him to see Davidson's point about truth; that "a being who could not be wrong about snow would not be a describer of snow, because she would not be a member of a norm-governed community of snow-describers."

The post-ontological distinction Rorty now accepts between the intentional and the non-intentional just *is* the distinction between the normativity constitutive of speaking a language and the causal relation obtaining between our thought and talk, and the world: between a 'logical space' governed by correctness conditions, and a 'logical space' which lacks such correctness conditions. Central to McDowell's view is Sellars's distinction between a non-causal, normative space of reasons and a non-normative, causal space of nature. However, now that Rorty has conceded Davidson's distinction between the intentional and the non-intentional, he cannot exclude McDowell's views from a rejuvenated 'new' pragmatism. Though there remains an *epistemological* disagreement between McDowell and Davidson about *where* to draw the distinction between the intentional and the non-intensional, McDowell closely follows Davidson's distinction between the intentional and the non-intentional, and Rorty has now conceded that this distinction is post- or non-ontological rather than ontological. This is how we should regard McDowell's view that thought is 'answerable' to the world: Not as an epistemological point, but as emphasizing the non-ontological distinction between the normative and the non-normative. By failing to see this, Rorty mistakes McDowell's yearning for answerability as perpetuating the "tedious controversy" between realism and anti-realism.

Rorty's mistake may be understandable, because McDowell's adherence to the basic post-ontological distinction between the normative and the non-normative is somewhat obscured by his epistemological attempt to include conceptually-informed perceptual experiences within the 'space of reasons'. Nevertheless, McDowell's claim that thought is answerable to the world is made within the Davidsonian framework. In the Introduction to *Mind and World* from 1996, McDowell identifies his "massive agreement" with Davidson that

> a normative context is necessary for the idea of being in touch with the world at all, whether knowledgeably or not. (McDowell 1996, *xiv*)

Thus the epistemological debate between Davidson and McDowell is best seen as an intramural debate about where to draw the post-ontological distinction between the intentional and the non-intentional. In defending the truth-conditional theory of meaning from the allegation that it harbors a realism, McDowell develops a view he dubs "anti-anti-realism," based on the fundamental distinction between the 'space of reason' and the 'space of nature'. Therefore McDowell's anti-anti-realism counts as a notational variant of the anti-metaphysical or post-ontological distinction between the intentional and the non-intentional that Ramberg attributes to Davidson.

4 CONCLUSION

A relentless critic of the notion of objectivity, Richard Rorty has long argued that we should redescribe talk of objectivity in terms of intersubjective agreement, contending that any more substantial notion of objectivity or answerability to the world involves metaphysical realism. Criticizing Davidson for failing to follow through on the anti-dualistic nature of his dismissal of the scheme-content distinction by holding that the intentional is irreducible to the non-intentional, Rorty charges that Davidson adheres to Quine's idea of a fact of the matter distinct from our intentional talk. However, some *new pragmatists* have argued that, contrary to Rorty, the notion of objectivity need not be incompatible with pragmatism, and that the resources for a rehabilitated notion of objectivity compatible with pragmatism may be found in Davidson's work. In particular, Bjørn Ramberg has argued recently that Davidson's point about the irreducibility of the intentional to the non-intentional is not only compatible with, but also congenial to Rorty's own pragmatism because it provides a post-ontological notion of objectivity.

Having accepted Ramberg's correction of his understanding of Davidson's distinction between the intentional and the non-intentional, Rorty has nevertheless denied that it requires him to accommodate McDowell's claim that thought is answerable to the world, which Rorty takes as a commitment to a reality beyond our access, to which our thought and talk must correspond. While denying that his truth-conditional theory of meaning requires realism, McDowell insists that our talk and thought is answerable to the world. This is McDowell's "anti-anti-realism," which adopts and adapts Davidson's post-ontological distinction between the intentional and the non-intentional, as the distinction between the normative, non-causal 'space of reasons' and the non-normative, causal 'space of nature'. This distinction is a version of the post-ontological distinction Rorty has recognized to be compatible with pragmatism. Hence McDowell's view counts as *pragmatist anti-anti-realism.*[10]

NOTES

1. Summing up his "On the Very Idea of a Conceptual Scheme," Davidson (2001, *xx*) argues that the essay "scouts the intelligibility of claims that different languages or conceptual schemes 'divide up', or 'cope with' reality in importantly different ways. . . . [I]f we reject the idea of an uninterpreted source of evidence no room is left for a dualism of scheme and content."
2. According to Hylton (2010, §5.2), Quine's physicalism amounts to the view that "the difference between a predicate's being true of a given object and its being false of it should, in all cases, be a physical difference: 'nothing happens in the world, not the flutter of an eyelid, not the flicker of a thought, without some redistribution of microphysical states. . . . physics can settle for nothing less' [Quine 1981, 98]. If a predicate such as '. . . . is thinking about Fermat's Last Theorem' picks out genuine events in the world then there is a physical difference between its being true of a person and its not being true of them, a fact of the matter."
3. Ramberg (2007, §2.3): "Indeed, that neither the concept of truth, nor those of objectivity and of reality, can be invoked to explain or legitimate our inferential practices and our standards of warrant, is the essence of Rorty's conversationalism, or epistemological behaviorism."
4. As Misak (2007, 2) puts it, "One of the pillars of the new pragmatism is the thought that standards of objectivity come into being and evolve over time, but that being historically situated in this way does not detract from their objectivity."
5. In the post-script to *Mind and World,* McDowell (1996, 155–156) writes, "I am claiming that Rorty's pragmatism is half-baked, according to standards set by his own account of what pragmatism is." And again, a page later: "language and traditions can figure not as '*tertia*' that would threaten to make our grip on

the world philosophically problematic, but as constitutive of our unproblematic openness to the world. . . . Rorty suggests (1986, 138) that 'intentionalistic notions' just as such breed unhealthy philosophical worries ('inserting imaginary barriers between you and the world'); in the context of a pragmatism less half-baked than Rorty achieves, the suggestion can stand exposed as absurd."

6. In the view in question, Miller (2001, 380) notes, "the sentence 'James the II suffered a migraine in 1665, on the afternoon of his 32nd birthday' has a truth-condition—James's suffering a migraine on the afternoon in question—and we can say that this condition either obtained or it did not, even though we may have no way, even in principle (because all the evidence appears to have vanished and time-travel is impossible) of determining which of these was the case."
7. As McDowell (1989, 355) argues, "If we reject the way Dummett generalizes his argument against platonism, we allow realism outside mathematics to refuse to model itself on platonism."
8. In *The Seas of Language*, Dummett (1993, 5) distinguishes a 'full-blooded' from a 'modest' theory of meaning thus: "to demand of the theory of meaning that it should explain new concepts to someone who does not already have them is to place too heavy a burden upon it, . . . all that we can require of such a theory is that it give the interpretation of the language to someone who already has the concepts required. Let us call a theory of meaning which purports to accomplish only this restricted task a *modest* theory of meaning, and one which seeks actually to explain the concepts expressed by the primitive terms of the language a *full-blooded* theory."
9. As McDowell (1977, 173) argues: "It was clear anyway that a truth-theory of the sort Davidson envisages does not, in saying what it does, *state* the senses of expressions. Why should we hanker after a theory that does that mysterious thing, if a theory that does some utterly unmysterious thing instead can be made to serve the purpose?"
10. A draft of this paper was delivered in the workshop, "Pragmatism and Realism," as part of the conference "Realism: A case of mere homonymy or identifiable common commitments?" (Helsinki, June 2011). I benefitted from comments from the audience, in particular from Bjørn Ramberg, who also read an early draft. The essay was completed during a research visit to UC-Berkeley, made possible by the generous support from the Fulbright Foundation and the Norwegian Research Council.

REFERENCES

Davidson, Donald, 1974. "Belief and the basis of meaning." *Synthese*, 27.3–4: 309–323.

———, 1986. "A coherence theory of truth and knowledge." In: E. Lepore, ed., *Truth and interpretation: Perspectives on the philosophy of Donald Davidson* (Oxford, Blackwell), 307–319.

———, 1999. "Reply to Rorty." In: L. E. Hahn, ed., *The Philosophy of Donald Davidson* (LaSalle, Il., Open Court), 595–600.

———, 2000. "Truth Rehabilitated." In: R. Brandom, ed., *Rorty and his Critics.* (Oxford, Blackwell), 65–74.

———, 2001. *Inquiries into Truth and Interpretation.* New York, Routledge.

Dummett, Michael, 1973. "The Philosophical Basis of Intuitinist Logic." Rpt. in: *idem., Truth and Other Enigmas* (Cambridge, Mass.: Harvard University Press, 1978), 215–247.

———, 1975. "What is a Theory of Meaning (I)." Rpt. in: *idem.*, *The Seas of Language* (New York: Oxford University Press, 1993), 1–33.
Hylton, Peter, 2010. "Willard van Orman Quine." In: E. N. Zalta, ed. *Stanford Encyclopaedia of Philosophy.* http://plato.stanford.edu/archives/fall2010/entries/quine/. Accessed: 09.08.2012.
McDowell, John, 1977. "On the Sense and Reference of a Proper Name." Rpt. in: *idem., Meaning, Knowledge, and Reality* (Cambridge, Mass.: Harvard University Press, 1998), 171–198.
———, 1981. "Anti-realism and the Epistemology of Understanding." Rpt. in: *idem., Meaning, Knowledge, and Reality* (Cambridge, Mass.: Harvard University Press, 1998), 314–343.
———, 1987. "In Defense of Modesty." Rpt. in: *idem., Meaning, Knowledge, and Reality* (Cambridge, Mass.: Harvard University Press, 1998), 87–107.
———, 1989. "Mathematical Platonism and Dummettian Anti-realism." Rpt. in: *idem., Meaning, Knowledge, and Reality* (Cambridge, Mass.: Harvard University Press, 1998), 344–365.
———, 1996. *Mind and World*, 2nd rev. ed. Cambridge, Mass., Harvard University Press.
———, 1997. "Another plea for modesty." Rpt. in: *idem., Meaning, Knowledge, and Reality* (Cambridge, Mass.: Harvard University Press, 1998), 108–107.
———, 1998. "The Constitutive Ideal of Rationality: Davidson and Sellars." Rpt. in: *idem.*, *Having the World in View.* (Cambridge, Mass.: Harvard University Press, 2009), 207–220.
———, 2000. "Towards Rehabilitating Objectivity." In: R. Brandom, ed., *Rorty and his Critics* (Oxford, Blackwell), 109–123.
Miller, Alexander, 2001. "Michael Dumemtt." In: A. P. Martinich and David Sosa, eds., *A Companion to Analytical Philosophy* (Oxford, Blackwell), 378–392.
Misak, Cheryl, 2007. "Introduction." In: C. Misak, ed. *New Pragmatists* (New York, Oxford University Press), 1–6.
Quine, W. V. O., 1951. "Two Dogmas of Empiricism." *The Philosophical Review* 60.1: 20–43.
———, 1981. *Theories and Things.* Cambridge, Mass., Harvard University Press.
Ramberg, Bjørn, 1989. *Donald Davidson's Philosophy of Language: An Introduction.* Oxford, Basil Blackwell.
———, 2000. "Post-ontological Philosophy of Mind: Rorty *versus* Davidson." In: R. Brandom, ed., *Rorty and his Critics* (Oxford, Blackwell), 351–370.
———, 2007. "Richard Rorty." In: E. N. Zalta, ed., *Stanford Encyclopaedia of Philosophy.* http://plato.stanford.edu/archives/spr2009/entries/rorty/. Accessed: 09.08. 2012.
Rorty, Richard, 1985. "Solidarity or Objectivity?" Rpt. in: *idem., Objectivity, Relativism, and Truth: Philosophical Papers, Vol. 1* (New York: Cambridge University Press, 1991), 21–34.
———, 1986. "Pragmatism, Davidson, and Truth." In: *idem., Objectivity, Relativism, and Truth: Philosophical Papers, Vol. 1* (New York: Cambridge University Press, 1991), 126–150.
———, 1991. "Introduction." In: *idem., Objectivity, Relativism, and Truth: Philosophical Papers, Vol. 1* (New York, Cambridge University Press), 1–20.
———, 1997. "Davidson Between Wittgenstein and Tarski." *Crítica. Revista hispanoamericana de filosofía*, 30.88: 49–71.
———, 1998a. "McDowell, Davidson, and Spontaneity." *Philosophy and Phenomenological Research*, 58.2: 389–394.
———, 1998b. "The Very Idea of Human Answerability to the World: John McDowell's Version of Empiricism." Rpt. in: *idem., Truth and Progress: Philosophical Papers, Volume 3* (New York, Cambridge University Press, 1998), 138–152.

———, 1999. "Davidson's Mental-Physical Distinction." In: L. E. Hahn, ed., *The Philosophy of Donald Davidson* (LaSalle, Il, Open Court), 575–594.
———, 2000. "Reply to Ramberg." In: R. Brandom, ed., *Rorty and his Critics.* (Oxford, Blackwell), 370–377.
Sellars, Wilfrid, 1956. "Empiricism and the Philosophy of Mind." In: H. Feigl and M. Scriven, eds., *Minnesota Studies in the Philosophy of Science* 1:253–329.
Stout, Jeffrey, 2007. "On Our Interests in Getting Things Right: Pragmatism without Narcissism." In: C. Misak, ed., *New Pragmatists* (New York, Oxford University Press), 7–31.

Contributors

Douglas R. Anderson is Professor of Philosophy at Southern Illinois University at Carbondale. He is author and editor of several books and numerous essays dealing with issues in American philosophy and culture. He is former Editor-in-Chief of the *Transactions of the Charles S. Peirce Society* and of the *Journal of Speculative Philosophy*. Presently he edits a book series in American thought for Fordham University Press. His recent work includes *Philosophy Americana* (2006) and, with Carl Hausman, *Conversations on Peirce* (2012).

Laurent Cesalli, Chargé de rechrerche at the CNRS (UMR 8163, *Savoirs, Textes, Langage,* Université de Lille 3), was born in 1968 in Vevey, Switzerland. He studied philosophy and musicology in Fribourg (1988–93), survived training as a mountain guide (1994–97), and bicycled from Geneva to Beijing (1995–96). Since 1998 he has worked at the universities of Geneva (PhD in Mediaeval philosophy, supervised by Alain de Libera), Freiburg in Breisgau, Lausanne, Fribourg (CH), and Cornell, NY. His field of research is semantics and ontology in mediaeval as well as in Austro-German philosophy. He is completing a monograph on Anton Marty's philosophy of language.

Jaakko Hintikka is Honorary Fellow in the Helsinki Collegium for Advanced Studies, and has received many international honors and awards. He was a Junior Fellow at Harvard (1956–59) and has held Professorships at the University of Helsinki, the Academy of Finland, Florida State University, and Boston University, and was affiliated with Stanford University (1965–82). His interests span philosophy of language, mathematical and philosophical logic, epistemology, philosophy of science, cognitive science, philosophy of mathematics, and history of philosophy, especially Aristotle, Descartes, Kant, Peirce, and Wittgenstein. He is the main architect of game-theoretical semantics and of the interrogative approach to inquiry, and one of the architects of distributive normal forms, possible-worlds semantics, tree methods, infinitely deep logics, and contemporary theory of inductive generalization. He has authored or co-authored over

30 books and monographs in nine languages. He has edited or co-edited 17 volumes and authored or co-authored over 300 scholarly papers, including five volumes of *Selected Papers* (1996–2003). *The Philosophy of Jaakko Hintikka* appears in the Library of Living Philosophers (2006, vol. 30).

Lauri Järvilehto, PhD, is a researcher and trainer at Filosofian Akatemia, Helsinki, and a Post-doctoral Affiliate at the Aalto University Systems Analysis Laboratory. His current research on 'Intuitive Thinking and Decision Making' draws upon recent developments in the philosophy of mind, cognitive psychology and neuroscience, and also the epistemology and semantics of C. I. Lewis, whose conception of *a priori* knowledge was the topic of Järvilehto's doctoral dissertation. Järvilehto has lectured on theoretical and applied philosophy at many Finnish universities, and at universities in Great Britain, France, Germany, Sweden, Denmark, and Estonia.

Antti Keskinen is a Post-doctoral Researcher at the University of Tampere, where he earned his PhD in 2010 with his dissertation, *Quine's Critique of Modal Logic and his Conception of Objects*. He has interpreted Quine's rejection of *de re* modalities on the basis of Quine's view of objects as posits. Currently he is researching the role of social cognition (empathy) in Quinean epistemology and philosophy of language, and problems related to the ascription of *de re* propositional attitudes. Part of the research on his contribution to this volume was done as a Post-doctoral Researcher at the University of Uppsala.

Jonathan Knowles is Professor of Philosophy at the Norwegian University of Science and Technology in Trondheim, Norway. He studied at the universities of Oxford, Edinburgh, and London, gaining his PhD in 1995 with a thesis on the philosophy of cognitive science and linguistics. He has written *Norms, Naturalism and Epistemology: The Case for Science Without Norms* (Palgrave 2003) and edited (with Henrik Rydenfelt) *Pragmatism, Science and Naturalism* (2011). He has published papers on folk psychology and explanation, knowledge of grammar and linguistic competence, reductionism in psychology, naturalistic theories of norms, the nature of naturalism, and realism and naturalized epistemology. He sits on the board of the Nordic Pragmatism Network.

Heikki A. Kovalainen is a Philosophy Researcher at the University of Tampere, Finland. His work centers on the intersections of American and Continental thought, including classical authors (especially Emerson and Nietzsche) as well as contemporary thinkers (especially Cavell and Rorty), focusing on philosophies of life richer than *Lebensphilosophie* and phenomenology, and encompassing both life's vulgarity and its sublimity. His publications include the monographs *Self as World—The New Emerson* (2010), *Emerson ja filosofia* [Emerson and Philosophy] (2007), and the articles "Emersonian Moral Perfectionism: An Alternative Ethics—But in What Sense?" (2010) and "Emersonian Self-Culture and Individual Growth: The American Appropriation of *Bildung*" (2013).

Dermot Moran holds the Chair of Philosophy (Metaphysics & Logic) at University College Dublin and is a Member of the Royal Irish Academy. A graduate of University College Dublin and Yale University, he has taught at Queen's University Belfast and Maynooth College, and held visiting professorships at Yale University, Connecticut College, Rice University, and Northwestern University. He has published widely on mediaeval philosophy (especially Christian Neoplatonism) and contemporary European philosophy (especially phenomenology). His books include *The Philosophy of John Scotus Eriugena* (1989, 2004), *Introduction to Phenomenology* (2000), and *Edmund Husserl: Founder of Phenomenology* (2005). He is Founding Editor of *The International Journal of Philosophical Studies.*

Ilkka Niiniluoto is, since 1977, Professor of Theoretical Philosophy at the University of Helsinki, and former Chair of the Department of Philosophy (1983–88, 1992, 1995–2000). He is former Rector of the University of Helsinki (2003–2008) and former Chancellor of the University (2008–2013). He has published widely on philosophy of science, and especially scientific realism. His books include *Is Science Progressive?* (1984), *Truthlikeness* (1987), and *Critical Scientific Realism* (1999). He was Associate Editor of *Synthese* (1973–76), then Editor (1977–79), and a member of its editorial board (1980–2007). He is Editor-in-Chief of *Acta Philosphica Fennica* (since 1980), and President of the Philosophical Society of Finland.

Mika Perälä holds an Academy of Finland Postdoctoral Research Fellowship in the University of Helsinki. Previously he was Postdoctoral Researcher in the Philosophical Psychology, Morality and Politics Research Unit, University of Jyväskylä, Academy of Finland. His doctoral dissertation examined Aristotle's account of perceiving that we see and hear. After obtaining his doctorate in Helsinki (2010), he pursued postdoctoral studies at Oriel College, Oxford, 2010–2011. He is currently writing articles on Aristotle's theory of perception, and a monograph on Aristotle's account of memory.

Sami Pihlström is, since 2006, Professor of Practical Philosophy at the University of Jyväskylä and, since 2009, Director of the Helsinki Collegium for Advanced Studies. He has published widely on the problem of realism and related matters, including the pragmatist tradition. His recent books include *Pragmatist Metaphysics: An Essay on the Ethical Grounds of Metaphysics* (2009), and *Transcendental Guilt: Reflections on Ethical Finitude* (2011); he edited the *Continuum Companion to Pragmatism* (2011) and is an Executive Editors of *SATS: North European Journal of Philosophy* and is the Book Review Editor of the *Transactions of the Charles S. Peirce Society.*

Panu Raatikainen is Adjunct Professor of Theoretical Philosophy at the University of Helsinki and the University of Tampere. His research embraces logic, philosophy of mathematics, philosophy of language, philosophy of mind, philosophy of science, and the history of analytic philosophy. He has held Research Fellowships in the Academy of Finland and in the Helsinki Collegium for Advanced Studies, and visiting research posts at the University of St. Andrews, the University of London, and the City University of New York. He publishes in such journals as *Analysis, Dialectica, Erkenntnis, Synthese, British Journal for the Philosophy of Science,* and *Journal of Symbolic Logic.*

Eirik Julius Risberg is a doctoral student in philosophy at the University of Oslo, researching his dissertation on meaning and objectivity, including whether pragmatism can accommodate a notion of objectivity. Risberg holds an MA in philosophy from the University of Oslo and has been a visiting student at the University of California, Berkeley, and at the University of British Columbia, Vancouver. With support from Fullbright Foundation (USA) and the Norwegian Research Council, he spent the academic year 2011–12 as a visiting research student at the University of California, Berkeley.

Peter Swirski is Professor of American Literature and Culture at the University of Missouri–St. Louis, Honorary Professor in American Studies at Jinan University (China), Honorary Professor in American Studies at the South China University of Technology, and former Research Fellow in the Helsinki Collegium for Advanced Studies. His research ranges from American Literature and American Studies to interdisciplinary studies in literature, philosophy, and science. He is the foremost critic on the late writer and philosopher, Stanislaw Lem. His books include *Between Literature and Science* (2000), *From Lowbrow to Nobrow* (2005), *Of Literature and Knowledge* (2007), *Literature, Analytically Speaking* (2010), and was a National Book Award finalist for his *Ars Americana, Ars Politica* (2010).

Kenneth R. Westphal is Professorial Fellow in the School of Philosophy, University of East Anglia, Norwich. He has held Humboldt Research Fellowships at the universities of Heidelberg, Bielefeld, and Göttingen; and been Visiting Professor at Northwestern University and at the Martin-Luther-Universität Halle-Wittenberg. His research on the character and scope of rational justification in non-formal domains integrates systematic with historical, and analytic with hermeneutical philosophy. His books include *Kant's Transcendental Proof of Realism* (2004); he edited *The Blackwell Guide to Hegel's Phenomenology of Spirit* (2009). His recent articles include: "Kant's *Critique of Pure Reason* and Analytic Philosophy" (2010); "Self-Consciousness, Anti-Cartesianism & Cognitive Semantics in Hegel's 1807 *Phenomenology*" (2011); "Norm Acquisition, Rational Judgment & Moral Particularism" (2012); "Hume, Empiricism & the Generality of Thought" (2013); "Natural Law, Social Contract & Moral Objectivity: Rousseau's Natural Law Constructivism" (2013); and "Rational Justification & Mutual Recognition in Substantive Domains" (2013).

Index

(Note: close cognate terms are all listed under one main form, without mentioning the latter variants; *e.g.*, 'actualism' also includes 'actualist'.)

For Product Safety Concerns and Information please contact our EU representative GPSR@taylorandfrancis.com Taylor & Francis Verlag GmbH, Kaufingerstraße 24, 80331 München, Germany